Algebra and Coding Theory

CONTEMPORARY MATHEMATICS

785

Algebra and Coding Theory

Virtual Conference in Honor of Tariq Rizvi
Noncommutative Rings and their Applications VII
July 5–7, 2021
Université d'Artois, Lens, France
Virtual Conference
Quadratic Forms, Rings and Codes
July 8, 2021
Université d'Artois, Lens, France

A. Leroy
S. K. Jain
Editors

AMS AMERICAN MATHEMATICAL SOCIETY

2020 *Mathematics Subject Classification.* Primary 16-XX, 16N40, 16S34, 16S36, 11T71, 94BXX, 94B05, 94B15, 94B60.

Library of Congress Cataloging-in-Publication Data

Names: Rizvi, S. Tariq, honoree. | Leroy, André (André Gerard), 1955– editor. | Jain, S. K. (Surender Kumar), 1938– editor. | International Conference on Noncommutative Rings and Their Applications (7th : 2021 : Lens, France), author. Virtual Conference on Quadratic Forms, Rings and Codes (2021 : Lens, France)

Title: Algebra and coding theory : virtual conference in honour of Tariq Rizvi, Noncommutative Rings and Their Applications, VII, July 5–7, 2021, Université d'Artois, Lens, France : Virtual Conference on Quadratic Forms, Rings and Codes, July 8, 2021, Université d'Artois, Lens, France / A. Leroy, S.K. Jain, editors.

Description: Providence, Rhode Island : American Mathematical Society, [2023] | Series: Contemporary mathematics, 0271-4132 ; volume 785 | Includes bibliographical references.

Identifiers: LCCN 2022048347 | ISBN 9781470468590 (paperback) | ISBN 9781470472986 (ebook)

Subjects: LCSH: Coding theory–Congresses. | Algebra–Congresses. | AMS: Associative rings and algebras. | Associative rings and algebras – Radicals and radical properties of rings – Nil and nilpotent radicals, sets, ideals, rings. | Associative rings and algebras – Rings and algebras arising under various constructions – Group rings. | Associative rings and algebras – Rings and algebras arising under various constructions – Ordinary and skew polynomial rings and semigroup rings. | Number theory – Finite fields and commutative rings (number-theoretic aspects) – Algebraic coding theory; cryptography. | Information and communication, circuits – Theory of error-correcting codes and error-detecting codes. | Information and communication, circuits – Theory of error-correcting codes and error-detecting codes – Linear codes, general. | Information and communication, circuits – Theory of error-correcting codes and error-detecting codes – Cyclic codes. | Information and communication, circuits – Theory of error-correcting codes and error-detecting codes – Other types of codes.

Classification: LCC QA268 .A388 2023 | DDC 512/.4601154–dc23/eng20230415

LC record available at https://lccn.loc.gov/2022048347

Contemporary Mathematics ISSN: 0271-4132 (print); ISSN: 1098-3627 (online)

DOI: https://doi.org/10.1090/conm/785

---

10 9 8 7 6 5 4 3 2 1 23 22 21 20 19 18

This volume is dedicated to Syed Tariq Rizvi for his outstanding contributions to Mathematics and in particular to Rings and Modules.

Tariq Rizvi, Ohio State University, Lima, Ohio.

Contents

Preface

As an outgrowth of the two conferences "Noncommutative rings and their Application (NCRA)" and "Quadratic forms, rings and codes(QFRC)" held online in July 2021 in Lens (France), this volume contains a variety of topics in algebra and exhibits their interplay as presented in the meetings by the participants.

The NCRA conference was dedicated to our colleague Tariq Rizvi. Tariq Rizvi is a prolific algebraist and his impact on rings and modules is tremendous. His influence is really impressive and many algebraists around the world benefited from his experience, expertise and generosity. He is frank, humble, generous and it is always a pleasure to meet him with his welcoming smile. We are lucky and proud to count him as our friend and we are very happy to give him, through this volume, a small token of our respect and gratitude. Some of the papers included in the volume reflect the influence of his work on the questions related to the direct sums of continuous and quasi-continuous modules, among others.

One of the aims of these meetings was to establish connections between ring theory and coding theory. In particular, the applications to coding theory of Frobenius rings, the skew group rings and iterated Ore extensions are the subject of some of the papers. The questions on the classical topics, such as Utumi rings, Baer rings, nil and nilpotent algebras, Brauer groups are discussed in some papers. Other articles deal with questions that relate to the elementwise study for rings and modules. Lastly, this volume includes papers dealing with questions in homological algebra and lattice theory. All these works show the vivacity of the research in the area of noncommutative rings and its influence on other subjects.

As usual, the papers went through a strict refereeing process and we would like to use this opportunity to thank all the anonymous referees who kindly reviewed the papers.

We thank Christine Thivierge who was so helpful and efficient in handling this publication.

André Leroy and S.K. Jain

List of Participants

Mehdi Aaghabali
University of Edinburg

Evrim Akalan
Hacettepe University, Ankara.

Mustafa Alkan
Akademiz University, Antalya.

Meltem Altun- Özarslan
Hacettepe University, Ankara.

Dilshad Alghazzawi
Université d'Artois.

María José Arroyo
Universidad Autónoma Metropolitana,
Iztapalapa (Mexico)

Noha Abdelghany
Colby College.

Mona Abdi
Shahrood University of Technology.

Sami Alabiad
King Saud University, Riyadh.

Yusuf Alagoz
Siirt University.

Dilshad Alghazzawi
Jeddah University, KAU.

Shakir Ali
Aligarh Muslim University, Aligarh.

Abeer Ali Rajaallah Alharbi
King Abdulaziz University

Branch Rabigh
Saudi Arabia.

Anani Alkathiry
Umm Al-Qura University.

Yousef Alkhamees
King Saud University.

Meltem Altun
Hacettepe University, Ankara.

Mohammad Ashraf
Alighar Muslim University.

Pinar Aydogdu
Hacettepe University, Ankara.

Per Bäck
Mälardalen University,.

Ayman Badawi
American University of Sharjah ,
Sharjah.

Mhammed Boulagouaz
Universite Sidi Mohammed ben
Abdellah Fes.

Ranya Djihad Boulanouar
Algiers University USTHB, Alger.

Ahmed Djamel Bouzidi
Algiers University USTHB, Alger.

Dipak Kumar Bhunia
Autonomous University of Barcelona.

Gary Birkenmeier
University of Louisiana at Lafayette.

Sinem Benli
Izmir Institute of Technology.

Delphine Boucher
IRMAR, Université de Rennes 1.

Engin Büyükasik
Izmir Institute of Technology.

Victor Camillo
University of Iowa.

Ahmed Cherchem
Algiers University USTHB, Alger.

Manuel Cortés-Izurdiaga
University of Málaga.

Soumitra Das
KPR Institute of Engineering and Technology, Coimbatore.

Avanish Kumar Chaturvedi
University of Allahabad, Prayagraj.

Wagner Cortes
Institute of Mathematics-UFRGS, Porto Alegre.

Najmeh Dehghani
Persian Gulf University, Bushehr.

Hai Dinh
Kent State University.

Steven T. Dougherty
University of Scranton.

Müge Diril
Izmir Institute of Technology.

Nguyen Viet Dung
Ohio University.

Susan El-Deken
Helwan University.

Fatma Ebrahim
Al-Azhar University, Cairo.

Lulwah M. Al-Essa
Imam Abdulrahman Bin Faisal University.

Gabriella d'Este
University of Milano.

Sergio Estrada
Universidad de Murcia.

Alberto Facchini
Università di Padova.

Walter Ferrer
Universidad de la República.

Cristina Fernandez-Cordoba
Autonomous University of Barcelona.

Surender Jain
Ohio University, Athens.

Àngel García Blàzquez
University of Murcia.

Diego García Lucas
Affiliation: Universidad de Murcia.

Deniz Gokalp
Istanbul University.

Franco Guerriero
Ohio University.

Pedro Antonio Guil Asensio
University of Murcia.

Ashok Ji Gupta
Indian Institue of technology(BHU), Varanasi.

Fatmanur Gürsoy
Yildiz Technical University.

Erik Hieta-aho
Aalto University.

Malgorzata Hryniewicka
Bialystok University.

Berke Kalebogaz
Hacettepe University.

Yeliz Kara, Bursa Uluda University (Turkey).

Fatma, Kaynarca, Afyon Kocatepe University, Afyonkarahisar.

Tamer Kosan
Gazi University.

Jan Krempa
Institute of Mathematics, Warsaw University.

Arda Kor
Gebze Technical University.

Adrian Korban
University of Chester.

Ahmed Laghribi
Université d'Artois.

Ganygong Lee
Chungnam National University.

André Leroy
Université d'Artois.

Yuanlin Li
Brock University.

Sergio Lopez Permouth
Ohio University, Athens.

Najib Mahdou
University S.M. Ben Abdellah, FEZ.

Omar Al-mallah Al-Balqa Applied University
Al-Salt.

Umberto Martines Penas
University of Toronto.

Xavier Mary
Université Paris Nanterre.

Jerzy Matczuk
Institute of Mathematics, Warsaw University.

Sanjeev Kumar Maurya
University of Allahabad.

Melaibari
University of Jeddah.

M.Kamal Mkhallalati
University of Istanbul.

Intan Muchtadi-Alamsyah
Institut Teknologi Bandung.

Ghulam Muhiuddin
University of Tabuk.

Diksha Mukhija
Université d'Artois.

Najat Muthana
University of Jeddah.

Figen TAKIL MUTLU
Eskisehir Technical University, Tepebasi.

Mehrdad Nasernejad
Institute for research in Fundamental sciences(IPM).

Zahra Nazemian, University of Graz (Austria).

Pace P. Nielsen
Brigham Young University.

Marta Nowakowska
University of Silesia in Katowice.

Ayse Cigdem Ozcan
Hacettepe University, Ankara.

Nil Orhan Ertas
Bursa Technical University, Bursa.

Salahattin Özdemir
Dokuz Eylül University,.

Jae Keol Park
Busan National University.

Cesar Polcino Milies
University of Sao Paulo.

Asiyeh Rafieipour
Ohio University.

Abeer Ali Rajaallah Alharbi
King Abdulaziz University Branch Rabigh.

Cosmin Roman
Ohio State University.

Louis Rowen
Bar-Ilan University.

Ángel del Río
Universidad de Murcia (Spain).

Tariq Rizvi, Ohio Sate University at Lima (USA).

Cosmin Roman The Ohio State University at Lima (USA).

Serap Sahinkaya
Tarsus University.

Martha Lizbeth Sandoval Miranda
Universidad Autónoma Metropolitana, Mexico City.

Bülent Saraç
Hacettepe University (Turkey).

Surjeet Singh
Chandigarh University.

Virgilio P. Sison
University of the Philippines Los Baños.

Daniel Smertnig
University of Graz.

Agata Smoktunowicz
University of Edinburgh.

Anuradha Sharma
IIIT-Delhi, New Delhi.

Patrick Solé
CNRS I2M Marseille.

Amin Soofiani
University of British Columbia.

Ashish K. Srivastava
Saint Louis University.

Steve Szabo
Eastern Kentucky University.

Sandeep Sharma
IIIT-Delhi, New Delhi.

Zubeyir Turkoglu
Dokuz Eyul University, Izmir

Sarra Talbi
U.S.T.H.B., Algiers.

Melis Tekin Akcin
Hacettepe University.

Sultan Eylem Toksoy
Hacettepe University.

Yaser Tolooei
Razi University.

Blas Torrecillas
Universidad de Almería.

Derya Keskin Tutuncu
Hacettepe University.

Gülsen Ulucak
Gebze Technical University.

Vaishali Varshney
department of Mathematics, AMU.

Maximiliano Vides
Universidad Nacional delLitoral.

Daniel Windisch
Graz University of Technology.

Jay Wood
University West Michigan.

Monika Yadav
IIIT-Delhi, New Delhi.

Tulay Yildirim
Karabuk University.

Eda Yildiz Yildiz Technical University, Istanbul, (Turkey).

Mohamed F. Yousif
The Ohio State University.

Yiqiang Zhou
Memorial University of Newfoundland.

Contemporary Mathematics
Volume **785**, 2023
https://doi.org/10.1090/conm/785/15770

A note on checkable codes over Frobenius and quasi-Frobenius rings

R. R. de Araujo, A. L. M. Pereira, and C. Polcino Milies

ABSTRACT. We study checkable codes in finite Frobenius and quasi-Frobenius rings and show that the Jacobson radical of a quasi-Frobenius ring is checkable if and only if the ring is Frobenius. Also, we show that in a quasi-Frobenius ring R every maximal left ideal is checkable and that R is code-checkable if and only if R is a principal right ideal ring.

Dedicated to Professor Syed Tariq Rizvi for his contribution to Ring and Module Theory.

1. Introduction

Cyclic codes over a finite field $\mathbb{F}$ were introduced by Prange [**14**] in 1957 and can be realized either as ideals in a factor ring $\mathbb{F}[X]/\langle X^n - 1\rangle$ or in its isomorphic image $\mathbb{F}C_n$, where C_n denotes a cyclic group of order n.

Extending this idea, S.D. Berman [**2**] in 1967 and F.J. MacWilliams [**13**] in 1970, independently, introduced the notion of a *group code*, which can be defined as a left (or right) ideal in a finite group algebra.

In 2008 J.A. Wood [**15**] showed that code equivalence characterizes finite Frobenius rings. More precisely, MacWilliams [**11, 12**] proved that linear codes over a finite field $\mathbb{F}$ have the following *extension property*: if $f : C_1 \to C_2$ is a linear isomorphism between linear codes C_1, C_2 in $\mathbb{F}^n$ and f preserves the Hamming weight, then f extends to a monomial transformation of $\mathbb{F}^n$. It is shown in [**15**] that if the extension property holds for linear codes over a finite ring R, then R is a finite Frobenius ring.

The concept of checkable codes, due to P. Hurley and T. Hurley [**7**], is related to the notion zero-divisors in group rings . If RG is a finite group ring, a code (left ideal) C of RG is said to be *checkable* if there exists $v \in RG$ such that $C = \{y \in RG \ : \ yv = 0\}$, i.e. if its left annihilator is a principal ideal.

A group ring RG is called *code-checkable* if every code in RG is checkable. Jitman et al. [**8**, Proposition 3.1] showed that if $\mathbb{F}$ is a finite field and G is a finite abelian group, then the group algebra $\mathbb{F}G$ is code-checkable if and only if $\mathbb{F}G$ is a

2020 *Mathematics Subject Classification.* Primary 11T71, 16S34, 20C05. Secondary 16L60, 94A24.

Key words and phrases. codes, checkable code, Frobenius ring, quasi-Frobenius ring.

The authors were partially supported by FAPESP, Proc 2015/09162-9.

principal ideal ring. In [**1**, Proposition 4.1], Abdelghany and Megahed extended this result to the case when coefficients are taken in a finite commutative ring.

Recently, Borello, de la Cruz and Willems showed that the same is true for Frobenius algebras:

[**3**, Corollary 2.3] *In a Frobenius algebra A, a right (resp. left) ideal I is checkable if and only if $Ann_l(I)$ resp. $Ann_r(I)$ is a principal left (resp. right) ideal.*

As pointed out in the same paper, there are many examples of checkable codes.

[**3**, Examples 2.4]

(a) Let $e = e^2$ be an idempotent element in A. Then the ideal eA is checkable.
(b) If A is a semisimple algebra, then all right and left ideals are generated by idempotents. Thus all right and left ideals are checkable.
(c) All cyclic codes are checkable, since the check equation is given by the check polynomial.
(d) LCD group codes C (that is, codes for which $C \cap C^{\perp} = \{0\}$) are checkable since $C = eKG$ with a self-adjoint idempotent e by [**4**].

The "double annihilator property" plays an important role in [**3**]. Actually, this property characterizes finite quasi-Frobenius rings [**10**]. Following this same approach, we study checkable codes also over quasi-Frobenius rings.

2. Finite Frobenius and quasi-Frobenius rings

In what follows we consider Frobenius rings and quasi-Frobenius rings only in the case of *finite* rings (rings are always associative with identity $1 \neq 0$). A general definition (including the infinite case) can be found in [**10**, Chapter 6].

For every left ideal $I \lhd_l R$, the right ideal $Ann_r(I) := \{x \in R : Ix = 0\}$ denotes the *right annihilator* of I. Similarly, the left ideal $Ann_l(I) := \{x \in R : xI = 0\}$ is the *left annihilator* of a right ideal $I \lhd_r R$.

DEFINITION 2.1. *A finite ring R is a* quasi-Frobenius ring *(QF ring) if it satisfies the two double annihilator properties:*

$$\text{(i) } Ann_r(Ann_l(I)) = I, \text{ for all right ideal } I \lhd_r R; \tag{1}$$

$$\text{(ii) } Ann_l(Ann_r(I)) = I, \text{ for all left ideal } I \lhd_l R. \tag{2}$$

The above definition indicates that, in QF rings, the maps $I \lhd_r R \mapsto Ann_l(I) \lhd_l R$ and $I \lhd_l R \mapsto Ann_r(I) \lhd_r R$ define mutually inverse lattice anti-isomorphisms [**10**, Corollary 15.6]. Also, denoting by $J(R)$ the Jacobson radical of R and by $Soc(R_R)$ and $Soc({}_RR)$ the socles of the R-modules R_R and ${}_RR$, i.e. the sum of all minimal left and of all minimal right ideals of R respectively, in [**10**, Corollary 15.7] it is shown that, in QF rings,

$$Ann_l(J(R)) = Soc(R_R) = Soc({}_RR) = Ann_r(J(R)). \tag{3}$$

If G is a finite group and R a finite QF ring, then the group ring RG is a QF ring. A direct sum of rings $R = \bigoplus_i^n R_i$ is a QF if and only if each R_i is a QF ring. In turn, if F is a finite field, $R = \begin{pmatrix} F & F \\ O & F \end{pmatrix}$ is an example of a ring that is not quasi-Frobenius [**10**].

A special class of QF rings are Frobenius rings which, in the finite case, can be defined as follows:

DEFINITION 2.2. *A finite ring R is said to be a* Frobenius ring *if $Soc(R_R)$ is isomorphic to $(R/J(R))_R$ and $Soc({}_RR)$ is isomorphic to ${}_R(R/J(R))$.*

Theorem 16.14 of [**10**] guarantees that Frobenius rings are QF rings. Honold, [**6**], pointed out important facts about finite Frobenius rings. In particular, in [**6**, Theorem 2] it is shown that only one of the isomorphisms in Definition 2.2 is needed to ensure that a *finite* ring R is Frobenius:

THEOREM 2.1. [**6**, Theorems 1 and 2] *Let R be a finite ring. Then, following properties are equivalent:*

(i) *R is Frobenius;*
(ii) *$Soc(R_R)$ is cyclic;*
(iii) *$Soc({}_RR)$ is cyclic;*
(iv) *$|I| \cdot |Ann_r(I)| = |R|$, for every left ideal $I \lhd_l R$;*
(v) *$|I| \cdot |Ann_l(I)| = |R|$, for every right ideal $I \lhd_r R$.*

Some examples of Frobenius rings are semisimple rings and finite chain rings. A direct sum $R = \bigoplus_i^n R_i$ is Frobenius if and only if every R_i is Frobenius. If R is a Frobenius ring and G is a finite group, then the matrix ring $M_n(R)$ and the group ring RG are Frobenius rings [**10**]. Examples of QF ring which are not Frobenius can be found in [**10**, Examples 16 and 19].

3. Checkable codes over finite QF rings

Let R be a finite ring. A set $C \subset R$ is a left (right) *code* in R if C is a left (right) ideal of R. In what follows, we always consider left ideals and hence will omit the word "left" in the sequel. Similar results hold for right ideals.

The following result gives a characterization of checkable codes in finite QF rings:

THEOREM 3.1. *Let R be a finite QF ring. A code C in R is checkable if and only if $Ann_r(C)$ is a principal right ideal of R.*

PROOF. A linear code C in R is checkable if and only if there exists $v \in R$ such that $C = Ann_l(vR)$. Since R is a QF ring, this is equivalent to the existence of some element $v \in R$ such that $Ann_r(C) = vR$, which is a principal right ideal. $\square$

Since the annihilator of the socle of a QF ring is its Jacobson radical, we have the following:

THEOREM 3.2. *Let R be a finite QF ring. Then:*

(i) *$Soc({}_RR)$ is checkable if and only if $J(R)$ is a principal right ideal of R.*

(ii) *If R is local and $J(R)$ is a principal right ideal of R, then there exists only one minimal left ideal I of R such that I is checkable.*

PROOF. The left ideal $Soc({}_RR)$ is checkable iff there exists $v \in R$ such that $Soc({}_RR) = Ann_l(vR)$. Since R is a QF ring, by (3), this is equivalent to $vR = Ann_r(Soc({}_RR)) = J(R)$, which proves (i).

To show (ii), suppose that R is local and $J(R)$ is a principal right ideal of R. If I is a minimal left ideal of R, then $I \subset Soc({}_RR)$, which implies $J(R) = Ann_r(Soc({}_RR)) \subset Ann_r(I)$. Since R is local, $J(R)$ is maximal and thus $J(R) = Ann_r(I)$. So, $Ann_r(I)$ is principal. By Theorem 3.1, I is checkable. Also, $I =$

$Ann_l(J(R)) = Soc({}_RR)$, which implies that I is the unique minimal left ideal of R. □

4. Code-checkable QF rings

In this section, we characterize code-checkable finite QF rings. Also, we prove that every code-checkable finite QF ring is a Frobenius ring.

We begin with an elementary result.

THEOREM 4.1. *Let R be Frobenius ring R of cardinality p^n such that $[R:I]=p$ and I an ideal such that $[R:I]=p$, where p is a prime number, $n \geq 2$. Then I is checkable. In particular, if $n=2$, the ring R is code-checkable.*

PROOF. Let I be a left ideal of R such that $[R:I]=p$. Then $|I| = |R|/[R:I] = p^{n-1}$. Since R is a Frobenius ring, it follows from Theorem 2.1 that $|I|\cdot|Ann_r(I)| = |R| = p^n$. So, $|Ann_r(I)| = p$. This means that $Ann_r(I)$ is a principal right ideal in R, implying that I is checkable. Now, suppose that $n=2$. Let I be a non-trivial left ideal of R (the cases $I=0$ and $I=R$ are trivial). Then $|I|$ divides $|R|=p^2$ and cannot be 1 or p^2, which means that $|I|=p$. So, $[R:I]=|R|/|I|=p$. It follows from the arguments above that I is checkable. Therefore, R is a code-checkable ring. □

Our next result shows a necessary and sufficient condition for a finite QF ring to be a Frobenius ring considering checkability of its Jacobson radical.

THEOREM 4.2. *Let R be a finite QF ring. The Jacobson radical $J(R)$ is checkable if and only if R is a Frobenius ring.*

PROOF. If $J(R)$ is checkable, it follows from Theorem 3.1 that $Ann_r(J(R))$ is a principal right ideal of R. Since R is a QF ring, it follows from (3) that $Soc(R_R) = Ann_r(J(R))$. So, $Soc(R_R)$ is a principal right ideal in R, that is, $Soc(R_R)$ is cyclic in the right R-module R_R. By theorem 2.1, this implies that R is Frobenius. Conversely, if R is a Frobenius ring, then R is a QF ring and $Soc(R_R) = Ann_r(J(R))$ is cyclic, that is, $Ann_r(J(R))$ is a principal right ideal in R. By Theorem 3.1, $J(R)$ is checkable. □

As an immediate consequence of Theorem 4.2 it follows that:

COROLLARY 4.1. *If a finite ring R is a QF ring and code-checkable, then R is a Frobenius ring.*

The next result extends [**3**, Corollary 2.11] to finite QF rings:

THEOREM 4.3. *Let R be a finite QF ring. Then every maximal left ideal of R is checkable.*

PROOF. For an R-module A, denote by $l(A)$ the composition length of A (that is, the number of nonzero submodules in a composition series for A). Let M be a maximal left ideal of R and $l = l(R)$. Then $l(M) = l-1$. Denote by M^* the right R-module $Hom_R(M,R)$ (the dual module of M) given by the action $\varphi(m)r = \varphi(rm)$, for all $\varphi \in M^*$, $r \in R$ and $m \in M$. It follows from duality properties (e.g., see [**10**, Equation 15.14]) that $R/Ann_r(M)$ is isomorphic to M^*. Since $l(M^*) = l(M)$, then $l(Ann_r(M)) = 1$, which means that $Ann_r(M)$ is a minimal right ideal of R. Therefore, $Ann_r(M)$ is a right principal ideal of R, since every minimal ideal is principal. It follows from Theorem 3.1 that M is checkable. □

The following theorem shows that, for QF rings, the class of code-checkable rings is precisely the class of principal right ideal rings. In particular, this means that a commutative QF ring is code-checkable iff it is a principal ideal ring extending [**3**, Theorem 3.1] and [**1**, Theorem 4.5].

THEOREM 4.4. *Let R be a finite QF ring. Then, R is code-checkable if and only if R is a principal right ideal ring.*

PROOF. Suppose that R is a code-checkable ring. Let I be a right ideal in R. Since $Ann_l(I)$ is a linear code (left ideal) and R is a QF ring, Theorem 3.1 implies that $I = Ann_r(Ann_l(I))$ is principal. Therefore, R is a principal right ideal ring. Conversely, let C be a linear code of R. As R is a principal right ideal ring, $Ann_r(C)$ is principal. It follows from Theorem 3.1 that C is checkable. Therefore, R is code-checkable. □

COROLLARY 4.2. *Consider the direct sum $R = \bigoplus_{i=1}^{n} R_i$ of finite commutative QF rings. Then, R is code-checkable if and only if R_i is code-checkable, for all $i, 1 \leq i \leq n$.*

PROOF. Since each R_i is a QF ring, for all $1 \leq i \leq n$, R is also a finite QF ring. From Theorem 4.4, R is a code-checkable ring iff R is a principal ideal ring. Also, since R is commutative, R is a principal ideal ring iff every R_i is a principal ideal ring, that is, every R_i is code-checkable and the result follows. □

COROLLARY 4.3. *Let R be a finite, commutative and code-checkable QF ring. Let C be a code in R. If $C = Ann_r(C)$ $(= Ann_l(C))$, then C is a nilpotent ideal of R.*

PROOF. By Theorem 4.4, R is a principal ideal ring. Then, for each code C in R, there exists $v \in R$ such that $Ann_r(C) = vR = Rv$, which implies $Cv = 0$. If $C = Ann_r(C)$, then $C = Rv$ and $v^2 = 0$. Therefore, C is nilpotent. □

References

[1] Noha Abdelghany and Nefertiti Megahed, *Code-checkable group rings*, J. Algebra Comb. Discrete Struct. Appl. **4** (2017), no. 2, 115–122, DOI 10.13069/jacodesmath.284939. MR3601343

[2] S.D. Berman, On the theory of group codes, *Kibernetika*,**1** (1967), 31-39.

[3] Martino Borello, Javier De La Cruz, and Wolfgang Willems, *On checkable codes in group algebras*, J. Algebra Appl. **21** (2022), no. 6, Paper No. 2250125, 11, DOI 10.1142/S0219498822501250. MR4429010

[4] Javier de la Cruz and Wolfgang Willems, *On group codes with complementary duals*, Des. Codes Cryptogr. **86** (2018), no. 9, 2065–2073, DOI 10.1007/s10623-017-0437-2. MR3816214

[5] Marcus Greferath, Alexandr Nechaev, and Robert Wisbauer, *Finite quasi-Frobenius modules and linear codes*, J. Algebra Appl. **3** (2004), no. 3, 247–272, DOI 10.1142/S0219498804000873. MR2096449

[6] Thomas Honold, *Characterization of finite Frobenius rings*, Arch. Math. (Basel) **76** (2001), no. 6, 406–415, DOI 10.1007/PL00000451. MR1831096

[7] P. Hurley and T. Hurley, Module codes in group rings. *Proc. Int. Symp. Information Theory (ISIT)*, 2007, pp. 1981-1985.

[8] S. Jitman; S. Ling; H. Liu; X. Xie. Checkable Codes from Group Rings, `arXiv:1012.5498`, 2010.

[9] Michael Klemm, *Selbstduale Codes über dem Ring der ganzen Zahlen modulo* 4 (German), Arch. Math. (Basel) **53** (1989), no. 2, 201–207, DOI 10.1007/BF01198572. MR1004279

[10] T. Y. Lam, *Lectures on modules and rings*, Graduate Texts in Mathematics, vol. 189, Springer-Verlag, New York, 1999, DOI 10.1007/978-1-4612-0525-8. MR1653294

[11] Jessie MacWilliams, *Error-correcting codes for multiple-level transmission*, Bell System Tech. J. **40** (1961), 281–308, DOI 10.1002/j.1538-7305.1961.tb03986.x. MR141541

[12] Florence Jessie MacWilliams, *Combinatorial Problems of Elementary Abelian Groups*, ProQuest LLC, Ann Arbor, MI, 1962. Thesis (Ph.D.)–Radcliffe College. MR2939359

[13] F. J. MacWilliams, *Binary codes which are ideals in the group algebra of an Abelian group*, Bell System Tech. J. **49** (1970), 987–1011, DOI 10.1002/j.1538-7305.1970.tb01812.x. MR265054

[14] E. Prange, Cyclic error-correcting codes in two symbols, AFCRC-TN-57-103, USAF, Cambridge Research Laboratories, New York (1957)

[15] Jay A. Wood, *Code equivalence characterizes finite Frobenius rings*, Proc. Amer. Math. Soc. **136** (2008), no. 2, 699–706, DOI 10.1090/S0002-9939-07-09164-2. MR2358511

Instituto Federal de Educação, Ciência e Tecnologia de São Paulo, CEP 11533-160, Brazil.

Email address: robson.ricardo@ifsp.edu.br

DEMAT, Universidade Federal Rural do Rio de Janeiro, Soropedica, RJ, Brazil, CEP Seropédica - RJ, 23890-000

Email address: almp1980@ufrrj.br

Instituto de Matemática e Estatística, Universidade de São Paulo, Caixa Postal 66281, CEP-05315-970, Brazil.

Email address: polcinomilies@gmail.com

Contemporary Mathematics
Volume **785**, 2023
https://doi.org/10.1090/conm/785/15771

Good classical and quantum codes from multi-twisted codes

Nuh Aydin, Thomas Guidotti, and Peihan Liu

Dedicated to Professor Syed Tariq Rizvi for his contribution to Ring and Module Theory

Abstract. Multi-twisted (MT) codes were introduced as a generalization of quasi-twisted (QT) codes. QT codes have been known to contain many good codes. In this work, we show that codes with good parameters and desirable properties can be obtained from MT codes. These include best known and optimal classical codes with additional properties such as reversibility and self-duality, and new and best known non-binary quantum codes obtained from special cases of MT codes. Often times best known quantum codes in the literature are obtained indirectly by considering extension rings. Our constructions have the advantage that we obtain these codes by more direct and simpler methods. Additionally, we found theoretical results about binomials over finite fields that are useful in our search.

1. Introduction and Motivation

A liner code C of length n over $\mathbb{F}_q$, the finite field of order q (also denoted by $GF(q)$), is a vector subspace of $\mathbb{F}_q^n$. The three important parameters of a linear code are: the length (n), the dimension (k), and the minimum distance (d). If a code C over $\mathbb{F}_q$ is of length n, dimension k, and minimum distance d, then we refer to it as an $[n,k,d]_q$-code. One of the most important and challenging problems in coding theory is to determine the optimal parameters of a linear code and to explicitly construct codes that attain them. For example, given the alphabet size q, length n, and dimension k, researchers try to determine the highest value of the minimum distance d, and also to give an explicit construction of such a code. Despite much work on this question, it still has many open cases. There are databases that contain information about best known linear codes (BKLC) and their constructions. Two of the well known databases are the maintained by M. Grassl ([**21**], codetables.de) and the database contained in the Magma software [**25**].

Given q, n and k, there are theoretical upper bounds on d that are given in the tables mentioned above. However, it is not guaranteed and it is not always possible that they can actually be attained. This is a challenging problem for two main reasons: firstly, computing the minimum distance of a linear code is an

2020 *Mathematics Subject Classification.* Primary 94B05, 94B60.

Key words and phrases. Multi-twisted Codes, best known linear codes, quantum codes, reversible codes, self-dual codes.

The authors were supported in part by a Kenyon Summer Science Scholars program.

NP-hard problem [**32**] and secondly, the number $\dfrac{(q^n-1)(q^n-q)\cdots(q^n-q^{k-1})}{(q^k-1)(q^k-q)\cdots(q^k-q^{k-1})}$ of linear codes over $\mathbb{F}_q$ of length n and dimension k grows fast when n, k and q increase. Therefore, it is impossible to do an exhaustive search except for very small parameters. Hence, researchers focus on sub-classes of linear codes with special algebraic properties. One very promising class has been the class of quasi-twisted (QT) codes which contains the important classes such as cyclic codes, constacyclic codes, and quasi-cyclic (QC) codes as special cases. Hundreds of BKLCs have been obtained, usually by computer searches, from the class of QT codes.

More recently, a generalization of QT codes called multi-twisted (MT) codes was introduced in [**4**]. Therefore, MT codes are a generalization of cyclic, negacyclic, constacyclic (CC), quasi-cyclic (QC), generalized quasi-cyclic (GQC) and quasi-twisted codes (QT). Compared to QT codes, the search space for MT codes is much larger. While this increases the chances of finding new codes, it also makes the search process computationally more expensive. In this work, we focus on a special type of MT codes whose generator matrix is of the form $G = [I_k \mid A]$ where I_k is the $k \times k$ identity matrix and A is a $k \times (n-k)$ circulant matrix. The idea of investigating these codes was first introduced in [**10**]. In addition to trying to find new linear codes from this type of MT codes, we also worked on obtaining codes that are reversible, self-dual, or self-orthogonal.

Finally, we also worked on obtaining quantum codes from classical codes, specifically from constacyclic, QC, QT, and MT codes. The idea of quantum codes was first introduced in [**28**]. Later, methods of constructing quantum codes from classical codes were introduced in [**15**] (known as the CSS construction) and [**30**] (the Steane construction). Since then, researchers have been working on various ways of obtaining quantum codes from classical codes. More recently, codes over extension rings have attracted a lot of attention. Many quantum codes in the literature have been obtained via the method of considering codes over an extension ring S of a base/ground ring (or field) R and using a map to eventually obtain codes over the ground ring R. We showed that in many cases, there is no need to obtain these codes in such an indirect way. We have obtained many of these best known codes presented in the literature, and even better ones, by using a direct approach. It is preferable to obtain codes in more direct ways with convenient algebraic structures.

2. Preliminaries

DEFINITION 2.1. [**4**] For each $i = 1, \dots, \ell$, let m_i be a positive integer and $a_i \in \mathbb{F}_q^* = \mathbb{F}_q \setminus \{0\}$. A multi-twisted (MT) module V is an $\mathbb{F}_q[x]$-module of the form

$$V = \prod_{i=1}^{\ell} \mathbb{F}_q[x]/\langle x^{m_i} - a_i \rangle.$$

A MT code is an $\mathbb{F}_q[x]$-submodule of a MT module V.

Equivalently, we can define a MT code in terms of the shift of a codeword. Namely, a linear code C is MT if for any codeword

$$\vec{c} = (c_{1,0}, \dots, c_{1,m_1-1}; c_{2,0}, \dots, c_{2,m_2-1}; \dots; c_{\ell,0}, \dots, c_{\ell,m_\ell-1}) \in C,$$

its multi-twisted shift

$$(a_1 c_{1,m_1-1}, c_{1,0}, \dots, c_{1,m_1-2}; a_2 c_{2,m_2-1}, c_{2,0}, \dots, c_{2,m_2-2}; \dots; a_\ell c_{\ell,m_\ell-1}, \dots, c_{\ell,m_\ell-2})$$

is also a codeword. If we identify a vector $\vec{c}$ with $C(x) = (c_1(x), c_2(x), \ldots, c_\ell(x))$ where $c_i(x) = c_{i,0} + c_{i,1}x + \cdots + c_{i,m_i-1}x^{m_i-1}$, then the MT shift corresponds to $xC(x) = (xc_1(x) \mod x^{m_1} - a_1, \ldots, xc_\ell(x) \mod x^{m_\ell} - a_\ell)$.

The following are some of the most important special cases of MT codes:

- $a_1 = a_2 = \cdots = a_\ell$ gives QT codes
- $a_1 = a_2 = \cdots = a_\ell = 1$ gives QC codes
- $\ell = 1, a_1 = 1$ gives cyclic codes
- $\ell = 1, a_1 = -1$ gives negacyclic codes
- $\ell = 1$ gives constacyclic codes

There has been much research on QT codes, a special case of MT codes. Hundreds of new codes have been obtained from QT codes by computer searches. An algorithm called ASR ([**3**]) has been particularly effective in this regard. The ASR algorithm is based on the following theorem.

THEOREM 2.2. [**3**] *Let C be a 1-generator QT code of length $n = m\ell$ over $\mathbb{F}_q$ with a generator of the form*

$$(g(x)f_1(x), g(x)f_2(x), ..., g(x)f_\ell(x))$$

where $x^m - a = g(x)h(x)$ and $\gcd(h(x), f_i(x)) = 1$ for all $i = 1, \ldots, \ell$. Then $dim(C) = m - deg(g(x))$, and $d(C) \geq \ell \cdot d$ where d is the minimum distance of the constacyclic code C_g generated by $g(x)$.

This algorithm has been refined and automatized in more recent works such as ([**5**], [**6**], [**7**], [**8**], [**9**]) and dozens of record breaking codes have been obtained over every finite field $\mathbb{F}_q$, for $q = 2, 3, 4, 5, 7, 8, 9$ through its implementation. Moreover, it has been further generalized in [**10**] and more new codes were discovered that would have been missed by its earlier versions.

We can consider MT analogues of the QT codes given in the previous theorem. The following theorem gives a basic fact about duals of such codes.

THEOREM 2.3. *Let $C = \langle (g_1(x), g_2(x), \ldots, g_\ell(x)) \rangle$ be an MT code such that the constacyclic codes generated by $g_i(x)$ have the same length n and the same dimension k for $i = 1, 2, \ldots, \ell$. Let D_i be the dual of $\langle g_i(x) \rangle$. Let $PD = D_1 \times D_2 \times \cdots \times D_\ell$. Then PD is contained in $C^\perp$.*

PROOF. Let G_i be a generator matrix of $C_i = \langle g_i(x) \rangle$, so a generator matrix G of C is given by

$$G = \begin{pmatrix} G_1 & G_2 & G_3 & \ldots & G_\ell \end{pmatrix}_{k \times \sum_{i=1}^{\ell} n_i}.$$

Let H_i be a generator matrix of $D_i = C_i^\perp$. Then a generator matrix GPD of PD is given by

$$GPD = \begin{pmatrix} H_1 & 0 & 0 & \ldots & 0 \\ 0 & H_2 & 0 & \ldots & 0 \\ 0 & 0 & H_3 & \ldots & 0 \\ \ddots & \ddots & \ddots & \ddots & \ddots \\ 0 & 0 & 0 & \ldots & H_\ell \end{pmatrix}_{(n-k)\ell \times \sum_{i=1}^{\ell} n_i}.$$

It is readily verified that $G \cdot GPD = 0$. Note that the dimension of $C^\perp$ is $\sum_{i=1}^{\ell} n_i - k$ and the dimension of PD is $\sum_{i=1}^{\ell}(n_i - k)$, so the dimension of PD is less than or equal to the dimension of $C^\perp$. Hence, $PD \subseteq C^\perp$. $\square$

3. A Special Type of MT Codes

We first introduced a special type of MT codes that we called ICY codes in [**10**]. The basic idea for this construction is inspired by the well known fact that any linear code is equivalent to a linear code whose generator matrix is of the form $G = [I_k \mid A]$ where I_k is the $k \times k$ identity matrix and A is a $k \times (n-k)$ matrix. A code with a generator matrix in this form is called a systematic code. In this search method, we begin with a $k \times k$ identity matrix on the left and append it with $k \times (n-k)$ circulant matrices on the right. The circulant matrix is constructed using a generator polynomial $g(x)|(x^{n-k} - a)$ of a constacyclic code (obtained using the partition program from [**9**]). For each polynomial $g(x)$, we consider a number of circulant matrices obtained from polynomials $g(x)\cdot f(x)$ where $\gcd(f(x), h(x)) = 1$, with the goal of finding a code with minimum distance equal to or greater than the BKLCs for parameters $[n, k]_q$. In [**10**] we were able to find many good codes over $GF(2)$ and $GF(5)$. In this paper we present a number of optimal codes over $GF(2)$ with parameters $[2\cdot k, k]$ obtained from an implementation of this method. A number of these codes we found are also self-dual, which makes them even more useful. Another property of a code that is useful in some applications, such as DNA codes is reversibility, is reversibility. A code is reversible if, whenever $(c_0, c_1, \ldots, c_{n-1})$ is a codeword, so is $(c_{n-1}, c_{n-2}, \ldots, c_0)$.

LEMMA 3.1. *If f(x) is a monomial and the block length of $\langle f(x)\rangle$ is the same as the block length of the identituy matrix, then the MT code C over GF(2) generated by $\langle 1, f(x)\rangle$ is always self-dual, reversible and of minimum distance 2. In particular, if the monomial is a constant, then C is a cyclic code.*

PROOF. Let $f(x) = x^i$ be a monomial and the block length of $\langle f(x)\rangle$ be the same as $k \times k$ identity matrix. Then the generator matrix of the MT code C over $GF(2)$ generated by $\langle 1, f(x)\rangle$ is given by:

$$G = \left(\begin{array}{cccc|ccccc} 1 & 0 & \ldots & 0 & 0 & \ldots & 1 & 0 & \ldots \\ 0 & 1 & \ldots & 0 & 0 & \ldots & 0 & 1 & \ldots \\ \ddots & \ddots & \ddots & \ddots & \ddots & \ddots & \ddots & \ddots & \ddots \\ 0 & 0 & \ldots & 1 & \ldots & 1 & 0 & 0 & \ldots \end{array}\right)_{k\times 2k}.$$

It is easy to verify that the reverse of the jth row is the $(k-i+1-j)^th$ row, so it follows that C is reversible. Also note that each row of G is orthogonal to itself and also orthogonal to every other row, so given $k = 2k/2$ we know C is also self-dual. Moreover, since each row is of weight 2, and both matrices are circulant, it follows that C is of minimum distance 2.

Specifically, if $f(x) = 1$, then the generator matrix G is given by

$$G = \left(\begin{array}{cccc|ccccc} 1 & 0 & \ldots & 0 & 1 & \ldots & 0 & \ldots \\ 0 & 1 & \ldots & 0 & 0 & 1 & 0 & \ldots \\ \ddots & \ddots & \ddots & \ddots & \ddots & \ddots & \ddots & \ddots \\ 0 & 0 & \ldots & 1 & \ldots & 0 & \ldots & 1 \end{array}\right)_{k\times 2k}.$$

Hence, the code is a cyclic code of length $2k$ generated by
$f(x) = (x^k - 1)|(x^{2k} - 1)$. □

The following codes in Table 1 are obtained by the ICY method where we chose $n = 2k$, so each one of these codes has parameters $[2k, k, d]$. All of these codes are optimal with additional properties of being reversible and/or self orthogonal. For the circulant matrix on the right, we used a random $m \times m$ circulant matrix by choosing a random vector and taking its cyclic shifts. This polynomial is given in the table. We also checked the conditions for reversibility and self-duality. Note that self-orthogonality is the same as self-duality since we set $n = 2k$. In this search, the binary field fields turned out to be most promising.

In this and every subsequent table presented in this paper the terms of the generator polynomials are listed in ascending order from left to right. For instance, the generator polynomial [111101] of the fourth code in this table is the vector representation of $g(x) = 1 + x + x^2 + x^3 + x^5$.

Table 1: Optimal codes from the ICY method with several desirable properties. Here, * delineates a reversible code, and ° delineates a self-dual code.

$[n,k,d]_q$	Polynomials
$[8,4,4]_2$ *°	$g = [111]$
$[10,5,4]_2$	$g = [10101]$
$[12,6,4]_2$ *°	$g = [110111]$
$[14,7,4]_2$	$g = [111101]$
$[20,10,6]_2$	$g = [11111011]$
$[24,12,8]_2$ *°	$g = [101111011]$
$[28,14,8]_2$	$g = [11101111001001]$
$[32,16,8]_2$ *°	$g = [1100100001100011]$
$[36,18,8]_2$ *°	$g = [111010010111001]$
$[42,21,10]_2$	$g = [1111001101101110100 1]$
$[48,24,12]_2$ *°	$g = [1011110100101111001110 1]$
$[16,8,6]_3$ *	$g = [212222]$
$[18,9,6]_3$ *	$g = [201002222]$
$[20,10,7]_3$ *	$g = [112012102]$
$[30,15,9]_3$ *	$g = [201212222200121]$
$[14,7,6]_4$ *°	$g = [\alpha\alpha\alpha^2\alpha\alpha^2\alpha^2]$
$[20,10,8]_4$ *°	$g = [\alpha\alpha 1\alpha 00\alpha^2 1\alpha^2\alpha^2]$
$[24,12,9]_4$	$g = [10\alpha\alpha^2\alpha\alpha^2\alpha 10\alpha 1]$
$[14,7,6]_5$	$g = [3343442]$
$[16,8,7]_5$	$g = [34404411]$
$[12,6,6]_7$	$g = [551641]$
$[14,7,7]_7$	$g = [5226211]$
$[16,8,7]_7$	$g = [42325344]$
$[18,9,8]_7$	$g = [524635101]$

The next table presents a list of optimal self-dual codes together with their weight enumerators obtained by this method. For codes over $\mathbb{F}_4 = \{0, 1, \alpha, \alpha + 1\}$ in this paper, α is a root of $x^2 + x + 1$ over $\mathbb{F}_2$. Again, * delineates a reversible code. The table also presents the weight enumerator of each code and for each binary code specifies whether it is Type I or Type II.

Table 2: Optimal self-dual codes from the ICY method.

$[n,k,d]_q$	Type	weight enumerator	Polynomials
$[8,4,4]_2$ *	II	A^8+14A^4+1	$g=[111]$
$[12,6,4]_2$	I	$A^{12}+15A^8+32A^6+15A^4+1$	$g=[110111]$
$[16,8,4]_2$ *	II	$A^{16}+28A^{12}+198A^8+28A^4+1$	$g=[110111]$
$[16,8,4]_2$ *	I	$A^{16}+12A^{12}+64A^{10}+102A^8+64A^6+12A^4+1$	$g=[1010111]$
$[18,9,4]_2$ *	I	$A^{18}+9A^{14}+75A^{12}+171A^{10}+171A^8+75A^6+9A^4+1$	$g=[11011001]$
$[20,10,4]_2$ *	I	$A^{20}+45A^{16}+210A^{12}+512A^{10}+210A^8+45A^4+1$	$g=[1011111111]$
$[22,11,6]_2$ *	I	$A^{22}+77A^{16}+330A^{14}+616A^{12}+616A^{10}+330A^8+77A^6+1$	$g=[1001011]$
$[26,13,6]_2$	I	$A^{26}+52A^{20}+390A^{18}+1313A^{16}+2340A^{14}+2340A^{12}+1313A^{10}+390A^8+52A^6+1$	$g=[10111110111]$
$[52,26,10]_2$	I	$A^{52}+442A^{42}+6188A^{40}+53040A^{38}+308958A^{36}+1270360A^{34}+3754569A^{32}+8065616A^{30}+12707500A^{28}+14775516A^{26}+12707500A^{24}+8065616A^{22}+3754569A^{20}+1270360A^{18}+308958A^{16}+53040A^{14}+6188A^{12}+442A^{10}+1$	$g=[1001111100101010100100101]$
$[14,7,6]_4$ *	*Euclidean*	$318A^{14}+1302A^{13}+2940A^{12}+3990A^{11}+3738A^{10}+2226A^9+1155A^8+546A^7+168A^6+1$	$g=[\alpha 1\alpha\alpha^2\alpha\alpha\alpha^2]$
$[16,8,6]_4$ *	*Euclidean*	$681A^{16}+3696A^{15}+7896A^{14}+14112A^{13}+16464A^{12}+9408A^{11}+7392A^{10}+4704A^9+678A^8+336A^7+168A^6+1$	$g=[\alpha\alpha 1\alpha^2\alpha\alpha^2]$
$[20,10,8]_4$ *	*Euclidean*	$171A^{12}+432A^{11}+864A^{10}+1440A^9+459A^8+432A^7+288A^6+9A^4+1$	$g=[\alpha\alpha 1\alpha 00\alpha^2\alpha^2\alpha^2]$
$[22,11,8]_4$ *	*Euclidean*	$171A^{12}+432A^{11}+864A^{10}+1440A^9+459A^8+432A^7+288A^6+9A^4+1$	$g=[\alpha^2\alpha\alpha\alpha^2 010\alpha\alpha^2\alpha^2\alpha]$

4. Reversible Codes

Reversible codes are essential in application of coding theory to DNA computing [**27**]. Each single DNA strand is composed of a sequence of four bases nucleotides (adenine (A), guanine (G), thymine (T), cytosine (C)) and it is paired up with a complementary strand to form a double helix [**1**]. Finding reversible codes is an essential requirement in order to find codes suitable for DNA computing. Also, reversible codes may be used in certain data storage applications [**26**]. For example, if the code is reversible then one can read the stored data from either end of a block. In this paper, we present a general method to construct reversible ℓ-block MT, QT and QC codes, and a number of good reversible QC codes over $GF(2)$ are obtained using this method.

The codes presented in Table 3 are reversible and self-orthogonal QC with the same parameters as BKLCs over $GF(2)$. Hence, they are more desirable than BKLCs that do not have these properties. In checking the reversibility of these codes, we used the conditions for reversibility of QC codes given in [**31**]. Theorem 1 in section IV of [**31**] presents three conditions for reversibility of QC codes with generator matrices of the form $[g(x), g(x)\cdot f(x)]$, where $g(x)|(x^m-1)$ and $\gcd(f(x), h(x))) = 1$. After finding all nonequivalent generators $g(x)$ for a given block length m, we check the first two conditions on $g(x)$ and $h(x)$ before proceeding. If these two checks pass, then we generate random polynomials $f(x)$ that satisfy the third condition in the theorem, resulting in a reversible QC code. Additionally, since these QC codes are constructed over $GF(2)$ they have the added property of being self-orthogonal. Theorem 1 also provides conditions for self-orthogonality of QC codes with such generator matrices and in the binary case these conditions are the same as those for reversibility. For a polynomial $f(x)$, it is customary to denote its reciprocal by $f(x)^*$.

THEOREM 4.1. *Let* $g_i|(x^{n_i} - a_i)$ *and* $\gcd(h_i, f_i) = 1$ *for* $i = 1, 2, \ldots, \ell$, *where* $h_i = \frac{x^{n_i}-a_i}{g_i}$. *If* $\deg(g_i f_i) = \deg(g_{\ell+1-i} f_{\ell+1-i})$, $n_i = n_{\ell+1-i}$ *and* $g_i f_i = (g_{\ell+1-i} f_{\ell+1-i})^*$ *for* $i = 1, 2, .., \ell$, *then the MT code* C *generated by* $(g_1 f_1, \ldots, g_\ell f_\ell)$ *is reversible.*

PROOF. Let $g_i f_i = a_{i,0} + a_{i,1}x + \cdots + a_{i,k_i}x^{k_i}$, so a generator matrix of this MT code with ℓ blocks is given by

$$G = \begin{pmatrix} G_1 & G_2 & \ldots & G_{\ell-1} & G_\ell \end{pmatrix},$$

where G_i is

$$G_i = \begin{pmatrix} a_{i,0} & a_{i,1} & a_{i,2} & \cdots & a_{i,k_i} & 0 & 0 & \cdots \\ 0 & a_{i,0} & a_{i,1} & \cdots & a_{i,k-1} & a_{i,k_i} & 0 & \cdots \\ \vdots & \ddots & \ddots & \ddots & \ddots & \ddots & \ddots & \ddots \\ 0 & 0 & \cdots & a_{i,0} & a_{i,1} & a_{i,2} & \cdots & a_{i,k_i} \end{pmatrix}_{(n_i-k_i)\times n_i}.$$

Hence, the reversed first row of G is given by

$$\begin{aligned} &(0, \ldots, 0, a_{\ell,k_\ell}, a_{\ell,k_\ell-1}, \ldots, a_{\ell,1}, a_{\ell,0};\\ &0, \ldots, 0, a_{\ell-1,k_{\ell-1}}, a_{\ell-1,k_{\ell-1}-1}, \ldots, a_{\ell-1,1}, a_{\ell-1,0};\\ &\vdots\\ &0, \ldots, 0, a_{1,k_1}, a_{1,k_1-1}, \ldots, a_{1,1}, a_{1,0}). \end{aligned}$$

Note that $g_i f_i = (g_{\ell+1-i} f_{\ell+1-i})^*$ for $i = 1, 2, .., \ell$, so $a_{i,j} = a_{\ell+1-i,k-j+1}$ for $i - 1, 2, .., \ell$ and $j = 1, 2, \ldots, k$. Also, since the degree of all $\deg(g_i f_i) = \deg(g_{\ell+1-i} f_{\ell+1-i})$, it is not hard to see that the reversed first row is the $(n-k)^{th}$ row, and in general the reversed j^{th} row is the same as $(n-k+1-j)^{th}$ row. Therefore, the reverse of each row in G is in G, so C is reversible. □

As mentioned previously all of these codes are constructed using the conditions for a 2-QC code to be reversible found in [**31**]. This paper gives conditions on the generator polynomials $g(x)$ and $f(x) \cdot g(x) \mod (x^m - 1)$ with $\gcd(h(x), f(x)) = 1$ so the first polynomial presented in this table is $g(x)$, and the second one is $f(x) \cdot g(x) \mod (x^m - 1)$.

Table 3: Reversible and self-orthogonal binary QC codes with best known parameters

$[n, k, d]$	Polynomials
$[60, 28, 12]$	[00101100111001101100101001101], [111]
$[60, 29, 12]$	[000111000101011000100100011111], [11]
$[72, 29, 16]$	[011111101110101111101101100111101011], [10111101]
$[72, 30, 16]$	[0110111100000111111100010010111001], [1101011]
$[80, 37, 16]$	[1010011110001011110000000111110001110001], [1111]
$[80, 38, 16]$	[001111111010101010010010110111010101011], [101]
$[84, 24, 24]$	[111100010111100111000111101110000000100101], [1010100000000010101]
$[84, 25, 24]$	[001111100001011100110011110100001101000111], [110011111111110011]
$[84, 26, 24]$	[001101010011000011101100000100001110001101], [11100000000000111]
$[84, 27, 24]$	[001001101000110101101001110110000011000001], [1011111111111101]
$[90, 28, 24]$	[011111010111100100001011100011000011110000001], [11011101111011011]
$[90, 34, 20]$	[000101110111111101100000000110101111111101101], [110001100011]
$[96, 36, 20]$	[100101001000000011000000011101110010011001110 01], [1010001000101]
$[96, 37, 20]$	[011001100110101010110111100100111011100100100 11], [100110011001]
$[96, 38, 20]$	[110011110011101110010010111100100111011010101], [11101110111]
$[100, 41, 20]$	[0011010001101110011000101010010010101000011100 1001], [1111111111]
$[102, 34, 24]$	[100111110000111000100100011000011000011011 10101], [10111011111011101]
$[102, 42, 20]$	[1010110010011100000100111101101100110001000000101], [1001111001]
$[104, 36, 24]$	[10101110001011100010010100101101110010011011101 10101], [11110000000001111]
$[108, 46, 20]$	[1100000010100000011011110001010100100110111010000 01], [111111111]
$[110, 40, 24]$	[11011111111110100001111111011001100001000101111100 0111], [1111100000011111]

5. Quantum Codes

In comparison to classical information theory, the field of quantum information theory is relatively young. The idea of quantum error correcting codes was first introduced in [**29**] and [**14**]. A method of constructing quantum error correcting codes (QECC) was given in [**15**]. Since then researchers have investigated various methods of using classical error correcting codes to construct new QECCs. The majority of the methods have been based on the CSS construction given in [**15**]. In this method, self-dual, self-orthogonal and dual-containing linear codes are used to construct quantum codes. The CSS construction requires two linear codes C_1 and C_2 such that $C_2^{\perp} \subseteq C_1$. Hence, if C_1 is a self-dual code, then we can construct a

CSS quantum code using C_1 alone since $C_1^{\perp} \subseteq C_1$. If C_1 is self-orthogonal, then we can construct a CSS quantum code with $C_1^{\perp}$ and C_1 since $C_1 \subseteq C_1^{\perp}$. Similarly in the case C_1 is a dual-containing code. The last few tables in this section present a number of good QECCs obtained from both the CSS method and the method of additive codes. Many of these codes have the same parameters as those found in the literature, however we obtained them using more direct and simpler constructions. Additionally, we have found other codes whose parameters do not appear in the literature.

The following codes are constructed using the method of quantum codes constructed from additive codes [**15**]. After generating all divisors $g(x)$ of $x^m - a$ that generate non-equivalent constacyclic codes and finding a generator matrix of the form $[C_{g(x)\cdot f_1(x)}, C_{g(x)\cdot f_2(x)}]$ where $\gcd(f_i(x), h(x))) = 1$ and $C_g(x)$ represents the circulant matrix from $g(x)$, we construct an additive code from this generator matrix and test it for symplectic self orthogonality. If the code passes this check, then we construct a quantum code from this additive code and compute its minimum distance. Then we can compare the minimum distances of these codes against the comparable best known QECCs given in the database codetables.de.

In Table 4, all of the codes are optimal quantum QT codes constructed over $GF(2^2)$. Note that α is a root of the irreducible polynomial x^2+x+1 over $GF(2)$.

Table 4 : Optimal quantum codes over GF(2^2) from QT Codes

$[n,k,d]$	Polynomials
$[[54,47,2]]$	$[00\alpha^2 1110\alpha 000\alpha^2 1110\alpha 000\alpha^2 1110\alpha]$, $[\alpha^2\alpha\alpha^2 01\alpha^2\alpha 0\alpha\alpha^2\alpha\alpha^2 01\alpha^2\alpha 0\alpha\alpha^2\alpha\alpha^2 01\alpha^2\alpha 0\alpha]$
$[[56,50,2]]$	$[10000\alpha^2\alpha\alpha 1\alpha^2 1\alpha^2\alpha 1\alpha^2\alpha\alpha\alpha 100\alpha^2 1\alpha^2 10\alpha^2]$, $[\alpha^2 10011\alpha\alpha\alpha^2 1\alpha 0\alpha^2\alpha^2 0\alpha\alpha^2\alpha^2\alpha\alpha 110\alpha 1\alpha^2]$
$[[70,64,2]]$	$[10\alpha^2\alpha 010\alpha 0\alpha\alpha 1\alpha\alpha\alpha 001\alpha\alpha\alpha^2 10\alpha 0\alpha 1\alpha^2 00\alpha^2 110\alpha]$, $[\alpha^2 01\alpha^2\alpha\alpha\alpha^2\alpha\alpha^2 101\alpha^2\alpha^2 01\alpha^2\alpha^2 0\alpha^2\alpha\alpha 1011\alpha 0\alpha^2\alpha^2\alpha^2 1\alpha 0\alpha]$
$[[88,80,2]]$	$[100\alpha^2 1\alpha\alpha 101\alpha^2\alpha\alpha\alpha^2 1\alpha 01\alpha^2\alpha^2\alpha^2 0011\alpha 0\alpha^2\alpha^2 010\alpha\alpha^2\alpha^2\alpha 0\alpha^2 10\alpha\alpha\alpha 1]$, $[0\alpha^2 11\alpha 00\alpha^2 000100\alpha\alpha^2\alpha^2 101\alpha^2 1\alpha 1\alpha^2\alpha^2 0\alpha\alpha 1\alpha\alpha\alpha^2\alpha\alpha 011\alpha^2\alpha\alpha^2 1\alpha^2]$
$[[98,91,2]]$	$[11\alpha^2 101011\alpha^2 101011\alpha^2 101011\alpha^2 101011\alpha^2 101011\alpha^2 101011\alpha^2 101]$, $[\alpha\alpha 000\alpha^2 0\alpha\alpha 000\alpha^2 0\alpha\alpha 000\alpha^2 0\alpha\alpha 000\alpha^2 0\alpha\alpha 000\alpha^2 0\alpha\alpha 000\alpha^2 0\alpha\alpha 000\alpha^2]$

We also found some QECCs whose parameters do not appear in the literature. Hence, we consider them new quantum codes. The inspiration for this idea came from [**17**]. These codes are obtained using direct sums of two cyclic codes, with generators $g_1(x), g_2(x)$ that are divisors of $x^m - 1$. Each $g_i(x)$ generates a dual containing cyclic code $C_i = \langle g_i(x)\rangle$ which implies that their direct sum $C_1 \times C_2$ is also dual containing. So, the CSS method can be used with the direct sum code and its dual. The resultant QECC will have parameters $[[2\cdot m, 2\cdot k - 2\cdot m, d]]_q$. Here, k is defined to be the dimension of the direct sum code which is equal to $k_1 + k_2$, the sum of the dimensions of the individual cyclic codes, and d is the minimum of the minimum distances of the cyclic codes generated by $g_1(x)$ and $g_2(x)$. Table 5 contains these new codes.

Table 5 : New QECCs

$[[n,k,d]]_q$	Polynomials
$[[60,54,2]]_3$	[21], [201]
$[[72,68,2]]_3$	[21], [21]
$[[84,80,2]]_3$	[21], [21]
$[[100,96,2]]_5$	[41], [41]

Next, we have a number of codes that have the same parameters as the best known QECCs presented in the literature. Our codes have the advantage that they are obtained via a more direct and simpler construction. They were all obtained using the same method as the codes in the previous table. Codes with the same parameters listed in Table 6 were found using more complex and indirect methods detailed in [**11**], [**2**], [**24**] , [**18**], [**19**], [**13**]. These constructions usually involve considering an extension ring S of a ground ring or field R, constructing a code over the extension ring, then coming back to R using some sort of Gray map. We argue that it is more desirable to construct codes with the same parameters using a more direct approach.

Table 6 : QECC ties

$[[n,k,d]]_q$	Polynomials	Source
$[[72,66,2]]_3$	[21], [111]	[**24**]
$[[84,78,2]]_3$	[21], [201]	[**11**]
$[[40,36,2]_5$	[41], [41]	[**19**]
$[[40,34,2]]_5$	[41], [131]	[**11**]
$[[40,28,3]]_5$	[3011], [3011]	[**19**]
$[[60,54,2]]_5$	[41], [131]	[**24**]
$[[90,84,2]]_5$	[41], [131]	[**18**]
$[[100,94,2]]_5$	[41], [401]	[**11**]
$[[140,134,2]]_5$	[41], [401]	[**13**]

5.1. New Quantum Codes from Constacyclic Codes and a Result About Binomials. In this section, we obtain new QECCs that are constructed by the method of CSS construction from constacyclic codes over finite fields. These codes are new in the sense that either there do not exist codes with these parameters in the literature (to the best of our knowledge) or the minimum distances of our codes are higher than the codes that are presented in the literature. Our work on this also led to a couple of theoretical results about the binomials of the form $x^n - a$ over $\mathbb{F}_q$. The first one (Lemma 2 below) is similar to Theorem 4.1 in [**4**]. We start with the results on binomials.

We will be using the fact that the Frobenius map $x \to x^{p^m}$ on a finite field $\mathbb{F}_q$ of characteristic p is a permutation of $\mathbb{F}_q$ for any positive integer m. In particular, for any $\alpha \in \mathbb{F}_q$ and for any $m \in \mathbb{Z}^+$, there exists a unique $\beta \in \mathbb{F}_q$ such that $\alpha = \beta^{p^m}$.

LEMMA 5.1. *Let $n \in \mathbb{Z}^+$, q be a prime power and $a, b \in \mathbb{F}_q$ such that $a \neq b$. Then* $\gcd(x^n - a, x^n - b) = 1$.

PROOF. It suffices to show that the polynomials $x^n - a$ and $x^n - b$ have no common roots. Let δ be a root of $x^n - a$. Then,

$$\delta^n - b = a - b \neq 0$$

It follows that $\gcd(x^n - a, x^n - b) = 1$. □

THEOREM 5.2. *Let q be a power of prime p, and let $g(x) = x^n - a \in \mathbb{F}_q[x]$. Then all irreducible factors of $x^n - a$ have the same multiplicity $p^z \| n$, where z is the largest positive integer such that p^z divides n.*

PROOF. Let $n = p^z \cdot b$, where $\gcd(b, p) = 1$. Then $x^n - a$ can be written as

$$\begin{aligned} x^n - a &= x^{p^z \cdot b} - a \\ &= (x^b - a')^{p^z}, \end{aligned}$$

where $a' \in \mathbb{F}_q$. Let's call $f(x) = x^b - a'$, and then consider $\gcd(f(x), f'(x))$. Note that $b \neq 0$ and $x^b - a'$ is not a multiple of x, so it follows that $\gcd(f(x), f'(x)) = \gcd(x^b - a', bx^{b-1}) = 1$. Hence, all irreducible factors of $f(x) = x^b - a'$ have the same multiplicity 1. Therefore, all irreducible factors of $g(x) = x^n - a = f^{p^z}(x)$ have the same multiplicity p^z. □

By the definition of CSS construction, we need two codes such that one is contained in the dual of the other one. By the well known fact about inclusion of ideals, we know that

$$\langle g(x)f(x)\rangle \subseteq \langle g(x)\rangle$$

and we consider $C_2^\perp = \langle g(x)f(x)\rangle \subseteq \langle g(x)\rangle = C_1$. Here $g(x)$ is a divisor of a binomial $x^n - a$ and it generates a constacyclic code.

Since for a given $n \in \mathbb{Z}^+$ and distinct elements $\alpha, \beta \in \mathbb{F}_q$, $\gcd(x^n - \alpha, x^n - \beta) = 1$, it follows that for any divisor $g(x)|(x^n - \alpha)$, we have $g(x)f(x) \nmid (x^n - \beta)$ for any $f(x) \in \mathbb{F}_q[x]$. In other words, for a given $C_1 = \langle g(x)\rangle$, where $g(x)|(x^n - \alpha)$, all possible constacyclic codes of the same length $C_2^\perp$ such that $C_2^\perp \subseteq C_1$ have the form $C_2^\perp = \langle g(x)f(x)\rangle$, where $g(x)f(x)|(x^n - \alpha)$. Hence, for given $\alpha \in \mathbb{F}_q$ and n, we only need to consider $g_i(x)|(x^n - \alpha)$ and $g_i(x)f_i(x)|(x^n - \alpha)$. By Lemma 5.1, any generator of $C_2^\perp$ will be a divisor of $x^n - \alpha$, but not of $x^n - \beta$ for $\beta \neq \alpha$. Also, based on the multiplicity of all irreducible factors of $x^n - \alpha$, for any $g(x)$, we can determine all $g(x)f(x)$ by enumerating all possible combinations of multiplicities of irreducible factors. Therefore, our algorithm goes through every possible pair of $g(x)$ and $g(x)f(x)$, and we view each $\langle g(x)\rangle$ as C_1 and each $\langle g(x)f(x)\rangle$ as $C_2^\perp$ in the CSS construction. This way, we obtained many new QECCs. The new codes we obtained are of two types: i) those that have higher minimum distances than the codes given in the literature, ii) those whose parameters do not appear in the literature. We present them in the following two tables. Table 7 contains references to the literature where a comparable code is given.

Table 7 : New QECCs that have higher minimum distances than the codes given in the literature

KnownQECC Ref	New $[[n,k,d]]_q$	$g(x)$ & $g(x)f(x)$
$[[11,1,4]]_5$ **[23]**	$[[11,1,5]]_5$	[124114], [1122031]
$[[30,20,3]]_5$ **[24]**	$[[30,20,4]]_5$	[142241], [14401243413033031434210441]
$[[40,32,2]]_5$ **[2]**	$[[40,32,3]]_5$	[12342], [12030222342024440142411123141333 40313]
$[[93,75,2]]_5$ **[12]**	$[[93,75,4]]_5$	[1032243334],[121431401222402220013421431143342442312423012012414312311210123302234142312110011121 1]
$[[60,52,2]]_5$ **[24]**	$[[60,52,3]]_5$	[10434],[10132120224322013313343134444001412441320244031203201432 1]
$[[78,60,3]]_3$ **[12]**	$[[78,60,4]]_5$	[1120201112],[120220102221110112120120121022021112001221211021021000010001210102002 2]
$[[18,2,3]]_7$ **[22]**	$[[18,2,6]]_7$	[143561634], [14631236125]
$[[12,4,3]]_{13}$ **[20]**	$[[12,4,5]]_{13}$	[18554], [184835595]
$[[24,18,2]]_{17}$ **[20]**	$[[24,18,3]]_{17}$	[1332], [95115211194105321084116 31]
$[[21,7,2]]_{19}$ **[16]**	$[[21,7,3]]_{19}$	[19999998], [1131313131313131319999991]

The final table shows the set of codes we have found whose parameters do not appear in the literature.

Table 8 : New QECCs

$[[n,k,d]]_q$	Polynomials
$[[61,41,5]]_3$	[10011111001], [1100101022020121010222110210221110202120101102 020022]
$[[52,7,6]]_3$	[12211120111], [100000202100102111]
$[[52,35,4]]_3$	[110000012], [11020221221200111012111020100112002220020221]
$[[52,15,5]]_3$	[12010202], [1220020210111022112120 1]
$[[40,11,6]]_3$	[112100201012], [1111200022110011102000 1]
$[[40,22,5]]_3$	[1222100011], [1222011110221122122002112121111 1]
$[[30,2,8]]_3$	[101101000202101], [10211201020112201]
$[[32,2,7]]_3$	[1211021211201002], [122010112001001202]
$[[30,10,4]]_3$	[120212022101], [121111110200121120220 1]
$[[30,16,4]]_3$	[11122111], [110112012020020210211011]
$[[33,1,10]]_3$	[1200002101202102 1], [111000222111222222]
$[[35,4,6]]_3$	[12122221121022], [101020002212012112]
$[[36,1,8]]_3$	[1112002010200211 1], [100110211221022002]
$[[36,2,8]]_3$	[1221210010012122 1], [1200122110112210021]
$[[36,18,4]]_3$	[1110220111], [1102010202122112212020102011]
$[[20,4,5]]_3$	[112212211], [1122201022211]
$[[23,1,8]]_3$	[100202011222], [1202121101001]
$[[26,9,6]]_3$	[111122121], [112200100222020102]
$[[28,0,9]]_3$	[12222100011212 2], [12222100011212 2]
$[[28,12,6]]_3$	[111000121], [1211001110201210021 11]
$[[14,2,5]]_3$	[1101011], [111202111]
$[[11,1,5]]_5$	[124114], [1122031]

$[[12, 4, 4]]_5$	[14102], [140124402]
$[[16, 1, 6]]_5$	[1320214], [11112222]
$[[20, 4, 6]]_5$	[121201212], [1231323013112]
$[[30, 20, 4]]_5$	[142241], [1440124341303303143421 0441]
$[[10, 2, 4]]_7$	[12134], [1230661]
$[[14, 6, 4]]_7$	[12056], [12462526421]
$[[40, 28, 4]]_7$	[1455541], [1115223606663003146303556 2341504546]
$[[12, 2, 5]]_{13}$	[154816], [1582412411]

5.2. New Quantum Codes from Constacyclic Random Search. In this section, we introduce another method to obtain quantum codes from classical codes that we call constacyclic random (CR) search. We have obtained several new quantum codes (codes with higher minimum distances than the ones in the literature) from this search.

There is a small difference between the CR search and the earlier method of constacyclic quantum search. In the CR search, we first find all $g(x)|(x^n - a)$, then find a large number of random polynomials $f(x)$, and construct CSS codes with $\langle g(x) \rangle$ and $\langle g(x)f(x) \rangle$.

By this method, we obtained a new $[[11, 1, 5]]_5$-code which was also found by the earlier search method and listed in Table 7. Additionally, we found a $[[19, 1, 9]]_{17}$-code from the generators $g = x^9+4x^8+15x^7+15x^6+6x^5+x^4+12x^3+2x^2+3x+16$ and $gf = x^{10}+3x^9+11x^8+8x^6+12x^5+11x^4+7x^3+x^2+13x+1$, which improves the minimum distance of the known QECC $[[19, 1, 5]]_{17}$-code in [**23**] by 4 units. Finally, we have found a $[[30, 10, 3]]_{19}$-code from the generators $g = x^4+11x^3+14x^2+3x+18$ and $gf = x^{14} + 11x^{12} + 6x^{11} + 8x^{10} + x^9 + 4x^8 + 17x^7 + 4x^6 + x^5 + 13x^4 + 7x^3 + 17x^2 + 15x + 8$, which improves the minimum distance of the known QECC code with parameters $[[30, 10, 2]]_{19}$ given in [**16**] by 1 unit.

Our computational results suggest that this is a promising search method that might yield even more new codes.

Acknowledgments

This work was supported by Kenyon College Summer Science Scholars program.

References

[1] Taher Abualrub, Ali Ghrayeb, and Xiang Nian Zeng, *Construction of cyclic codes over* GF(4) *for DNA computing*, J. Franklin Inst. **343** (2006), no. 4-5, 448–457, DOI 10.1016/j.jfranklin.2006.02.009. MR2262559

[2] Mohammad Ashraf and Ghulam Mohammad, *Quantum codes from cyclic codes over* $F_q + uF_q + vF_q + uvF_q$, Quantum Inf. Process. **15** (2016), no. 10, 4089–4098, DOI 10.1007/s11128-016-1379-8. MR3547872

[3] Nuh Aydin, Irfan Siap, and Dijen K. Ray-Chaudhuri, *The structure of 1-generator quasi-twisted codes and new linear codes*, Des. Codes Cryptogr. **24** (2001), no. 3, 313–326, DOI 10.1023/A:1011283523000. MR1857145

[4] Nuh Aydin and Ajdin Halilović, *A generalization of quasi-twisted codes: multi-twisted codes*, Finite Fields Appl. **45** (2017), 96–106, DOI 10.1016/j.ffa.2016.12.002. MR3631356

[5] Nuh Aydin, Nicholas Connolly, and John Murphree, *New binary linear codes from quasi-cyclic codes and an augmentation algorithm*, Appl. Algebra Engrg. Comm. Comput. **28** (2017), no. 4, 339–350, DOI 10.1007/s00200-017-0327-x. MR3679335

[6] Nuh Aydin, Nicholas Connolly, and Markus Grassl, *Some results on the structure of constacyclic codes and new linear codes over* $GF(7)$ *from quasi-twisted codes*, Adv. Math. Commun. **11** (2017), no. 1, 245–258, DOI 10.3934/amc.2017016. MR3613297

[7] Nuh Aydin and Derek Foret, *New linear codes over $GF(3)$, $GF(11)$, and $GF(13)$*, J. Algebra Comb. Discrete Struct. Appl. **6** (2019), no. 1, 13–20, DOI 10.13069/jacodesmath.508968. MR3904733

[8] Nuh Aydin, Ghada Bakbouk, and Jonathan G. G. Lambrinos, *New linear codes over non-prime fields*, Cryptogr. Commun. **11** (2019), no. 5, 891–902, DOI 10.1007/s12095-018-0333-2. MR3990970

[9] Nuh Aydin, Jonathan Lambrinos, and Oliver VandenBerg, *On equivalence of cyclic codes, generalization of a quasi-twisted search algorithm, and new linear codes*, Des. Codes Cryptogr. **87** (2019), no. 10, 2199–2212, DOI 10.1007/s10623-019-00613-0. MR4001788

[10] Nuh Aydin, Thomas H. Guidotti, Peihan Liu, Armiya S. Shaikh, and Robert O. VandenBerg, *Some generalizations of the ASR search algorithm for quasitwisted codes*, Involve **13** (2020), no. 1, 137–148, DOI 10.2140/involve.2020.13.137. MR4059947

[11] T. Bag, H. Q. Dinh, A. K. Upadhyay, W. Yamaka, *New non-binary quantum codes from cyclic codes over product rings*, IEEE Commun. Lett. **24** (2020), 486–490.

[12] Tushar Bag, Ashish K. Upadhyay, Mohammad Ashraf, and Ghulam Mohammad, *Quantum codes from cyclic codes over the ring $F_p[u]/\langle u^3 - u\rangle$*, Asian-Eur. J. Math. **12** (2019), no. 7, 2050008, 10, DOI 10.1142/s1793557120500084. MR4032245

[13] J. Borges, C. Fernández-Córdoba, J. Pujol, J. Rifà, and M. Villanueva, *$\mathbb{Z}_2\mathbb{Z}_4$-linear codes: generator matrices and duality*, Des. Codes Cryptogr. **54** (2010), no. 2, 167–179, DOI 10.1007/s10623-009-9316-9. MR2576874

[14] A. R. Calderbank, P. W. Shor, *Good quantum error-correcting codes exist*, Phys. Rev. Lett. **54** (1996), 1098–1106.

[15] A. Robert Calderbank, Eric M. Rains, P. W. Shor, and Neil J. A. Sloane, *Quantum error correction via codes over* GF(4), IEEE Trans. Inform. Theory **44** (1998), no. 4, 1369–1387, DOI 10.1109/18.681315. MR1665774

[16] Y. Cengellenmis, A. Dertli, *Quantum codes over F_q and quantum quasi-cyclic codes over F_p*. MSAEN. **7** (2019), 87–93.

[17] H. Q. Dinh, T. Bag, S. Pathak, A. K. Upadhyay, W. Chinnakum, *Quantum codes obtained from constacyclic codes over a family of finite rings $\mathbb{F}_p[u_1, u_2, \ldots, u_s]$*, IEEE Access **8** (2020), 194082–194091.

[18] Hai Q. Dinh, Tushar Bag, Ashish K. Upadhyay, Mohammad Ashraf, Ghulam Mohammad, and Warattaya Chinnakum, *Quantum codes from a class of constacyclic codes over finite commutative rings*, J. Algebra Appl. **19** (2020), no. 12, 2150003, 19, DOI 10.1142/S0219498821500031. MR4160995

[19] H.Q. Dinh, T. Bag, A. K. Upadhyay, R. Bandi, W. Chinnakum, *On the structure of cyclic codes over $\mathbb{F}_qRS$ and applications in quantum and LCD codes constructions*, IEEE Access. **8** (2020), 18902–18914.

[20] Yun Gao, Jian Gao, and Fang-Wei Fu, *Quantum codes from cyclic codes over the ring $\mathbb{F}_q + v_1\mathbb{F}_q + \cdots + v_r\mathbb{F}_q$*, Appl. Algebra Engrg. Comm. Comput. **30** (2019), no. 2, 161–174, DOI 10.1007/s00200-018-0366-y. MR3914594

[21] M. Grassl, *Code Tables: Bounds on the parameters of of codes*, online, available at, `http://www.codetables.de/`

[22] M. E. Koroglu and I. Siap, *Quantum codes from a class of constacyclic codes over group algebras*, Malays. J. Math. Sci. **11** (2017), no. 2, 289–301. MR3669560

[23] G. G. La Guardia, *Quantum codes derived from cyclic codes*, Int. J. Theor. Phys. **56** (2017), 2479–2484.

[24] Fanghui Ma, Jian Gao, and Fang-Wei Fu, *New non-binary quantum codes from constacyclic codes over $\mathbb{F}_q[u, v]/\langle u^2 - 1, v^2 - v, uv - vu\rangle$*, Adv. Math. Commun. **13** (2019), no. 3, 421–434, DOI 10.3934/amc.2019027. MR3959338

[25] Magma computer algebra system, online, available at, `http://magma.maths.usyd.edu.au/`

[26] James L. Massey, *Reversible codes*, Information and Control **7** (1964), 369–380. MR168397

[27] Elif Segah Oztas, Bahattin Yildiz, and Irfan Siap, *A novel approach for constructing reversible codes and applications to DNA codes over the ring $\mathbb{F}_2[u]/(u^{2k} - 1)$*, Finite Fields Appl. **46** (2017), 217–234, DOI 10.1016/j.ffa.2017.04.001. MR3655756

[28] P. W. Shor, *Scheme for reducing decoherence in quantum memory*, Phys. Rev. A, **52** (1995), 2493.

[29] A. M. Steane, *Error correcting codes in quantum theory*, Phys. Rev. Lett. **77** (1996), no. 5, 793–797, DOI 10.1103/PhysRevLett.77.793. MR1398854

[30] Andrew M. Steane, *Enlargement of Calderbank-Shor-Steane quantum codes*, IEEE Trans. Inform. Theory **45** (1999), no. 7, 2492–2495, DOI 10.1109/18.796388. MR1725135

[31] R. Takieldin, H. Matsui, *On reversibility and self-duality for some classes of quasi-cyclic codes*, IEEE Access. **8** (2020), 143285–143293.

[32] Alexander Vardy, *The intractability of computing the minimum distance of a code*, IEEE Trans. Inform. Theory **43** (1997), no. 6, 1757–1766, DOI 10.1109/18.641542. MR1481035

Department of Mathematics and Statistics, Kenyon College, Gambier, Ohio, 43022
Email address: aydinn@kenyon.edu

Department of Mathematics and Statistics, Kenyon College, Gambier, Ohio, 43022
Email address: guidotti1@kenyon.edu

Department of Mathematics and Statistics, Kenyon College, Gambier, Ohio, 43022
Current address: Department of Mathematics, University of Michigan, Ann Arbor, Michigan, 48104
Email address: paulliu@umich.edu

Contemporary Mathematics
Volume **785**, 2023
https://doi.org/10.1090/conm/785/15772

Evaluation of iterated Ore polynomials and skew Reed-Muller codes

Nabil Bennenni and André Leroy

Dedicated to Professor Syed Tariq Rizvi for his contribution to Ring and Module Theory.

ABSTRACT. In this paper, we study two ways of evaluating iterated Ore polynomials. We provide many examples and compare these evaluations. We use the evaluation maps to construct Reed-Muller codes and compute explicitly some of the data that are associated to such codes.

Introduction

The evaluation of polynomials is a keystone in many branches of mathematics. In noncommutative algebra one of the most attractive notion of polynomials is the Ore polynomials (a.k.a. skew polynomials). There are plenty of works dealing with the evaluations of these polynomials with coefficients in division rings or prime rings [**7**],[**8**]. In coding theory the evaluation of polynomials is used, in particular, for constructing Reed-Muller codes [**10**], [**5**], [**4**].

Let us recall the general definition of the skew polynomial ring. Let K be a ring, $\sigma \in End(K)$, and δ a skew σ-derivation on K (i.e. δ is an additive map from K to K such that for any $a, b \in K$, $\delta(ab) = \sigma(a)\delta(b)+\delta(a)b$). The elements of a skew polynomial ring $R = K[t;\sigma,\delta]$ are polynomials $\sum_{i=0}^{n} a_i t^i$ with coefficients $a_i \in K$ written on the left. The addition of these polynomials is the usual one and the multiplication is based on the commutation rule $ta = \sigma(a)t+\delta(a)$, for any $a \in K$. With these two operations, R becomes a ring. When K is a division ring, R is a left principal ideal domain. If σ is an automorphism of the division ring K then R is also a right principal ideal domain. This construction can be iterated: we consider a ring K, a sequence of skew polynomial rings $R_1 = K[t_1;\sigma_1,\delta_1]$, where $\sigma_1 \in End(K)$ and δ_1 is a σ_1-derivation of K. We then build $R_2 = R_1[t_1;\sigma_2,\delta_2] = K[t_1;\sigma_1,\delta_1][t_2;\sigma_2,\delta_2]$, where $\sigma_2 \in End(R_1)$ and δ_2 is a σ_2-derivation of R_1. Continuing this construction we arrived at the iterated polynomial ring in n noncommutative variables $R = R_n = K[t_1;\sigma_1,\delta_1][t_2;\sigma_2,\delta_2]\cdots[t_n;\sigma_n,\delta_n]$. The evaluation of skew polynomials has a very different behavior than the usual one for commutative variables. Most of the literature is related to right evaluation of univariate (i.e. $n = 1$) polynomials with coefficients in a division ring (e.g. [**8**], [**7**]). For the case when the base ring K is not necessarily a division ring, the pseudo-linear transformations are very useful [**7**]. When K is a division ring the left evaluation was

2020 *Mathematics Subject Classification.* Primary 94B05, 11T71, 14G50; Secondary 16L30.
Key words and phrases. Evaluation of iterated Ore polynomials, skew Reed-Muller codes.

also considered even when σ is not onto [**9**]. While considering the evaluation of a polynomial $f(t_1,\dots,t_n) \in R = K[t_1;\sigma_1,\delta_1][t_2;\sigma_2,\delta_2]\cdots[t_n;\sigma_n,\delta_n]$ at a point $(a_1,\dots,a_n) \in K^n$, we are quickly facing a "noncommutative phenomenon": the ideal $I = R(t_1-a_1)+\cdots+R(t_2-a_2)$ can be equal to R. We will show how it is possible to avoid this problem.

Reed-Muller codes over $\mathbb{F}_2$ were introduced by Reed in 1954 [**10**]. Kasami, Lin and Peterson in [**6**] extended the definition to any finite field $\mathbb{F}_q$. Reed-Muller codes are very useful in cryptography. Indeed, they show good resistance to linear cryptanalysis. The interested reader may consult [**4**], for more details. Reed-Muller codes are amongst the simplest examples of the class of geometrical codes, which also includes Euclidean geometry and projective geometry codes.

The paper is organized as follows. Section 2 presents two ways of evaluating iterated polynomials in a general context. We provide many examples and compare these evaluations. In Section 3, we apply the evaluation maps to construct Reed-Muller codes and compute some of the data that are associated to such codes.

1. Polynomial maps in n variables

In this section, we define a notion of evaluation for polynomials in an iterated Ore extension $R = K[t_1;\sigma_1,\delta_1][t_2;\sigma_2,\delta_2]\cdots[t_n;\sigma_n,\delta_n]$ over an arbitrary ring K (with unity). Throughout the paper we require that $\sigma_i(K)\subset K$ and $\delta_i(K)\subset K$ for any $1\le i\le n$. We will need some facts about iterated polynomial rings.

REMARK 1.1. (1) When $\delta = 0$ we denote the Ore extension as $R = K[t;\sigma]$. Similarly when $\sigma = id$, we write $R = K[t;\delta]$.

(2) We remark that if $u\in K$ an invertible element and, for any $x\in K$, we have $\sigma(x) = I_u(x) = uxu^{-1}$ then $K[t;\sigma;\delta] = K[u^{-1}t;u^{-1}\delta]$. Notice that if $\delta = 0$, this gives a usual polynomial ring.

(3) If σ is an endomorphism of the ring K and x,a are elements in K, we define $\delta_a(x) = ax-\sigma(x)a$. This is a σ-derivation on K called the inner derivation on K. Let us notice that in this case we have, for any $x\in K$ that $(t-a)x = \sigma(x)(t-a)$ so that $R = K[t;\sigma,\delta] = K[t-a;\sigma]$.

(4) Combining the two previous remarks we have that, for any $u,v\in K$ with u an invertible element, $K[t;I_u,\delta_v] = K[u^{-1}(t-v)]$ is a usual polynomial ring.

Let us briefly recall the definition of the evaluation of a polynomial $f(t)\in R = K[t;\sigma,\delta]$ at an element $a\in K$. We define $f(a)\in K$ to be the only element of K such that $f(t)-f(a)\in R(t-a)$. This means, in particular, that $a\in K$ is a right root of $f(t)$ when $t-a$ is right factor of $f(t)$. We then introduce, for any $i\ge 0$, a map N_i defined by induction as follows:

$$N_0(a) = 1, \quad N_{i+1}(a) = \sigma(N_i(a))a+\delta(N_i(a)).$$

This leads to a concrete formula for the evaluation of any polynomial $f(t) = \sum_{i=0}^n b_it^i \in R = K[t;\sigma,\delta]$ at an element $a\in K$:

$$f(a) = \sum_{i=0}^n b_iN_i(a).$$

Before defining the evaluation of iterated polynomials, we now provide a few classical examples.

EXAMPLE 1.2.

(1) Consider the complex number $\mathbb{C}$ and the conjugation denoted by "$-$". We then construct the Ore extension: $R = \mathbb{C}[t; -]$; the commutation rule is here $t(a + ib) = (a - ib)t$, where $a, b \in \mathbb{R}$. An element $a \in \mathbb{C}$ is a (right) root of $t^2 + 1$ if $N_2(a) + 1 = 0$ i.e. $\overline{a}a + 1 = 0$. From this it is clear that $t^2 + 1$ is an irreducible polynomial in R. On the other hand, it is easy to check that $t^2 + 1$ is a central polynomial and we get that the quotient $R/(t^2 + 1)$ is isomorphic to the quaternion algebra $\mathbb{H}$. On the other hand, the roots of the polynomial $t^2 - 1$ are exactly the complex numbers of norm 1.

(2) Another important kind of Ore extension is obtained by presenting the Weyl algebra $A_1 = k[X][Y; id., \frac{d}{dx}]$, where k is a field. The commutation rule that comes up is thus $YX = XY + 1$. In characteristic zero this algebra is simple (it doesn't have any two sided ideal except 0 and A_1). If $char(k) = p > 0$, A_1 is a p^2 dimensional algebra over its center $k[X^p, Y^p]$. Evaluating powers of Y at X we get for instance $Y^2(X) = X^2 + 1$, $Y^3(X) = X^3 + 3X$, and $Y^4(X) = X^4 + 6X^2 + 3$.

    We can iterate the procedure and get the Weyl algebra $A_n(k)$.

(3) In coding theory, Ore extensions of the form $\mathbb{F}_q[t; \theta]$, where $q = p^n$ and θ is the Frobenius automorphism defined by $\theta(a) = a^p$, for $a \in \mathbb{F}_q$, have been extensively used (cf. [**2**]). In [**7**], both the evaluation and the factorization for these extensions were described in term of classical (untwisted) evaluation of polynomials. General Ore extensions have also been used for constructing codes [**3**].

Let $f(t_1, t_2, \ldots, t_n) \in R = K[t_1; \sigma_1, \delta_1][t_2; \sigma_2, \delta_2] \cdots [t_n; \sigma_n, \delta_n]$ and an n-tuple $(a_1, a_2, \ldots, a_n) \in K^n$ be given (K is a ring which is assumed to be stable by the maps σ_i and δ_i, with$1 \leq i \leq n$). We want to define the value of $f(a_1, a_2, \cdots, a_n)$. It will be convenient to have notations for intermediate Ore extensions. So we put $R_0 = K$, and, for $i \in \{1, \ldots, n-1\}$, $R_i = R_{i-1}[t_i; \sigma_i, \delta_i]$. Notice that $R_n = R$.

Let $(a_1, a_2, \cdots, a_n) \in K^n$ and consider $f_0 = f(t_1, \ldots, t_n) \in R = R_{n-1}[t_n; \sigma_n, \delta_n]$. We can define $f_1 = f(t_1, \ldots, t_{n-1}, a_n) \in R_{n-1}$ as the remainder of the division of f_0 on the right by $t_n - a_n$. This is

$$f_0 = q(t_1, \ldots, t_{n-1})(t_n - a_n) + f_1.$$

By the same procedure we divide $f_1 = f(t_1, \ldots, t_{n-1}, a_n)$ by $t_{n-1} - a_{n-1}$ and get the remainder $f_2 = f(t_1, \ldots, t_{n-2}, a_{n-1}, a_n) \in R_{n-2}$. This is

$$f_1 = f(t_1, \ldots, t_{n-1}, a_n) = q_2(t_1, t_2, \ldots, t_{n-1})(t_{n-1} - a_{n-1}) + f_2.$$

Continuing this process, we define $f_3 \in R_{n-3}, \cdots f_{n-1} \in R_1, f_n \in R_0 = K$. The evaluation of $f(t)$ at $(a_1, \ldots, a_n)$ is $f_n \in K$.

This leads to the following definition.

DEFINITION 1.3. For $f(t_1, \ldots, t_n) \in R = K[t_1; \sigma_1, \delta_1][t_2; \sigma_2, \delta_2] \cdots [t_n; \sigma_n, \delta_n]$ and $(a_1, \ldots, a_n) \in K^n$, we define $f(a_1, \ldots, a_n)$, the evaluation of $f(t_1, t_2, \ldots, t_n)$ at $(a_1, \ldots, a_n)$, as the representative in K of $f(t_1, t_2, \ldots, t_n)$ modulo $I_n(a_1, \ldots, a_n) = R_1(t_1 - a_1) + \cdots + R_{n-1}(t_{n-1} - a_{n-1}) + R(t_n - a_n)$, where, for $1 \leq i \leq n$, R_i stands for $R_i = K[t_1; \sigma_1, \delta_1] \cdots [t_i; \sigma_i, \delta_i]$.

Remark that $I_n = I_n(a_1, \ldots, a_n)$ is an additive subgroup of $R, +$ and is not, in general, an ideal in $R = R_n$.

Another way to compute the evaluation is to remark that the polynomials in $R = K[t_1;\sigma_1,\delta_1][t_2;\sigma_2,\delta_2]\cdots[t_n;\sigma_n,\delta_n]$ can be written in a unique way as sums of monomials of the form $\alpha_{l_1,\ldots,l_n}t_1^{l_1}t_2^{l_2}\cdots t_n^{l_n}$, for some $0 \le l_1, l_2, \ldots, l_n \le n$ and $\alpha_{l_1,\ldots,l_n} \in K$. The sum $\sum_{i=1}^n l_i$ is called the degree of the monomial and the degree of a polynomial is given by the degree of the monomials with higher degree. First remark that if a monomial $m = m(t_1,\ldots,t_n)$ is of degree l, then for any $a \in K^*$, ma is a polynomial of degree l as well. We define the evaluation by induction on the degree. In practice, it is sufficient to define the evaluation of a monomial.

For $(a_1, a_2, \ldots, a_n) \in K^n$ and $m = m(t_1,\ldots,t_n)$ we define $m(a_1,\ldots,a_n)$ as follows:

If $deg(m) = 1$ and $m = \alpha t_i$ for some $1 \le i \le n$ and $\alpha \in K$ then $m(a_1,\ldots,a_n) = a_i$. So, assume that the evaluation of the monomials of degree $< l$ has been defined and consider a monomial $m = m(t)$ such that $deg(m) = l$ and $m = m'(t_1,\ldots,t_j)t_j$ for some $1 \le j \le n$, then $m(a_1,\ldots,a_n) = m''(a_1,\ldots,a_j)$ where the polynomial $m''(t_1,\ldots,t_j) = m'(t_1,\ldots,t_j)a_j$ is of degree smaller than l.

Let us make some remarks about this evaluation.

REMARK 1.4.

(1) First let us remark that, before we evaluate a polynomial, we must write it with the variables appearing in the precise order $t_1, t_2, \ldots, t_n$ (from left to right). In other words, before evaluating a polynomial we must write it as a sum of monomials of the form $t_1^{l_1}t_2^{l_2}\cdots t_n^{l_n}$.

(2) Of course, this is not the only possible definition, but although it might look a bit strange, still it is natural if we want the zeros being right roots. Let us look more closely to the case of two variables. In other words it is natural for (a_1, a_2) to annihilate a polynomial of the form $g(t_1)(t_2 - a_2)$; but we don't necessarily expect (a_1, a_2) to be a zero of a polynomial of the form $(t_1 - a_1)h(t_1)$.

(3) Since the base ring is not assumed to be commutative we must be very cautious while evaluating a polynomial, even when the variables commute. With this definition, $t_1t_2 \in K[t_1, t_2]$ evaluated at (a, b) gives $(t_1t_2)(a, b) = ba$. This might look very strange but if we think of evaluation in terms of "operators" via the right multiplication by b followed by the right multiplication by a this evaluation looks perfectly fine and the apparent awkwardness disappears.

(4) We assumme that the different endomorphisms σ_is are such that $\sigma_i(K) \subseteq K$. In other words, for $i > 1$, σ_i is an extension of $\sigma_i|_K$ to R_{i-1}. Some commutation relations exist between the different endomorphisms σ_i. For instance, consider the polynomial ring extension $R = K[t_1;\sigma_1][t_2;\sigma_2]$. If we put $\sigma_2(t_1) = \sum_{i=0}^l a_i t^i$ computing $\sigma_2(\sigma_1(a)t_1) = \sigma_2(t_1 a) = \sigma_2(t_1)\sigma_2(a)$, leads to the following equations

$$\forall\, 0 \le i \le l, \quad a_i\sigma_1^i(\sigma_2(a)) = \sigma_2(\sigma_1(a))a_i.$$

Let us now give some examples.

EXAMPLE 1.5.

(1) Let $A_1(k) = k[X][Y; id., \frac{d}{dX}]$ and $(a, b) \in K^2$, then
   - $YX = XY + 1 = X(Y - b) + bX + 1 = X(Y - b) + b(X - a) + ba + 1$ and hence $(YX)(a, b) = ba + 1$.

- $YX^2 = X^2Y + 2X = X^2(Y-b) + bX^2 + 2X = X^2(Y-b) + bX(X-a) + baX + 2(X-a) + 2a$ and hence $(YX^2)(a,b) = ba^2 + 2a$.
- $Y^2X = XY^2 + 2Y = XY(Y-b) + bX(Y-b) + bXb + 2(Y-b) + 2b$ and hence $(Y^2X)(a,b) = b^2a + 2b$.

(2) Consider the double Ore extension $R = \mathbb{F}_q[t_1;\theta][t_2;\overline{\theta}]$, where $q = p^n$, $\overline{\theta}(a) = a^p$, and $\overline{\theta}(t_1) = t_1$. A polynomial $p(t_1,t_2) \in R$ can be written as $p(t_1,t_2) = \sum_{i=0}^n p_i(t_1)t_2^i = \sum_{i,j} a_{i,j}t_1^jt_2^i$ and we can easily check that

$$p(t_1,t_2)(a,b) = \sum_{i,j} \theta^j(N_i(b))N_j(a) = b^{\frac{(p^i-1)p^j}{p-1}} a^{\frac{p^i-1}{p-1}}.$$

Let us now turn to another possible way of evaluating a polynomial $f(t_1,\ldots,t_n) \in R = K[t_1;\sigma_1,\delta_1][t_2;\sigma_2,\delta_2]\cdots[t_n;\sigma_n,\delta_n]$ at $(a_1,a_2,\ldots,a_n) \in K^n$. We consider the element of K representing f in the quotient R/I where $I = R(t-a_1) + R(t-a_2) + \cdots + R(t-a_n)$. The set I is the usual left ideal of R and this evaluation looks more classical. We now compare the two evaluations by comparing $I_n = R_1(t_1-a_1) + \cdots + R_{n-1}(t_{n-1}-a_{n-1}) + R(t_n-a_n)$ and I.

THEOREM 1.6. *Let K be a ring and $R = K[t_1;\sigma_1,\delta_1][t_2;\sigma_2,\delta_2]\cdots[t_n;\sigma_n,\delta_n]$. We consider $(a_1,a_2,\ldots,a_n) \in K^n$ and put $I = R(t-a_1)+R(t-a_2)+\cdots+R(t-a_n)$ and $I_n = R_1(t_1-a_1)+\cdots+R_{n-1}(t_{n-1}-a_{n-1})+R(t_n-a_n)$, where, for $1 \le i \le n$, R_i, $R_i = K[t_1;\sigma_1,\delta_1]\cdots[t_i;\sigma_i,\delta_i]$. With these notations, the following are equivalent:*

1. *$I_n = I$;*
2. *$R(t_i - a_i) \subseteq I_n$;*
3. *$I \neq R$;*
4. *For $1 \le i < j \le n$, we have $t_j(t_i - a_i) \in I_n$;*
5. *For $1 \le i < j \le n$, we have $\sigma_j(t_i - a_i)a_j + \delta_j(t_i - a_i) \in I_n$;*
6. *For $1 \le i < j \le n$, we have $(t_jt_i)(a_1,\ldots,a_n) = \sigma_j(a_i)a_j + \delta_j(a_i)$.*

PROOF. 1) $\Leftrightarrow$ 2) is clear.
2) $\Rightarrow$ 3) If $I = R$, then $I_n = R$ and, for $1 \le i \le n$, there exist polynomials $g_i(t_1,t_2\cdots,t_i) \in R_i$ such that $1 = \sum_{i=1}^n g_i(t_1,\ldots,t_i)(t_i - a_i)$. The change of variables defined by putting $y_i = t_i - a_i$ gives that, for some $h_i \in R_i$, we have $1 = \sum_{i=1}^n h_i(y_1,\ldots,y_i)y_i$. A comparison of the coefficients of degree zero of this equality leads to a contradiction.
3) $\Rightarrow$ 4) There exists $c \in K$ such that $t_j(t_i - a_i) - c \in I_n \subseteq I$. Since $t_j(t_i-a_i) \in I$ we obtain $c \in I$ hence, by 3), we must have $c = 0$. This shows that $t_j(t_i - a_i) \in I_n$.
4) $\Rightarrow$ 5) The fact that $t_j(t_i - a_i) \in I_n$, implies that $\sigma_j(t_i - a_i)t_j + \delta_j(t_i - a_i) \in I_n$. Since $(t_j - a_j) \in I_n$, we obtain that $\sigma_j(t_i - a_i)a_j + \delta_j(t_i - a_i) \in I_n$.
5) $\Rightarrow$ 6) This is clear since 5) implies that $t_jt_i - \sigma_j(a_i)a_j - \delta_j(a_i) = \sigma_j(t_i)t_j + \delta_j(t_i) - \sigma_j(a_i)a_j - \delta_j(a_i) \in I_n$.
6) $\Rightarrow$ 1) The equality in 6) implies that $t_jt_i - \sigma_j(a_i)a_j - \delta_j(a_i) \in I_n$ for $1 \le i < j \le n$. This gives that $t_j(t_i - a_i) \in I_n$. Since also we have that $R_i(t_i - a_i) \subseteq I_n$ we conclude that $t_j(t_i - a_i) \in I_n$ for any integer $1 \le i \le j \le n$ and hence $R(t_i - a_i) \subseteq I_n$ for any $1 \le i \le n$. This yields $I_n = I$, as required. □

REMARK 1.7.

(1) If $n = 1$, we obviously have $I = I_1$ and all points are good.
(2) In general, the two additive subset $I_n \subset I$ are different. As mentioned above, we will use I_n for our evaluation. The reason is that while evaluating with respect to I we often face the following problem: the left

R-module I can be the entire ring. So that the evaluation of any polynomial at $(a_1, a_2 \ldots, a_n) \in K^n$ with respect to I is zero. This the case in the Weyl algebra $R = A_1(K) = K[t_1][t_2; id, \frac{d}{dt_1}]$ for the piont $(0,0)$ since we then have $t_2t_1 - t_1t_2 = 1$ and hence $Rt_1 + Rt_2 = R$. This is not the case with our evaluation since, for instance, $t_2t_1 = t_1t_2 + 1$, so that $t_2t_1 + I_2(0,0) = 1 + I_2(0,0)$ and hence the evaluation of t_2t_1 at $(0,0)$ is just 1.

(3) In fact, it is quite often the case that $I = R$, even if we are using Ore polynomials with zero derivations. Consider for instance the Ore extension $R = K[t_1; \sigma_1][t_2; \sigma_2]$ where K is a field and σ_2 is an endomorphism of $K[t_1; \sigma_1]$ such that $\sigma_2(t_1) = t_1$. It is easy to check that for any $(a_1, a_2) \in K^2$ we have $(t_2 - \sigma_1(a_2))(t_1 - a_1) + (-t_1 + \sigma_2(a_1))(t_2 - a_2) = \sigma_1(a_2)a_1 - \sigma_2(a_1)a_2$. So that if $\sigma_1(a_2)a_1 - \sigma_2(a_1)a_2 \neq 0$, then the left ideal $I(a_1, a_2) = R$. This shows that very often the evaluation modulo I turns out to be trivial. Once again, this is not the case with our evaluation, since we have that $t_2(t_1 - a_1)$ is represented by $\sigma_1(a_2)a_1 - \sigma_2(a_1)a_2$ modulo $I_2(a_1, a_2)$.

DEFINITION 1.8. A point $(a_1, \ldots, a_n) \in K^n$ will be called a good point if the two ways of evaluating a polynomial $inK[t_1; \sigma_1, \delta_1][t_2; \sigma_2, \delta_2] \cdots [t_n; \sigma_n, \delta_n]$ at $(a_1, \ldots, a_2)$ coincide i.e. if $I_n(a_1, \ldots, a_n) = I$.

The advantage of the good points is that in this case the evaluations via a left ideal I and I_n coincide. But of course, we can still evaluate a polynomial at any points via our additive subset $I_n = R_1(t_1 - a_1) + \cdots + R_{n-1}(t_{n-1} - a_{n-1}) + R(t_n - a_n)$.

EXAMPLE 1.9. (1) In the classical case ($\sigma_i = id_K$ and $\delta_i = 0$, for every $1 \leq i \leq n$), every point $(a_1, \ldots, a_n) \in K^n$ is good.

(2) If K is a division ring $\sigma_1 = id_K$, $\delta_1 = 0$ and $\sigma_2 = id, \delta_2 = d/dt_1$, we have for any $a, b \in K$, $(t_2 - b)(t_1 - a) = (t_1 - a)(t_2 - b) + 1$. This shows that in this case there are no good points.

Working with I_n instead of I we avoid the problem of having points that are zeroes of every polynomial in the ring $R = K[t_1; \sigma_1, \delta_1][t_2; \sigma_2, \delta_2] \cdots [t_n; \sigma_n, \delta_n]$.

We will consider Reed-Muller codes and hence we will need to evaluate polynomials in several variables. We consider the case of iterated Ore polynomials defined on a finite base ring (field) K. In classical Reed-Muller coding the polynomials that are used to make evaluations are restricted to be of some bounded degree in each variables. Indeed polynomial maps associated to $X^q - X \in \mathbb{F}_q[X]$ are identically zero. So our first task is to consider what is the analogue of this polynomial in an Ore polynomial ring. We will need a few definitions.

DEFINITION 1.10. Let $a \in K$ be an element of a division ring K, σ and δ an endomorphism of K and a σ-derivation of K, respectively.

(1) For a nonzero $x \in K$, we denote $a^x = \sigma(x)ax^{-1} + \delta(x)x^{-1}$ and put $\Delta(a) = \{a^x \mid x \in K^*\}$. This set is called the (σ, δ) conjugacy class of a.
(2) The map $T_a : K \longrightarrow K$ defined by $T_a(x) = \sigma(x)a + \delta(x)$ is called the (σ, δ) pseudo-linear map associated to a.
(3) The (σ, δ) centralizer of a is the set $C^{\sigma,\delta}(a) = \{x \in K^* \mid a^x = a\} \cup \{0\}$.

Let us mention that the notion of σ, δ conjugation appears naturally due to the fact that while evaluating a product fg we have the nice formula $fg(a) = 0$ if $g(a) = 0$ and $(fg)(a) = f(a^{g(a)})g(a)$ if $g(a) \neq 0$. Let us also mention that the evaluation at an element $a \in K$ is strongly related to the (σ,δ)-pseudo linear map T_a via the equality $f(a) = f(T_a)(1)$ or more generally that, for any $h \in R = K[t;\sigma,\delta]$, $a \in K$ and $x \in K^*$, $h(T_a)(x) = h(a^x)x$. We first need the following lemma which is part of folklore.

THEOREM 1.11. *Let a be an element of a division ring K and $f(t) \in R = K[t;\sigma,\delta]$ be a polynomial of degree n. Then:*

(1) *$C^{\sigma,\delta}(a)$ is a subdivision ring of K.*
(2) *The map T_a is a right $C^{\sigma,\delta}(a)$ linear map.*
(3) *$f(t)$ has roots in at most n (σ,δ)-conjugacy classes, say $\{\Delta(a_1),\dots,\Delta(a_r)\}$, $r \le n$;*
(4) *$\sum_{i=1}^{r} dim_{C(a_i)} \ker(f(T_{a_i})) \le n$, where $C(a_i) := C^{\sigma,\delta}(a_i)$ for $1 \le i \le r$.*

PROOF. We refer the reader to [**8**] and [**11**]. □

Notice that point (4) above generalizes several classical theorems, in particular, the theorem by Gordon-Motzkin that states that a polynomial $f(X) \in K[X]$ with coefficients in a division ring K can have roots in at most $n = deg(f)$ classical $(\sigma = id., \delta = 0)$-conjugacy classes. Theorem 1.12 below can be found in [**7**].

THEOREM 1.12. *Let p be a prime number and $\mathbb{F}_q$ be the finite field with $q = p^n$ elements. Denote by θ the Frobenius automorphism. Then:*

(a) *There are p distinct θ-conjugacy classes in $\mathbb{F}_q$.*
(b) *$C^\theta(0) = \mathbb{F}_q$ and, for $0 \ne a \in \mathbb{F}_q$, we have $C^\theta(a) = \mathbb{F}_p$.*
(c) *In $\mathbb{F}_q[t;\theta]$, the least left common multiple of all the elements of the form $t - a$ for $a \in \mathbb{F}_q$ is the polynomial $G(t) := t^{(p-1)n+1} - t$. In other words, $G(t) \in \mathbb{F}_q[t;\theta]$ is of minimal degree such that $G(a) = 0$ for all $a \in \mathbb{F}_q$.*
(d) *The polynomial $G(t)$ obtained in c) above is invariant, i.e. $RG(t) = G(t)R$.*

PROOF.

(a) Let us denote by g a generator of the cyclic group $\mathbb{F}_q^* := \mathbb{F}_q \setminus \{0\}$. The θ-conjugacy class determined by the zero element is reduced to $\{0\}$ i.e. $\Delta(0) = \{0\}$. The θ-conjugacy class determined by 1 is a subgroup of $\mathbb{F}_q^*$: $\Delta(1) = \{\theta(x)x^{-1} \mid 0 \ne x \in \mathbb{F}_q\} = \{x^{p-1} \mid 0 \ne x \in \mathbb{F}_q\}$. It is easy to check that $\Delta(1)$ is cyclic generated by g^{p-1} and has order $\frac{p^n-1}{p-1}$. Its index is $(\mathbb{F}_q^* : \Delta(1)) = p - 1$. Since two nonzero elements a, b are θ-conjugate if and only if $ab^{-1} \in \Delta(1)$, we indeed get that the number of different nonzero θ-conjugacy classes is $p - 1$. This yields the result.

(b) If $a \in \mathbb{F}_q$ is nonzero, then $C^\theta(a) = \{x \in \mathbb{F}_q \mid \theta(x)a = ax\}$ i.e. $C^\theta(a) = \mathbb{F}_p$.

(c) We have, for any $x \in \mathbb{F}_q$, $(t^{(p-1)n+1} - t)(x) = \theta^{(p-1)n}(x)\dots\theta(x)x - x$. Since $\theta^n = id$, and $N_n(x) := \theta^{n-1}(x)\dots\theta(x)x \in \mathbb{F}_p$, we get $(t^{(p-1)n+1} - t)(x) = x(\theta^{n-1}(x)\dots\theta(x)x)^{p-1} - x = xN_n(x)^{p-1} - x = 0$. This shows that indeed $G(t)$ annihilates all the elements of $\mathbb{F}_q$ and hence $G(t)$ is a left common multiple of the linear polynomials $\{(t-a) \mid a \in \mathbb{F}_q\}$. Let $h(t) := [t - a \mid a \in \mathbb{F}_q]_l$ denote their least left common multiple. It remains to show that $\deg h(t) \ge n(p-1)+1$. Let $0 = a_0, a_1, \dots, a_{p-1} \in \mathbb{F}_q$ be elements representing the θ-conjugacy classes (Cf. a) above). Denote by $C_0, C_1, \dots, C_{p-1}$ their respective θ-centralizer. The formulas recalled in the paragraph before Theorem 1.11 shows that $h(T_a)(x) = h(a^x)x = 0$ for any nonzero element $x \in \mathbb{F}_q$ and any element $a \in \{a_0,\dots,a_{p-1}\}$. Hence $\ker h(T_{a_i}) = \mathbb{F}_q$ for $0 \le i \le p-1$. Using Inequality (4) in Theorem 1.11 and the statement b) above, we get $\deg h(t) \geq \sum_{i=0}^{p-1} dim_{C_i} \ker h(T_{a_i}) = dim_{\mathbb{F}_q}\mathbb{F}_q + \sum_{i=1}^{p-1} dim_{\mathbb{F}_p}\mathbb{F}_q = 1 + (p-1)n$, as required.

(d) Since $\theta^n = id.$, we have immediately that $G(t)x = \theta(x)G(t)$ and obviously $G(t)t = tG(t)$.

□

Let us now consider an iterated Ore extension $K[t_1;\sigma_1,\delta_1][t_2;\sigma_2,\delta_2]\cdots[t_n;\sigma_n,\delta_n]$, where K is a finite ring. Using classical notations from the world of algebraic geometry, we consider

$$I(K^n) = \{f(t_1,\ldots,t_n) \in R \mid \forall (a_1,\ldots,a_n) \in K^n, f(a_1,\ldots,a_n) = 0\}.$$

In other words: $I(K^n) = \cap_{(a_1,\ldots,a_n)} I_n(a_1,\ldots,a_n)$. In the case of an Ore extension $\mathbb{F}_q[t_1;\theta]\ldots[t_n;\theta]$, where θ is the Frobenius automorphism of $\mathbb{F}_q = \mathbb{F}_{p^n}$, the above theorem shows that, for any $1 \leq i \leq n$, $t_i^{(n-1)p} - t_i \in I(K^n)$. Having this in mind we introduce, for any $1 \leq i \leq n$, the monic polynomials $G_i = G_i(t_1,\ldots t_i)$ of minimal degree in $R_i = K[t_1;\sigma_1,\delta_1]\ldots[t_i;\sigma_i,\delta_i]$ such that $G_i \in I(K^n)$. With these notations, we can now state the following proposition.

PROPOSITION 1.13. *Let K be a finite ring and, for $1 \leq i \leq n$, consider the polynomials $G_i \in R_i = K[t_1,\sigma_1,\delta_1]\ldots[t_i,\sigma_i,\delta_i]$ defined above. Then*

$$I(K^n) = R_1G_1(t_1) + R_2G_2(t_1,t_2) + \cdots + R_nG_n(t_1,\ldots,t_n).$$

PROOF. It is clear that $R_1G_1(t_1) + R_2G_2(t_1,t_2) + \cdots + R_nG_n(t_n) \subseteq I(K^n)$. Now, if a polynomial $f \in K[t_1;\sigma_1,\delta_1][t_2;\sigma_2,\delta_2]\cdots[t_n;\sigma_n,\delta_n]$ belongs to $I(K^n)$, then, for any $(a_1,\ldots,a_n) \in K^n$, $f(a_1,a_2,\ldots,a_n) = 0$. Let us write the consecutive remainders obtained during the evaluation process as follows. For any $(a_1,\ldots,a_n) \in K^n$, we have

$f(t_1,\ldots,t_n) = q_1(t_1,\ldots,t_n)(t_n - a_n) + f_1(t_1,\ldots,t_{n-1},a_n)$.

$f_1(t_1,\ldots,t_{n-1},a_n) = q_2(t_1,\ldots,t_{n-1})(t_{n-1} - a_{n-1}) + f_2(t_1,\ldots,t_{n-2},a_{n-1},a_n)$.

Continuing this process, we get

$f_{n-2}(t_1,t_2,a_3,\ldots,a_n) = q_{n-1}(t_1,t_2)(t_2 - a_2) + f_{n-1}(t_1,a_2\ldots,a_n)$ and

$f_{n-1}(t_1,a_2\ldots,a_n) = q_n(t_1)(t_1 - a_1) + f_n(a_1,\ldots,a_{n-1},a_n)$.

Since $f(a_1,a_2,\ldots,a_n) = f_n(a_1,\ldots,a_{n-1},a_n) = 0$, we conclude that for any for any $(a_1,\ldots,a_n \in K^n$, $f_{n-1}(t_1,a_2\ldots,a_n) \in R_1(t_1 - a_1)$, i.e. that $f_{n-1}(t_1,a_2\ldots,a_n) \in R_1G_1(t_1)$. Similarly, we have that $f_{n-2}(t_1,t_2,a_3,\ldots,a_n) - f_{n-1}(t_1,a_2\ldots,a_n) \in R_2 \cap I(K^n) = R_2G_2(t_1,t_2)$. An easy induction yields that for any $1 \leq i \leq n$, we have $f_{n-i}(t_1,\ldots t_i,a_{i+1},\ldots,a_n) - f_{n-i+1}(t_1,\ldots,t_{i-1},a_i,\ldots,a_n) \in R_iG_i$. From this we get successively that $f \in R_nG_n + f_1 \subseteq R_nG_n + R_{n-1}G_{n-1} + f_2 \subseteq R_nG_n + R_{n-1}G_{n-1} + R_{n-3}G_{n-3} + f_3 \subseteq R_1G_1(t_1) + R_2G_2(t_1,t_2) + \cdots + R_nG_n(t_1,\ldots,t_n)$. This yields the desired result. □

2. Skew Reed-Muller codes using iterated skew polynomial rings

In the first part of this section, we give some preliminaries on Reed-Muller codes over the commutative polynomial ring.

Let m be a positive integer and let $P_1,P_2,\ldots,P_n$ be the $n = q^m$ points in the affine space $\mathbb{A}^m(\mathbb{F}_q)$. For any integer r with $0 \leq r \leq m(q-1)$, let $\mathcal{R}_m \subseteq \mathbb{F}_q[x_1,x_2,\ldots,x_m]/(x_1^p - x_1, x_2^p - x_2,\cdots,x_m^p - x_m)$ be the set representing polynomials of degree less or equal than r.

DEFINITION 2.1. Let $f = \sum_{i=0}^{n} a_I X^I$ be a polynomial in $\mathbb{F}_q[x_1, x_2, \ldots, x_m]$, where for $I = (i_1, \ldots, i_n)$ we define X^I to be $x^{i_1} x^{i_2} \cdots x^{i_n}$. The operator evaluation $\mathrm{ev}(f)$ at $b = (b_1, b_2 \ldots b_n) \in \mathbb{F}_q^n$ is defined as

$$\mathcal{R}_m \to \mathbb{F}_q$$
$$\mathrm{ev}: \quad f = \sum_{i \in I} a_i X^i \mapsto f(b) = \sum_{i \in I} a_i (b^i).$$

The Reed-Muller codes of the length $n = q^m$ can be obtained as evaluation of the polynomial f in R_m in all points $P_1, P_2, \cdots, P_n$ in the affine sapce $\mathbb{A}^m(\mathbb{F}_q)$. The Reed-Muller codes of order r is defined as

$$\mathcal{R}_q(r, m) = \{(f(P_1), f(P_2), \ldots, f(P_n)) \mid f \in R_m\} = \{\mathrm{ev}(f) : f \in R_m, \deg f \leqslant r\}.$$

The function $\mathrm{ev} : \mathcal{R}_m \to \mathbb{F}_q$ is an homomorphism of algebras (see [**1**]). Hence the Reed-Muller codes is generated by the codewords $\mathrm{ev}(f)$ where f is a monomial of degree less than or equal to r in the set of monomials in m variables. It will be denoted

$$\mathcal{M} = \{X_1^{a_1} X_2^{a_2} \ldots X_n^{a_n}\}.$$

2.1. Iterated skew polynomial and Reed-Muller codes. In a recent paper, skew Reed-Muller codes were defined via the use of iterated Ore extensions with three variables over a finite field [**5**]. The authors used inner derivations and inner automorphisms, and (cf. Remarks 1.1) these can be erased by a change of variables. Their idea was to used Gröbner bases to make computations. They used the available software on the subject. But their methods used the ideal I in our notations and hence they faced the problem that I might be equal to the whole iterated Ore extension R. Some of their computations were wrong and for the sake of future works, we will briefly correct them. In the lines below the variables X_i are the one used in their paper.

1. Let $\mathbb{F}_4 = \mathbb{F}_2(\alpha)$, where $\alpha^2 = \alpha + 1$. In the ring $R_1 = \mathbb{F}_4[Y_1; \theta]$ where θ is the Frobenius automorphism, we put $X_1 = Y_1 + 1$ and we have the commutation relation

$$X_1 \alpha = (Y_1 + 1)\alpha = \theta(\alpha) Y_1 + \alpha = \theta(\alpha) X_1 + \theta(\alpha) + \alpha = \alpha^2 X_1 + 1.$$

2. In the ring $R_2 = R_1[Y_2]$ we put $X_2 = \alpha^2 Y_2 + Y_1 + \alpha^2$ so that we have $R_2 = R_1[X_2; \theta_\alpha, \delta_{X_1+\alpha}]$, where $\theta_\alpha(x) = \alpha^{-1} x \alpha$ for any $x \in R_1$. In R_2 we have the above commutation relation $X_1 \alpha = \alpha^2 X_1 + 1$ together with

$$X_2 \alpha = \alpha X_2 + X_1 + 1 \quad \text{and} \quad X_2 X_1 = \alpha X_1 X_2 + \alpha^2 X_2 + \alpha^2 X_1^2 + \alpha^2.$$

3. In the ring $R_3 = R_2[Y_3]$, we put $X_3 = \alpha^2 Y_3 + \alpha Y_1 + \alpha$. We then have $R_3 = R_2[X_3; \theta_\alpha, \delta_{\alpha X_1}]$ where, as above $\theta_\alpha(x) = \alpha^{-1} x \alpha$, for any $x \in \mathbb{F}_4$. We have the above commutation relations together with

$$\begin{aligned} X_3 \alpha &= \alpha X_3 + \alpha X_1 + \alpha, \\ X_3 X_1 &= \alpha X_1 X_3 + \alpha^2 X_3 + \alpha^2 X_1^2 + \alpha^2 X_1, \\ X_3 X_2 &= X_2 X_3 + \alpha^2 X_1 X_3 + \alpha^2 X_3 + X_1 X_2 + \alpha X_1^2 + \alpha^2 X_1 + X_2 + 1. \end{aligned}$$

Of course, the monomials in X_1, X_2, X_3 are not monomials in Y_1, Y_2, Y_3 and conversely. The advantage of the variables Y_1, Y_2, Y_3 is clear. We continue with these variables and consider,

$$R_n = R_{n-1}[Y_n] = R_1[Y_2, Y_3, \dots, Y_n] = \mathbb{F}_4[Y_1, \theta][Y_2, Y_3, \dots, Y_n].$$

As in the commutative case, there are polynomials that annihilate any element of $\mathbb{F}_q^n$. Thanks to Theorem 1.12 and 1.13, we have that, in our situation $I(K^n) = R_1G_1 + R_2G_2 + \cdots + R_nG_n$ where $G_1 = Y_1^3 - Y_1$ and $G_i = Y_i^4 - Y_i$ for any $2 \le i \le n$ (notice that the form of G_1 is due to the theorem 1.12). This iterated skew polynomial ring leads to a new family of skew Reed-Muller code generated by the evaluations of a base of monomials in m variables and of degree less than or equal to r given by

$$\mathcal{M} = \left\{ Y_1^{l_1} Y_2^{l_2} \dots Y_n^{l_n} \right\}.$$

By iterated skew polynomial rings given above we have the evaluation of $f \in R_m$ at $b = (b_1, b_2 \dots b_n) \in \mathbb{F}_q^n$ is defined as

$$\begin{aligned} R_m &\to \mathbb{F}_q \\ \text{ev}: \quad f(X_1, X_2, \dots, X_n) &\mapsto f(b_1, \dots, b_n) = \sum_{i=0}^{n} P(b_2, \dots, b_n)\theta^{\frac{i(i+1)}{2}}(b_1). \end{aligned}$$

Where $f(X_1, X_2, \dots, X_n) = g(Y_1)P(Y_2, \dots, Y_n) \in R_1G_1 + R_2G_2 + \cdots + R_nG_n$. By the iterated skew polynomial ring and the evaluation we have the new new family of the skew Reed-Muller codes in the following proposition.

PROPOSITION 2.2. *The skew Reed-Muller codes of parameters r and m over $R_m = \mathbb{F}_q[X_1; \theta_1, \delta_1] \dots [X_m; \theta_m, \delta_m]$ are generated as*

$$\mathcal{R}(r, m) \stackrel{\text{def}}{=} \left\{ \sum_{i=0}^{n} P(b_2, \dots, b_n)\theta^{\frac{i(i+1)}{2}}(b_1) : P \in R_m \textit{ such that } \deg P \leqslant \deg G_n \right\}.$$

The choice of the rows used to form the matrix generating the Reed-Muller codes $\mathcal{R}(r, m)$ consists in selecting the monomials $Y_1^{l_1} Y_2^{l_2} \dots Y_n^{l_n}$ of degree less than or equal to $r = (n-1)p+1$. This choice is based on the order of the monomials. The group of permutations of Reed-Muller codes $\mathcal{R}(r, m)$ is the set of transformations affines $x \to Ax + b$ where $A \in \mathbb{F}_q^{m \times m}$ is an invertible matrix and $b \in \mathbb{F}_q^m$.

2.2. Skew Reed-Muller codes using polynomial maps in n variables. In this section we give the construction of skew Reed-Muller codes using polynomial maps in n variables. We know that the Reed-Muller codes are defined as the codes generated by the evaluations of a base of monomials in m variables and of degree less than or equal to r, the set of monomials in m variables will be denoted

$$\mathcal{M} = \left\{ t_1^{l_1} t_2^{l_2} \dots t_n^{l_n} \right\}.$$

Notice that in the case of iterated Ore extensions $K[t_1; \sigma_1, \delta_1][t_2; \sigma_2, \delta_2] \cdots [t_n; \sigma_n, \delta_n]$ it is always possible to order the indeterminates of a word in $t_1, \dots, t_n$ so that the only words we have to consider are the one of the form $t_1^{l_1} t_2^{l_2} \dots t_n^{l_n}$. We now give the evaluation of some monomials in $R = K[t_1, \sigma_1, \delta_1][x_2, \sigma_2, \delta_2], \cdots, [x_n, \sigma_n, \delta_n]$ at $(a_1, a_2, \cdots, a_n) \in K^n$.

(1) Consider the polynomial ring $R = K[t_1; \sigma_1, \delta_1][t_2; \sigma_2, \delta_2]$ and let us evaluate the polynomials t_1t_2 at $(a_1, a_2) \in K^2$. We have $t_1t_2 = t_1(t_2 - a_2) + t_1a_2 = t_1(t_2 - a_2) + \sigma_1(a_2)t_1 + \delta_1(a_2)$. This leads to $(t_1t_2)(a_1, a_2) = \sigma_1(a_2)a_1 + \delta_1(a_2)$.

(2) Consider the polynomial ring $R = K[t_1; \sigma_1, \delta_1][t_2; \sigma_2, \delta_2][t_3; \sigma_3, \delta_3]$ and let us evaluate the polynomials $t_1t_2t_3$ at $(a_1, a_2, a_3) \in K^3$. We have $t_1t_2t_3 = t_1t_2(t_3 - a_3) + t_1t_2a_3 = t_1t_2(t_3 - a_3) + t_1(\sigma_2(a_3)t_2 + \delta_2(a_3)) = t_1t_2(t_3 - a_3) + t_1\sigma_2(a_3)t_2 + t_1\delta_2(a_3) = t_1t_2(t_3 - a_3) + \sigma_1(\sigma_2(a_3))t_1t_2 + \sigma_1(\delta_2(a_3))t_1 + \delta_1(\delta_2(a_3))$. This leads to
$(t_1t_2t_3)(a_1, a_2, a_3) = \sigma_1(\sigma_2(a_3))a_1a_2 + \sigma_1(\delta_2(a_3))a_1 + \delta_1(\delta_2(a_3))$.

(3) Consider the polynomial ring

$$R = K[t_1; \sigma_1, \delta_1][t_2; \sigma_2, \delta_2][t_3; \sigma_3, \delta_3][t_4; \sigma_4, \delta_4].$$

We have
$(t_1t_2t_3t_4)(a_1, a_2, a_3, a_4) = \sigma_1(\sigma_2(\sigma_3(a_4)a_3)a_2)a_1 + \delta_1(\sigma_2(\sigma_3(a_4)a_2)a_3) + \sigma_1(\delta_2(\sigma_3(a_4)a_3))a_1 + \delta_1(\delta_2(\sigma_3(a_4)a_3)) + \sigma_1(\sigma_2(\delta_3(a_4))a_2)a_1 + \delta_1(\sigma_2(\delta_3(a_4))a_2) + \sigma_1(\delta_2(\delta_3(a_4)))a_1 + \delta_1(\delta_2(\delta_3(a_4))))$.

Polynomial maps in n variables and the evaluation leads to the new family of skew Reed-Muller codes in the following example.

EXAMPLE 2.3. In this example, we give the construction of the Reed-Muller codes with polynomial maps in 2 or 3 variables. In these examples we used the Frobenius automorphsim for σ_1 and the identity for σ_2 and σ_3. The derivations are all zeroes.

(1) The evaluation of the set monomial $\{1, t_1, t_2, t_1t_2\}$ over $R = \mathbb{F}_4[t_1; \sigma_1, \delta_1][t_2; \sigma_2, \delta_2]$ gives the Reed-Muller code with parameters $[16, 4, 8]$.
(2) The evaluation of the set monomial $\{1, t_1, t_2, t_2t_1\}$ over $R = \mathbb{F}_4[t_1; \sigma_1, \delta_1][t_2; \sigma_2, \delta_2]$ gives the Reed-Muller code with parmeters $[16, 4, 7]$.
(3) The evaluation of the set monomial $\{1, t_1, t_2, t_1, t_3, t_1t_3, t_2t_3, t_1t_2t_3\}$ over $R = \mathbb{F}_4[t_1; \sigma_1, \delta_1][t_2; \sigma_2, \delta_2][t_3; \sigma_3, \delta_3]$ gives the Reed-Muller code with parameters $[64, 8]$.

References

[1] M. Bardet, V. Dragoi, A. Otmani, and J.P. Tillich, *Algebraic properties of polar codes from a new polynomial formalism*, IEEE International Symposium on Information Theory (ISIT), 230–234, 2016.

[2] Delphine Boucher and Felix Ulmer, *Coding with skew polynomial rings*, J. Symbolic Comput. **44** (2009), no. 12, 1644–1656, DOI 10.1016/j.jsc.2007.11.008. MR2553570

[3] M'Hammed Boulagouaz and André Leroy, *(σ, δ)-codes*, Adv. Math. Commun. **7** (2013), no. 4, 463–474, DOI 10.3934/amc.2013.7.463. MR3119685

[4] I. Dumer, R. Fourquet, and C. Tavernier, *Cryptanalysis of Block Ciphers via Decoding of Long Reed-Muller Codes*, Algebraic and Combinatorial Coding Theory (ACCT 2010), Novosibirsk, Russia September 2010.

[5] Willi Geiselmann and Felix Ulmer, *Skew Reed-Muller codes*, Rings, modules and codes, Contemp. Math., vol. 727, Amer. Math. Soc., [Providence], RI, [2019] ©2019, pp. 107–116, DOI 10.1090/conm/727/14628. MR3938143

[6] Tadao Kasami, Shu Lin, and W. Wesley Peterson, *New generalizations of the Reed-Muller codes. I. Primitive codes*, IEEE Trans. Inform. Theory **IT-14** (1968), 189–199, DOI 10.1109/tit.1968.1054127. MR275989

[7] André Leroy, *Noncommutative polynomial maps*, J. Algebra Appl. **11** (2012), no. 4, 1250076, 16, DOI 10.1142/S0219498812500764. MR2959425

[8] T. Y. Lam, A. Leroy, and A. Ozturk, *Wedderburn polynomials over division rings. II*, Noncommutative rings, group rings, diagram algebras and their applications, Contemp. Math., vol. 456, Amer. Math. Soc., Providence, RI, 2008, pp. 73–98, DOI 10.1090/conm/456/08885. MR2416145

[9] T. Y. Lam and André Leroy, *Principal one-sided ideals in Ore polynomial rings*, Algebra and its applications (Athens, OH, 1999), Contemp. Math., vol. 259, Amer. Math. Soc., Providence, RI, 2000, pp. 333–352, DOI 10.1090/conm/259/04106. MR1780532

[10] D.E. Muller, *Application of Boolean algebra to switching circuit design and to error detection*, Transactions of the IRE professional group on electronic computers, vol.**3** (1954), 6–12.

[11] André Leroy and Adem Ozturk, *Algebraic and F-independent sets in 2-firs*, Comm. Algebra **32** (2004), no. 5, 1763–1792, DOI 10.1081/AGB-120029901. MR2099700

USTHB, Faculté de Mathématiques, BP 32 El Alia Bab Ezzouar, Algiers, Algeria
Email address: nabil.bennenni@gmail.com

Université Artois, UR 2462, Laboratoire de Mathématiques de Lens (LML), F-62300 Lens, France
Email address: andre.leroy@univ-artois.fr

Contemporary Mathematics
Volume **785**, 2023
https://doi.org/10.1090/conm/785/15773

Identities of the algebra $\mathbb{O} \otimes \mathbb{O}$

Guy Blachar, Louis H. Rowen, and Uzi Vishne

To Professor Syed Tariq Rizvi, for his contribution to Ring and Module Theory

ABSTRACT. We find nonassociative polynomial identities of minimal degree, which is 7, for the algebras $\mathbb{O}\otimes\mathbb{O}$ and $\mathrm{M}_2(\mathbb{O})$, where $\mathbb{O}$ is the octonion algebra.

1. Introduction

One of the basic invariants of an algebra A satisfying a polynomial identity (PI) is the minimal degree of an identity, which we call the **PI degree**. A commutative algebra has PI degree 2, because it satisfies $[x_1, x_2]$. That the PI degree of the matrix algebra $\mathrm{M}_n(F)$ is $2n$ is the Amitsur-Levitzki theorem. The Grassmann algebra satisfies $[x_1, [x_2, x_3]]$, so its PI degree is 3.

The same notion applies to nonassociative algebras. The substitution in an associative polynomial is not well defined if the algebra is not associative, so an identity of A would be an element in the free nonassociative algebra, which vanishes under every substitution from A. Again we are concerned with the minimal degree of an identity. The associator $(x_1, x_2, x_3) = (x_1x_2)x_3 - x_1(x_2x_3)$ is an identity in the nonassociative variety precisely when the algebra is associative. The notion of polynomial identities, and thus the PI degree, depends on the variety. If $\mathcal{V}$ is a variety of algebras and $A \in \mathcal{V}$, the nontrivial polynomial identities of A with respect to $\mathcal{V}$ are nonzero elements of the free algebra in $\mathcal{V}$, so one excludes consequences of the defining identities of $\mathcal{V}$.

Let $\mathbb{O}$ denote the split octonion algebra over a base field F. The alternative PI degree of the octonion algebra is 5 (see Section 3 below), because the identities of lower degree, namely the consequences of (x, x, y) and (y, x, x), hold in any alternative algebra. The purpose of this note is to attract attention to the following question:

QUESTION 1.1. What is the (nonassociative) PI degree of the matrix algebras $\mathrm{M}_n(\mathbb{O})$? What is the PI degree of the tensor power $\mathbb{O}^{\otimes n}$?

Here we are concerned with the (general) nonassociative identities, as the central simple algebras $\mathrm{M}_n(\mathbb{O})$ and $\mathbb{O}^{\otimes n}$ are not alternative (for $n > 1$); indeed, the only simple alternative algebras are associative or the octonion algebras.

2020 *Mathematics Subject Classification.* Primary 17A30; Secondary 17D05, 16R99.
Partially supported by ISF grant #1994/20.

We assume throughout this paper that char$F \neq 2$. Bremner and Hentzel [**3**] showed that the PI degree of $\mathrm{M}_2(\mathbb{O})$ is 7, thereby answering the matrix part of Question 1.1 for $n = 2$. We answer the tensor product part for $n = 2$:

THEOREM 1.2. *The PI degree of* $\mathbb{O}\otimes\mathbb{O}$ *is* 7.

Since polynomial identities over infinite fields are stable under extension of scalars, the results apply to all octonion algebras. Section 2 presents this problem in the general framework of the nonassociative PI degree.

Our proof of the identities for $\mathbb{O}\otimes\mathbb{O}$ uses the decomposition $\mathbb{O} = F \oplus \mathbb{O}_0$ where $\mathbb{O}_0$ is the subspace of the traceless elements. We spell out some identities of $\mathbb{O}$ and $\mathbb{O}_0$ in Section 3, where we also indicate which subspaces of $\mathbb{O}\otimes\mathbb{O}$ are of interest to us, see Figure 1. Other than Theorem 1.2, our results include:

(1) Identities of degree 3 for $\mathbb{O}_0\otimes\mathbb{O}_0$ and $\mathrm{M}_2(F)_0\otimes\mathbb{O}$.
(2) Identities of degree 4 (in a single variable) for $\mathrm{M}_2(F)\otimes\mathbb{O}_0$.
(3) Identities of degree 7 for $\mathbb{O}\otimes\mathbb{O}$ (and its subalgebra $\mathrm{M}_2(\mathbb{O})$).

The identities are described and fully proved in Sections 4, 5 and 6, respectively. The fact that in each case there are no identities of smaller degree is verified by a computer using standard linear-algebra techniques, as in [**12**] or [**13**].

2. Varieties and matrices

If A is an associative PI algebra, then by Regev's theorem [**9**] on tensor product of PI algebras, the matrix algebra $\mathrm{M}_n(A)$ is also PI. This raises a natural question:

QUESTION 2.1. Find an optimal bound $f_n(d)$ for the PI degree of $\mathrm{M}_n(A)$ in terms of n and $d = \mathrm{PIdeg}\, A$.

For example, $f_2(3) \geq 8$ by [**12**], where minimal identities are found for the algebra of 2×2 matrices over the Grassmann algebra.

However, as Regev showed in [**10**], the tensor product of nonassociative PI algebras is not necessarily PI. The following special case is open as far as we know:

QUESTION 2.2. If B is a nonassociative PI algebra, is the matrix algebra $\mathrm{M}_n(B)$ PI as well?

Let us say that a variety $\mathcal{V}$ of nonassociative algebras is **matricial** if $\mathrm{M}_n(A) \in \mathcal{V}$ for any $A \in \mathcal{V}$ and any n (equivalently for $n = 2$, by iteration). Matricial varieties can easily be characterized in terms of the defining identities. We say that a nonassociative polynomial is **sorted** if its monomials all reduce to the same monomial modulo associators (for example, $((x_1x_2)x_3)x_4 - 2(x_1x_2)(x_3x_4) + x_1(x_2(x_3x_4))$ is sorted). For example, the sorted identities

$$(x_1, x_2, (x_3, x_4, x_5)), \quad (x_1, (x_2, x_3, x_4), x_5), \quad ((x_1, x_2, x_3), x_4, x_5)$$

define the matricial variety of algebras in which all associators are in the nucleus.

REMARK 2.3. Recall that a polynomial is Spechtian if it vanishes under any substitution $x_i \mapsto 1$ [**1**, Section 6.2].

Any associator polynomial (in the sense of [**2**]) is Specht. But not every sorted Specht polynomial is an associator polynomial, as illustrated by

$$(x_1, x_2, (x_3, x_4, x_5)x_6) - (x_1, x_2, (x_3, x_4, x_5))x_6.$$

PROPOSITION 2.4. *A variety $\mathcal{V}$ is matricial if and only if it is defined by sorted identities.*

PROOF. An identity can be written as a sum of a minimal number of sorted polynomials. Each sorted summand of a multilinear identity of degree n is an identity in $\mathcal{V}$ by substituting $(w_1e_{12}, w_2e_{23}, \dots, w_ne_{n,n+1})$. □

It follows that a matricial unital variety must contain all associative algebras. We can now state Question 2.1 more generally.

More concretely, let A be a nonassociative algebra. Substituting elements from $\bigcup e_{ij}A$, we obtain:

PROPOSITION 2.5. *A multilinear identity holds in all the matrix algebras $\mathrm{M}_n(A)$ if and only if it is a sum of sorted identities of A.*

QUESTION 2.6. Given a matricial variety $\mathcal{V}$, find an optimal bound $f_n^{\mathcal{V}}(d)$ for the PI degree of $\mathrm{M}_n(A)$ (with respect to $\mathcal{V}$) in terms of n and $d = \mathrm{PIdeg}^{\mathcal{V}} A$.

(where $\mathrm{PIdeg}^{\mathcal{V}} A$ is the minimal degree of an identity of A which does not hold for the free algebra in $\mathcal{V}$). The PI degree of $\mathrm{M}_n(\mathbb{O})$ will thus be a lower bound for $f_n^{\mathcal{N}}(3)$, where $\mathcal{N}$ is the variety of all nonassociative algebras. In this language, our results here on $\mathrm{M}_2(\mathbb{O})$ show that $f_2^{\mathcal{N}}(3) \geq 7$.

3. Identities of $\mathbb{O}$ and $\mathbb{O}_0$

The proof of identities for $\mathbb{O}\otimes\mathbb{O}$ in Section 6 leads us to consider identities of subspaces, resulting from the inclusion of the matrix algebra $\mathrm{M}_2(F)$ in the split octonions $\mathbb{O}$, as well as the decompositions $\mathbb{O} = F \oplus \mathbb{O}_0$ and $\mathrm{M}_2(F) = F \oplus \mathrm{M}_2(F)_0$. Polynomials are always evaluated in some algebra A. Nevertheless, we say that a polynomial f is a **PI of a subspace** $V \subseteq A$ if f vanishes under all substitutions of values from V. We may thus speak of the PI degree of a subspace, as well. More precisely, we present minimal identities for each of the subspaces in Figure 1.

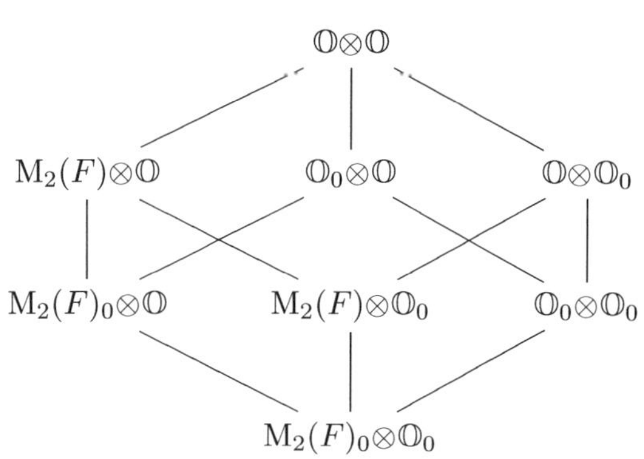

FIGURE 1. Some subspaces of $\mathbb{O}\otimes\mathbb{O}$

REMARK 3.1. The algebra $\mathbb{O}\otimes\mathbb{O}$ contains a copy of $\mathrm{M}_2(F)\otimes\mathrm{M}_2(F) = \mathrm{M}_4(F)$, whose associative PI degree is 8. Being associative, the nonassociative PI degree of this algebra is of course 3.

Recall the associator $\mathfrak{a} = \mathfrak{a}(x_1, x_2, x_3) = (x_1x_2)x_3 - x_1(x_2x_3)$. The polynomials

$$\mathfrak{l} = \mathfrak{a} + (12)\mathfrak{a} = (x_1, x_2, x_3) + (x_2, x_1, x_3) \tag{3.1}$$
$$\mathfrak{r} = \mathfrak{a} + (23)\mathfrak{a} = (x_1, x_2, x_3) + (x_1, x_3, x_2) \tag{3.2}$$
$$\mathfrak{f} = \mathfrak{a} + (13)\mathfrak{a} = (x_1, x_2, x_3) + (x_3, x_2, x_1) \tag{3.3}$$

linearize the identities (x, x, y), (y, x, x) and (x, y, x) of left-alternativity, right-alternativity and flexibility, respectively. Any two of those imply the third. An alternative algebra satisfies all three.

The space of multilinear identities of degree 3 of $\mathbb{O}$ is spanned as an S_3-module by $\mathfrak{l}$ and $\mathfrak{r}$. The space of multilinear identities of degree 3 of $\mathbb{O}_0$ is spanned by $\mathfrak{l}$, $\mathfrak{r}$, and the identity

$$\mathfrak{c} = [x_1, x_2 \circ x_3], \tag{3.4}$$

which linearizes $[x, x'^2]$. The latter is a baby-version of the Wagner-Hall identity $[x, [x', x'']^2]$, which holds in $\mathbb{O}$.

Iltyakov solved the Specht problem, on a finite basis of identities for affine alternative algebras [**6**]. Racine [**8**] shows that all the identities of degree at most 5 of $\mathbb{O}$ follow from alternativity, the Hall identity, and a Jordan algebra identity of degree 5. Two central identities of degree 5 (hence identities of degree 6) were found in [**7**] and [**11**]. In [**4**] it is shown that these known identities, together with a new identity of degree 6, induce all identities of degree at most 7. Henry [**5**] described the graded identities of $\mathbb{O}$, a direction we do not pursue here.

4. Identities of degree 3

We first present multilinear identities of degree 3 for the spaces $\mathbb{O}_0 \otimes \mathbb{O}_0$ and $\mathrm{M}_2(F)_0 \otimes \mathbb{O}$. Since we are considering identities of a tensor product, it would be convenient to employ two further sets of variables, y_i and z_i, referring to the first and second component of the product, respectively. The symmetric group acts on all sets simultaneously.

4.1. Preparations.

LEMMA 4.1. *Let A and B be any nonassociative algebras. For any $y_1, y_2, y_3 \in A$ and $z_1, z_2, z_3 \in B$ we have that*

$$(y_1 \otimes z_1, y_2 \otimes z_2, y_3 \otimes z_3) = (y_1, y_2, y_3) \otimes (z_1z_2)z_3 + y_1(y_2y_3) \otimes (z_1, z_2, z_3).$$

COROLLARY 4.2. *Let $\mathfrak{D}(\cdot)$ denote the associator ideal of the algebra. Then $\mathfrak{D}(A \otimes B) \subseteq \mathfrak{D}(A) \otimes B + A \otimes \mathfrak{D}(B)$.*

REMARK 4.3. Assume that A and B are (say) left-alternative. For simple tensors $y_1 \otimes z_1$ and $y_3 \otimes z_3 \in A \otimes B$,

$$(y_1 \otimes z_1, y_1 \otimes z_1, y_3 \otimes z_3) = 0 \tag{4.1}$$

by Lemma 4.1. However $A \otimes B$ is not left-alternative in general. (Linearizing (4.1) results in an identity of $A \otimes B$ which is strictly weaker than left-alternativity).

Summing up monomials by Lemma 4.1, we obtain:

LEMMA 4.4. *Let A and B be alternative algebras. For $y_1, y_2, y_3 \in A$ and $z_1, z_2, z_3 \in B$, we have that*

$$\begin{aligned}
\mathfrak{l}(y_1\otimes z_1, y_2\otimes z_2, y_3\otimes z_3) &= (y_1,y_2,y_3)\otimes[z_1,z_2]z_3 \\
&\quad + (y_1(y_2y_3) - (y_2y_1)y_3)\otimes(z_1,z_2,z_3); \\
\mathfrak{r}(y_1\otimes z_1, y_2\otimes z_2, y_3\otimes z_3) &= (y_1,y_2,y_3)\otimes((z_1z_2)z_3 - (z_1z_3)z_2) \\
&\quad + y_1[y_2,y_3]\otimes(z_1,z_2,z_3); \\
\mathfrak{f}(y_1\otimes z_1, y_2\otimes z_2, y_3\otimes z_3) &= (y_1,y_2,y_3)\otimes((z_1z_2)z_3 - (z_3z_2)z_1) \\
&\quad +(y_1(y_2y_3) - y_3(y_2y_1))\otimes(z_1,z_2,z_3).
\end{aligned}$$

4.2. The irreducible component of $F[S_3]\mathfrak{f}$. The natural action of the symmetric group S_n on an n-dimensional space shows up in our context. The S_3-module spanned by $\mathfrak{f}$ is isomorphic to the natural representation, as seen from the action on the intermediate variable in (3.3). This module decomposes into a direct sum of the trivial module (spanned by the sum of all associators (x_i, x_j, x_k), which is contained in $F[S_3]\mathfrak{l}$ and $F[S_3]\mathfrak{r}$ as well), and the 2-dimensional irreducible component

$$N = \{\alpha_0\mathfrak{f} + \alpha_1(123)\mathfrak{f} + \alpha_2(123)^2\mathfrak{f}\colon\ \alpha_0+\alpha_1+\alpha_2 = 0\}.$$

Let

$$\phi(x_1,x_2,x_3) = \mathfrak{f}(x_1,x_2,x_3) - \mathfrak{f}(x_2,x_3,x_1)\ \in N. \tag{4.2}$$

Let us define

$$\begin{aligned}
\mathfrak{p}(x_1,x_2,x_3) &= \mathfrak{c}+\mathfrak{r} = (x_1x_2)x_3 + (x_1x_3)x_2 - (x_2x_3+x_3x_2)x_1, \\
\mathfrak{q}(x_1,x_2,x_3) &= \mathfrak{c}+(123)\mathfrak{l} = x_1(x_2x_3+x_3x_2) - x_2(x_3x_1) - x_3(x_2x_1),
\end{aligned}$$

where $\mathfrak{l}$, $\mathfrak{r}$ and $\mathfrak{c}$ are defined in (3.1), (3.2) and (3.4), respectively. These are identities of $\mathbb{O}_0$, symmetric under switching $x_2 \leftrightarrow x_3$.

PROPOSITION 4.5. *Assume A, B are alternative. Substituting $x_i = y_i\otimes z_i$ in $\phi(x_1,x_2,x_3)$ and taking $\mathbf{y} = (y_1,y_2,y_3)$ and $\mathbf{z} = (z_1,z_2,z_3)$ yields*

$$\mathfrak{a}(\mathbf{y})\otimes\mathfrak{p}(\mathbf{z}) + \mathfrak{q}(\mathbf{y})\otimes\mathfrak{a}(\mathbf{z}).$$

PROOF. Direct computation using Lemma 4.4. □

THEOREM 4.6. *The polynomial ϕ is an identity of both $\mathbb{O}_0\otimes\mathbb{O}_0$ and $\mathrm{M}_2(F)_0\otimes\mathbb{O}$.*

PROOF. The first claim follows since $\mathfrak{p}(z) = \mathfrak{q}(y) = 0$ for any $y_i, z_i \in \mathbb{O}_0$. Similarly, $\mathfrak{a}(y) = \mathfrak{q}(y) = 0$ for any $y_i \in \mathrm{M}_2(F)_0$, because $\mathrm{M}_2(F)_0 \subseteq \mathbb{O}_0$ and is associative. □

REMARK 4.7. We have proved that ϕ is an identity of $\mathrm{M}_2(F)_0\otimes B$ for any alternative algebra B.

Being a linear combination of substitutions in $\mathfrak{f}$, the identity ϕ clearly holds in any flexible algebra. In fact, in characteristic prime to 6 an algebra is flexible if and only if it satisfies ϕ and $(x,x,x) = 0$ (whose linearization is the sum of associators). However:

REMARK 4.8.

(1) The space $\mathrm{M}_2(F)_0\otimes\mathbb{O}_0$ is not flexible, because $(x,x,x) \neq 0$ for generic $x \in \operatorname{span}\{i\otimes i, j\otimes j, ij\otimes k\}$.

(2) In particular, neither $\mathbb{O}_0 \otimes \mathbb{O}_0$ nor $\mathrm{M}_2(F)_0 \otimes \mathbb{O}$ are flexible.
(3) $\mathrm{M}_2(F) \otimes \mathbb{O}_0$ does not satisfy ϕ (because for any space W, if $\mathrm{M}_2(F) \otimes W$ satisfies any linear combination of permutations of the associator then W must be associative, by substituting $e_{11}w_1$, $e_{12}w_2$ and $e_{22}w_3$ for $w_1, w_2, w_3 \in W$).

5. Identities of degree 4

The matrix algebra $\mathrm{M}_2(F) \otimes \mathbb{O}_0 = \mathrm{M}_2(\mathbb{O}_0)$ is the only space participating in Figure 1 whose PI degree is 4. In fact, this space satisfies a one-variable identity of degree 4. In this section we prove this identity, and report on the single-variable identities of degree 4 of the spaces discussed in the previous section.

5.1. An identity of $\mathrm{M}_2(\mathbb{O}_0)$. We start with the following observation:

PROPOSITION 5.1. *The polynomial*

$$p_4(x_1, x_2, x_3, x_4) = ((x_1x_2)x_3)x_4 - x_1(x_2(x_3x_4)) \tag{5.1}$$

is alternating on the space $\mathbb{O}_0$.

PROOF. Since $\mathbb{O}$ is a quadratic algebra, $x^2 \in F$ for every $x \in \mathbb{O}_0$. It is easy to verify that p_4 vanishes under every substitution $x_{i+1} \mapsto x_i$ using the fact that any two-generated subspace of $\mathbb{O}_0$ is associative. □

THEOREM 5.2. *Let V be a subspace of an alternative algebra. Let $f(x_1, \dots, x_4)$ be an alternating form of degree 4 on V. Then $f(w, \dots, w) = 0$ for any $w \in \mathrm{M}_2(F) \otimes V$.*

PROOF. Write $w = \sum a_{ij}e_{ij}$ for $a_{ij} \in V$. Then

$$f(w, \dots, w) = \sum_{i_0, \dots, i_4} f(a_{i_0i_1}, \dots, a_{i_3i_4})e_{i_0i_4},$$

summing over all $i_0, \dots, i_4 \in \{1, 2\}$. But since f is alternating over V, it suffices to sum over $(i_0, \dots, i_4)$ for which the pairs $(i_0, i_1), \dots, (i_3, i_4)$ are distinct. Since there are only four pairs, $(1, 1)$, $(1, 2)$, $(2, 1)$, and $(2, 2)$, each of the pairs must appear once, forming a Hamiltonian path. There are four Hamiltonian paths on the complete directed graph (with loops) on $\{1, 2\}$, summing up to

$$\begin{aligned} f(w, w, w, w) &= (f(a_{11}, a_{12}, a_{22}, a_{21}) + f(a_{12}, a_{22}, a_{21}, a_{11}))e_{11} \\ &+ (f(a_{22}, a_{21}, a_{11}, a_{12}) + f(a_{21}, a_{11}, a_{12}, a_{22}))e_{22}. \end{aligned}$$

But this is zero because $f(a, b, c, d) + f(d, a, b, c) = 0$ over V by the alternating property. □

Taking the alternating form p_4 of (5.1) in the theorem, we obtain that $p_4(x, x, x, x)$ is an identity for $\mathrm{M}_2(F) \otimes \mathbb{O}_0$:

COROLLARY 5.3. *The space of matrices $\mathrm{M}_2(F) \otimes \mathbb{O}_0$ satisfies the identity*

$$((xx)x)x - x(x(xx)).$$

REMARK 5.4. The identity $p = ((xx)x)x - x(x(xx))$ follows from $(x, x, x) = 0$ because $2p = (x, x, x^2) + (x, x^2, x) + (x^2, x, x) + x \circ (x, x, x)$.

(However, $\mathrm{M}_2(F) \otimes \mathbb{O}_0$ does not satisfy other power-associative identities such as $((xx)x)x = (xx)(xx)$).

5.2. Symmetric identities of degree 4 of spaces of PI degree 3. The multilinearization of a polynomial of degree n in one variable is symmetric under the action of S_n; namely it belongs to the trivial submodule of the module V_n^{na} of nonassociative polynomials. For $n = 4$, this trivial submodule is spanned by $(x^2x)x$, $(xx^2)x$, $(xx)^2$, $x(x^2x)$, and $x(xx^2)$.

Let $\mathrm{id}_n^{(1)}(V)$ denote the space of multilinear identities of degree n of V which are elementwise-symmetric under the action of the symmetric group S_n. We describe this space for the three spaces of PI degree 3 which were discussed in Section 4. Let

$$\begin{aligned} f_1 &= 4[(x,x,x),x] + 3(x,x,x^2) - 3(x^2,x,x) \\ &= (x^2x)x - 4(xx^2)x + 6(x^2)^2 - 4x(x^2x) + x(xx^2). \end{aligned}$$

There is an action $f \mapsto f^*$ of $\mathbb{Z}/2\mathbb{Z}$ on the space of nonassociative multilinear polynomials of degree n by reversing the order of monomials; for example $x_1((x_2x_3)x_4) \mapsto (x_4(x_3x_2))x_1$. Since $\mathbb{O}$ and thus $\mathbb{O}\otimes\mathbb{O}$ are algebras with involution, if f is an identity then so is f^*. In particular, the spaces $\mathrm{id}_4^{(1)}(V)$ decompose as a direct sum of symmetric and antisymmetric identities. For example, f_1 is symmetric. Fact 5.5 provides the space of identities of a single variable, as a direct sum of the symmetric and antisymmetric subspaces.

FACT 5.5. The spaces of identities of degree 4 in a single variable are:

$$\begin{aligned} \mathrm{id}_4^{(1)}(\mathrm{M}_2(F)_0\otimes\mathbb{O}_0) &= \langle f_1\rangle \oplus \langle (x,x,x)\circ x,\ (x,x^2,x)\rangle, \\ \mathrm{id}_4^{(1)}(\mathbb{O}_0\otimes\mathbb{O}_0) &= \langle f_1\rangle \oplus \langle (x,x,x)\circ x - (x,x^2,x)\rangle, \\ \mathrm{id}_4^{(1)}(\mathrm{M}_2(F)_0\otimes\mathbb{O}) &= \langle f_1\rangle \oplus \langle (x,x,x)\circ x + 3(x,x^2,x)\rangle, \\ \mathrm{id}_4^{(1)}(\mathrm{M}_2(F)\otimes\mathbb{O}_0) &= 0 \oplus \langle (x,x,x)\circ x + (x,x^2,x)\rangle. \end{aligned}$$

(the latter is the identity of Corollary 5.3).

This was verified by substituting basis elements in the generic linear combination of the 5 monomials on a single variables.

6. Identities of degree 7

We now treat $\mathbb{O}\otimes\mathbb{O}$ and its subalgebra $\mathrm{M}_2(F)\otimes\mathbb{O}$. Again our computer verified that there are no identities of lower degree:

FACT 6.1. The algebra $\mathbb{O}\otimes\mathbb{O}$ has no identities of degree 6. (This is true for the subspaces $\mathrm{M}_2(F)\otimes\mathbb{O}$ and $\mathbb{O}\otimes\mathbb{O}_0$ as well).

What we show in this section is that $\mathbb{O}\otimes\mathbb{O}$ does satisfy identities of degree 7. We exhibit identities of the form

$$h''(x_1,x_2,h'(x_3,x_4,h(x_5,x_6,x_7))),$$

where each of h, h', h'' is multilinear of degree 3. In fact, for every $x, x' \in \mathbb{O}\otimes\mathbb{O}$ we have that

$$\begin{aligned} h(x,x',y) &\in \mathbb{O}_0\otimes\mathbb{O}_0 && \text{for any } y \in \mathbb{O}\otimes\mathbb{O}; \\ h'(x,x',y) &\in \mathbb{O}\otimes F + F\otimes\mathbb{O} && \text{for any } y \in \mathbb{O}_0\otimes\mathbb{O}_0; \\ h''(x,x',y) &= 0 && \text{for any } y \in \mathbb{O}\otimes F + F\otimes\mathbb{O}. \end{aligned}$$

6.1. From $\mathbb{O}\otimes\mathbb{O}$ to $\mathbb{O}_0\otimes\mathbb{O}_0$. The algebra $\mathbb{O}$ is endowed with a trace function $\mathrm{tr}:\mathbb{O}\to F$. The trace vanishes on associators, so we treat expressions under the trace as associative. We define

$$\mathrm{tr}_1:\mathbb{O}\otimes\mathbb{O}\to F\otimes\mathbb{O} \qquad \text{by} \qquad \mathrm{tr}_1(y\otimes z)=\mathrm{tr}(y)\otimes z,$$

and

$$\mathrm{tr}_2:\mathbb{O}\otimes\mathbb{O}\to\mathbb{O}\otimes F \qquad \text{by} \qquad \mathrm{tr}_2(y\otimes z)=y\otimes\mathrm{tr}(z).$$

LEMMA 6.2. *For $y_i, z_i\in\mathbb{O}$, let $x_i=y_i\otimes z_i$. Then*

$$\mathrm{tr}_1(\mathfrak{a}(x_1,x_2,x_3))=\mathrm{tr}(y_1y_2y_3)\otimes\mathfrak{a}(z)$$

and

$$\mathrm{tr}_2(\mathfrak{a}(x_1,x_2,x_3))=\mathfrak{a}(y)\otimes\mathrm{tr}(z_1z_2z_3).$$

PROOF. Immediate from Lemma 4.1. □

LEMMA 6.3. *Recall $\mathfrak{l}$, $\mathfrak{r}$ and $\mathfrak{f}$ from (3.1)–(3.3). Write $x_i=y_i\otimes z_i$ and let $\alpha=\mathrm{tr}(y_1y_2y_3)$ and $\alpha'=\mathrm{tr}(y_1y_3y_2)$. Then*

$$\mathrm{tr}_1(\mathfrak{l})=\mathrm{tr}_1(\mathfrak{r})=\mathrm{tr}_1(\mathfrak{f})=(\alpha-\alpha')\otimes\mathfrak{a}(z).$$

PROOF. Immediate from Lemma 4.4, as the trace of an associator is zero. □

Since $\ker(\mathrm{tr}_1)\cap\ker(\mathrm{tr}_2)=\mathbb{O}_0\otimes\mathbb{O}_0$, we have proved:

COROLLARY 6.4. *If $\lambda_1,\lambda_2,\lambda_3\in F$ sum to zero, then the image of*

$$\lambda_1\mathfrak{l}+\lambda_2\mathfrak{f}+\lambda_3\mathfrak{r}$$

under substitution of $x_1,x_2,x_3\in\mathbb{O}\otimes\mathbb{O}$ is in $\mathbb{O}_0\otimes\mathbb{O}_0$.

6.2. From $\mathbb{O}_0\otimes\mathbb{O}_0$ to $\mathbb{O}\otimes F+F\otimes\mathbb{O}$. Consider the polynomial

$$s(x,x',w)=\mathfrak{l}(x,x',w)+\mathfrak{r}(w,x,x'),$$

which is a linearization of $(x,x,w)+(w,x,x)$.

Opening parenthesis, one checks that in any alternative algebra,

$$(xa)b-(xb)a \;=\; x[a,b]+2(a,b,x), \tag{6.1}$$

$$a(bx)-b(ax) \;=\; [a,b]x-2(a,b,x). \tag{6.2}$$

LEMMA 6.5. *If A,B are alternative, then for any $y_1,y_2,y_3\in A$ and $z_1,z_2,z_3\in B$ we have that*

$$s(y_1\otimes z_1,y_2\otimes z_2,y_3\otimes z_3)=\mathfrak{a}(y)\otimes(z_3\circ[z_1,z_2])+(y_3\circ[y_1,y_2])\otimes\mathfrak{a}(z). \tag{6.3}$$

PROOF. Use Lemma 4.4 and (6.1)–(6.2) to get

$$\begin{aligned} s(y_1\otimes z_1,y_2\otimes z_2,y_3\otimes z_3) &= (y_1\otimes z_1,y_2\otimes z_2,y_3\otimes z_3)+(y_2\otimes z_2,y_1\otimes z_1,y_3\otimes z_3)\\ &\quad +(y_3\otimes z_3,y_1\otimes z_1,y_2\otimes z_2)+(y_3\otimes z_3,y_2\otimes z_2,y_1\otimes z_1)\\ &= (y_1,y_2,y_3)\otimes[(z_1z_2)z_3-(z_2z_1)z_3+(z_3z_1)z_2-(z_3z_2)z_1]\\ &\quad +[y_1(y_2y_3)-y_2(y_1y_3)+y_3(y_1y_2)-y_3(y_2y_1)]\otimes(z_1,z_2,z_3)\\ &= (y_1,y_2,y_3)\otimes(z_3\circ[z_1,z_2]+2(z_1,z_2,z_3))\\ &\quad +(y_3\circ[y_1,y_2]-2(y_1,y_2,y_3))\otimes(z_1,z_2,z_3), \end{aligned}$$

and cancel. □

COROLLARY 6.6. *Let $y_1,y_2,y_3\in\mathbb{O}$ and $z_1,z_2,z_3\in\mathbb{O}$, and write $x_i=y_i\otimes z_i$. Assume $y_3,z_3\in\mathbb{O}_0$. Then $s(x_1,x_2,x_3)\in\mathbb{O}\otimes F+F\otimes\mathbb{O}$.*

PROOF. By linearity, we may assume y_i and z_i are elements in the standard basis of $\mathbb{O}$, whose elements either commute or anticommute. The value of $s(x_1, x_2, x_3)$ is then a monomial (namely a scalar multiple of a tensor product of basis elements). By assumption $y_3, z_3 \notin F$. If $y_1y_2y_3 \in F$ or $z_1z_2z_3 \in F$ then we are done as the value of s is in the desired space. So we assume that $y_3 \notin Fy_1y_2$ and $z_3 \notin Fz_1z_2$.

If $y_1y_2 \in F$ then $(y_1, y_2, y_3) = [y_1, y_2] = 0$; likewise if $z_1z_2 \in F$ then $(z_1, z_2, z_3) = [z_1, z_2] = 0$; in both cases $s(x_1, x_2, x_3) = 0$ by the lemma.

Otherwise, the conditions imply that y_3 does not commute with y_1y_2, so they anticommute, and $y_3 \circ [y_1, y_2] = 0$; and z_3 does not commute with z_1z_2, so similarly $z_3 \circ [z_1, z_2] = 0$. Now (6.3) gives $s(x_1, x_2, x_3) = 0$. □

6.3. Annihilating $\mathbb{O}\otimes F + F\otimes\mathbb{O}$. Let A, B be alternative algebras over F. Write $x_i = y_i\otimes z_i$, where $y_1, y_2, y_3 \in A$ and $z_1, z_2, z_3 \in B$. By Lemma 4.4 (noting the cyclic rotation for $\mathfrak{r}$ and $\mathfrak{f}$), we have that if $y_3 = 1$ then

$$\begin{aligned}
\mathfrak{l}(y_1\otimes z_1, y_2\otimes z_2, y_3\otimes z_3) &= [y_1, y_2]\otimes(z_1, z_2, z_3);\\
\mathfrak{f}(y_2\otimes z_2, y_3\otimes z_3, y_1\otimes z_1) &= [y_2, y_1]\otimes(z_2, z_3, z_1);\\
\mathfrak{r}(y_3\otimes z_3, y_1\otimes z_1, y_2\otimes z_2) &= [y_1, y_2]\otimes(z_3, z_1, z_2);
\end{aligned}$$

and similarly if $z_3 = 1$ then

$$\begin{aligned}
\mathfrak{l}(y_1\otimes z_1, y_2\otimes z_2, y_3\otimes z_3) &= (y_1, y_2, y_3)\otimes[z_1, z_2];\\
\mathfrak{f}(y_2\otimes z_2, y_3\otimes z_3, y_1\otimes z_1) &= (y_2, y_3, y_1)\otimes[z_2, z_1];\\
\mathfrak{r}(y_3\otimes z_3, y_1\otimes z_1, y_2\otimes z_2) &= (y_3, y_1, y_2)\otimes[z_1, z_2].
\end{aligned}$$

COROLLARY 6.7. *For any $\lambda_1, \lambda_2, \lambda_3 \in F$, if $\lambda_1 - \lambda_2 + \lambda_3 = 0$ then*

$$h(x_1, x_2, x_3) = \lambda_1\mathfrak{l}(x_1, x_2, x_3) + \lambda_2\mathfrak{f}(x_2, x_3, x_1) + \lambda_3\mathfrak{r}(x_3, x_1, x_2)$$

satisfies $h(x_1, x_2, x_3) = 0$ for any elements $x_1, x_2 \in \mathbb{O}\otimes\mathbb{O}$ and $x_3 \in \mathbb{O}\otimes F + F\otimes\mathbb{O}$.

COROLLARY 6.8. *We have that $(x, x, x) = 0$ for any $x \in \mathbb{O}\otimes F + F\otimes\mathbb{O}$.*

6.4. Conclusion. We now prove Theorem 1.2.

THEOREM 6.9. *The algebra $\mathbb{O}\otimes\mathbb{O}$ satisfies nonassociative polynomial identities of degree 7.*

PROOF. The quadratic operators $\mathcal{L}_x = L_{x^2} - L_x^2$, $\mathcal{C}_x = [R_x, L_x]$ and $\mathcal{R}_x = R_x^2 - R_{x^2}$ linearize to $\mathfrak{l}(x, x', \cdot)$, $\mathfrak{f}(x, \cdot, x')$ and $\mathfrak{r}(\cdot, x, x')$, respectively. Notice that these functions are symmetric with respect to switching x, x'. In Corollary 6.4 we saw that any operator from $\mathrm{span}\{\mathcal{L}_x - \mathcal{R}_x, \mathcal{L}_x - \mathcal{C}_x\}$ carries $\mathbb{O}\otimes\mathbb{O}$ to $\mathbb{O}_0\otimes\mathbb{O}_0$. In Corollary 6.7 we saw that any operator from $\mathrm{span}\{\mathcal{L}_x - \mathcal{R}_x, \mathcal{L}_x + \mathcal{C}_x\}$ annihilates the space $\mathbb{O}\otimes F+F\otimes\mathbb{O}$. By Corollary 6.6, we have proved that for any $x_1, x_2, x_3, y \in \mathbb{O}\otimes\mathbb{O}$,

$$\begin{aligned}
(\mathcal{L}_{x_1} - \mathcal{R}_{x_1})(\mathcal{L}_{x_2} + \mathcal{R}_{x_2})(\mathcal{L}_{x_3} - \mathcal{R}_{x_3})(y) &= 0;\\
(\mathcal{L}_{x_1} - \mathcal{R}_{x_1})(\mathcal{L}_{x_2} + \mathcal{R}_{x_2})(\mathcal{L}_{x_3} - \mathcal{C}_{x_3})(y) &= 0;\\
(\mathcal{L}_{x_1} + \mathcal{C}_{x_1})(\mathcal{L}_{x_2} + \mathcal{R}_{x_2})(\mathcal{L}_{x_3} - \mathcal{R}_{x_3})(y) &= 0;\\
(\mathcal{L}_{x_1} + \mathcal{C}_{x_1})(\mathcal{L}_{x_2} + \mathcal{R}_{x_2})(\mathcal{L}_{x_3} - \mathcal{C}_{x_3})(y) &= 0.
\end{aligned}$$

□

In particular, $\mathbb{O}\otimes\mathbb{O}$ satisfies an identity in two variables which is linear in one variable and has degree 6 in the other, namely

$$(\mathcal{L}_x - \mathcal{R}_x)(\mathcal{L}_x + \mathcal{R}_x)(\mathcal{L}_x - \mathcal{R}_x) = 0.$$

This shows that $\mathbb{O}\otimes\mathbb{O}$ has PI degree 7, as does $\mathrm{M}_2(\mathbb{O})$.

REMARK 6.10. In [**3**] Bremner and Hentzel find the nonassociative identities of degree 7 satisfied by $\mathrm{M}_2(F)\otimes A$ when A is the free alternative algebra, and in particular by $\mathrm{M}_2(F)\otimes\mathbb{O}$. The identities of Theorem 6.9 coincide with the ones given in Theorem 1 of [**3**].

Along these lines, one might look for identities satisfied by $\mathbb{O}\otimes A$ for the free alternative algebra; however, such identities would hold in $\mathbb{O}\otimes\mathrm{M}_n(F)$ for any n, so must be sorted by Proposition 2.5.

QUESTION 6.11. Does $\mathbb{O}$ satisfy any sorted identities?

References

[1] Alexei Kanel-Belov, Yakov Karasik, and Louis Halle Rowen, *Computational aspects of polynomial identities. Vol. 1: Kemer's theorems*, 2nd ed., Monographs and Research Notes in Mathematics, CRC Press, Boca Raton, FL, 2016. MR3445601

[2] Murray Bremner and Irvin Hentzel, *Identities for the associator in alternative algebras*, J. Symbolic Comput. **33** (2002), no. 3, 255–273, DOI 10.1006/jsco.2001.0510. MR1882229

[3] Murray Bremner and Irvin Hentzel, *Identities for algebras of matrices over the octonions*, J. Algebra **277** (2004), no. 1, 73–95, DOI 10.1016/j.jalgebra.2004.03.001. MR2059621

[4] Murray R. Bremner, Sara Madariaga, and Luiz A. Peresi, *Structure theory for the group algebra of the symmetric group, with applications to polynomial identities for the octonions*, Comment. Math. Univ. Carolin. **57** (2016), no. 4, 413–452, DOI 10.14712/1213-7243.2015.188. MR3583300

[5] Fernando Henry, *Some graded identities of the Cayley-Dickson algebra*, Internat. J. Algebra Comput. **28** (2018), no. 5, 697–718, DOI 10.1142/S0218196718500327. MR3844699

[6] A. V. Il′tyakov, *Finiteness of the basis of identities of a finitely generated alternative* PI-*algebra over a field of characteristic zero* (Russian), Sibirsk. Mat. Zh. **32** (1991), no. 6, 61–76, 204, DOI 10.1007/BF00971199; English transl., Siberian Math. J. **32** (1991), no. 6, 948–961 (1992). MR1156745

[7] Irvin R. Hentzel and Luiz A. Peresi, *Identities of Cayley-Dickson algebras*, J. Algebra **188** (1997), no. 1, 292–309, DOI 10.1006/jabr.1996.6814. MR1432358

[8] Michel L. Racine, *Minimal identities of octonion algebras*, J. Algebra **115** (1988), no. 1, 251–260, DOI 10.1016/0021-8693(88)90294-3. MR937613

[9] Amitai Regev, *Existence of polynomial identities in $A\otimes_F B$*, Bull. Amer. Math. Soc. **77** (1971), 1067–1069, DOI 10.1090/S0002-9904-1971-12869-0. MR284468

[10] Amitai Regev, *Nonassociative P.I. algebras*, J. Algebra **28** (1974), 247–252, DOI 10.1016/0021-8693(74)90036-2. MR332904

[11] Ivan Shestakov and Natalia Zhukavets, *Skew-symmetric identities of octonions*, J. Pure Appl. Algebra **213** (2009), no. 4, 479–492, DOI 10.1016/j.jpaa.2008.07.012. MR2483833

[12] Uzi Vishne, *Polynomial identities of $M_2(G)$*, Comm. Algebra **30** (2002), no. 1, 443–454, DOI 10.1081/AGB-120006502. MR1880684

[13] Uzi Vishne, *Polynomial identities of $M_{2,1}(G)$*, Comm. Algebra **39** (2011), no. 6, 2044–2050, DOI 10.1080/00927871003667486. MR2813162

[14] Yu. P. Razmyslov, *Identities of algebras and their representations*, Translations of Mathematical Monographs, vol. 138, American Mathematical Society, Providence, RI, 1994. Translated from the 1989 Russian original by A. M. Shtern, DOI 10.1090/mmono/138. MR1291603

[15] K. A. Zhevlakov, A. M. Slin′ko, I. P. Shestakov, and A. I. Shirshov, *Rings that are nearly associative*, Pure and Applied Mathematics, vol. 104, Academic Press, Inc. [Harcourt Brace Jovanovich, Publishers], New York-London, 1982. Translated from the Russian by Harry F. Smith. MR668355

Department of Mathematics, Bar Ilan University, Ramat Gan, Israel
Email address: guy.blachar@gmail.com

Department of Mathematics, Bar Ilan University, Ramat Gan, Israel
Email address: rowen@math.biu.ac.il

Department of Mathematics, Bar Ilan University, Ramat Gan, Israel
Email address: vishne@math.biu.ac.il

Contemporary Mathematics
Volume **785**, 2023
https://doi.org/10.1090/conm/785/15774

When Is Every Non Central-Unit A Sum of Two Nilpotents?

Simion Breaz and Yiqiang Zhou

Dedicated to Professor Syed Tariq Rizvi for his contribution to Ring and Module Theory

ABSTRACT. A ring is said to satisfy the 2-nil-sum property if every non central-unit is the sum of two nilpotents. We prove that a ring satisfies the 2-nil-sum property iff it is either a simple ring with the 2-nil-sum property or a commutative local ring with nil Jacobson radical, and we provide an example of a simple ring with the 2-nil-sum property that is not commutative. Moreover, a simple right Goldie ring has the 2-nil-sum property iff it is a field.

1. Introduction

Throughout, rings are associative with identity. We start by recalling three special types of elements in a ring. An element in a ring is 2-good if it is a sum of two units (see [**18**]), is fine if it is a sum of a unit and a nilpotent (see [**4**]), and is 2-nilgood if it is a sum of two nilpotents (see [**3**]). By the terminology of Vámos [**18**], a ring is 2-good if each element is 2-good. The study of 2-good rings (also called rings with the 2-sum property in the literature), initiated by Wolfson [**19**] and Zelinsky [**20**], has attracted considerable interest (see, for example, [**2**], [**11**], [**14**], [**17**], [**18**] and the references there). A ring is called a fine ring if each nonzero element is fine (the zero element being fine implies that the ring is trivial). Fine rings were introduced and extensively investigated by Călugăreanu and Lam [**4**].

Motivated by the notions of 2-good rings and fine rings, we define a ring to satisfy the 2-*nil-sum property* if every element that is not a central unit is 2-nilgood (a central unit in a ring being 2-nilgood implies that the ring is trivial). These rings belong to a larger class of rings for which every element is either a unit or a sum of two units. One easily sees that a ring without nonzero nilpotents satisfies the 2-nil-sum property iff it is a field. The rings which are additively generated by nilpotents have been studied by several authors with connections to rings additively generated by commutators. We refer to [**5**] and [**10**] for more details. As far as we are aware, the 2-nil-sum property of rings has not been discussed in the literature. Here a

2020 *Mathematics Subject Classification.* Primary 16U99; 16S50.

Key words and phrases. Nilpotent, unit, 2-nil-sum property, matrix ring, simple ring.

The first author was supported by a grant of the Ministry of Research, Innovation and Digitization, CNCS/CCCDI – UEFISCDI, project number PN-III-P4-ID-PCE-2020-0454, within PNCDI III.

The second author was supported by a Discovery Grant (RGPIN-2016-04706) from NSERC of Canada.

study of this topic is conducted. The main results proved here are the following: A ring satisfies the 2-nil-sum property iff it is either a simple ring with the 2-nil-sum property or a commutative local ring with nil Jacobson radical (Theorem 2.2); there exists simple rings with the 2-nil-sum property that are not commutative (Example 3.1); a simple right Goldie ring has the 2-nil-sum property iff it is a field (Theorem 3.8).

For a ring R, we denote by $C(R)$, $J(R)$, $U(R)$, and $\mathrm{nil}(R)$, the center, the Jacobson radical, the group of units of R, and the set of nilpotents of R, respectively. We write $\mathbb{M}_n(R)$ for the ring of $n \times n$ matrices over R whose identity is denoted by I_n. By a *non central-unit* we mean an element that is not a central unit. For $i, j \in \{1, 2, \ldots, n\}$, we denote by E_{ij} the $n \times n$ matrix whose (i, j)-entry is 1 and all other entries are 0.

2. Basic properties and the reduction theorem

We start by recording some basic properties.

LEMMA 2.1. *Let R be a ring with the 2-nil-sum property. The following hold:*

(1) *$C(R)\backslash U(R) \subseteq \mathrm{nil}(R)$. Consequently, $C(R)$ is a local ring with nil Jacobson radical.*
(2) *For every proper ideal I of R we have $I \subseteq C(R) \cap \mathrm{nil}(R)$.*
(3) *All proper ideals of R are contained in $J(R)$.*
(4) *$J(R)$ is nil and $J(R) \subseteq C(R)$.*
(5) *For any ideal I of R, central units of R/I can be lifted to central units of R.*

PROOF. (1) Assume that r is a central element that is not a unit. Then $r = b_1 + b_2$, where $b_1, b_2 \in \mathrm{nil}(R)$. Since r is central, b_1 and b_2 commute. So $r = b_1 + b_2 \in \mathrm{nil}(R)$.

(2) Assume that $I\backslash C(R) \neq \emptyset$. Let $a \in I\backslash C(R)$. Then $1 + a$ is a sum of two nilpotents, so $\bar{1} \in R/I$ is a sum of two nilpotents, a contradiction. Hence $I \subseteq C(R)\backslash U(R)$, so $I \subseteq \mathrm{nil}(R)$ by (1).

(3) and (4) are clear by (2).

(5) Let $\bar{u} := u + I$ be a central unit of R/I. We claim that $u \in R$ is a central unit. Otherwise, u is a sum of two nilpotents in R. Hence $\bar{u}$ is a sum of two nilpotents in R/I, a contradiction. □

THEOREM 2.2 (Reduction Theorem). *A ring R satisfies the 2-nil-sum property iff R is either a simple ring with the 2-nil-sum property or a commutative local ring with nil Jacobson radical.*

PROOF. The sufficiency is clear. For the necessity, suppose that R satisfies the 2-nil-sum property but R is not simple. Then, by Lemma 2.1(3), $J(R) \neq 0$. Let $0 \neq x \in J(R)$. If $a, b \in R$ then, noting $J(R) \subseteq C(R)$, we have

$$(ab)x = a(bx) = (bx)a = b(xa) = b(ax) = (ba)x,$$

hence $(ab - ba)x = 0$. It follows that $ab - ba$ belongs to the ideal $Ann(x) = \{r \in R \mid rx = 0\}$. Since $Ann(x) \neq R$, $Ann(x) \subseteq J(R)$. It follows that $R/J(R)$ is a commutative ring with the 2-nil-sum property, and hence is local by Lemma 2.1(1). It follows that $R/J(R)$ is a field. Thus, by Lemma 2.1(5), for any $a \in R\backslash J(R)$, $a = u + j$ where u is a central unit and $j \in J(R)$. It follows that a is central. Hence, R is commutative, and so R is a commutative local ring with $J(R)$ nil. □

If R is a commutative ring with the 2-nil-sum property then every non central-unit $a \in R$ is nilpotent, so it is a sum of the form $a = b + c$ with $b^1 = 0$ and $c = a$ is nilpotent. Thus, R is a ring with the 2-nil-sum property of type $(1, \infty)$ as defined below. The 2-nil-sum property is refined as follows: Let p, q be integers with $p \geq 1$ and $q \geq 1$. We say that a ring R has the 2-nil-sum property of type (p, q) if, for each non central-unit a in R, $a = b + c$ where $b^p = c^q = 0$. The ring R has the 2-nil-sum property of type (p, ∞) if, for each non central-unit a in R, $a = b + c$ where $b^p = 0$ and c is nilpotent. Clearly, a ring R is a field iff R has the 2-nil-sum property of type $(1, 1)$.

PROPOSITION 2.3. *Suppose that R has the 2-nil-sum property of type $(2, \infty)$. Then R is commutative.*

PROOF. We observe that R has no non-trivial idempotents: If $1 \neq e^2 = e \in R$ is a sum of a square-zero element and a nilpotent it follows by [**7**, Proposition 2] that $e = 0$.

Suppose that R is not commutative. Then R is a simple ring by Theorem 2.2, hence 0 is the only central nilpotent. But $\mathrm{nil}(R) \neq 0$, hence there exists a square-zero element x that is not central. Since R is of type $(2, \infty)$, there exist $y, z \in R$ such that

$$1 - x = y + z,$$

where $y^2 = 0$ and $z^n = 0$ for some $n \geq 1$. Then $(1 - x - y)^n = 0$. It follows that

$$1 = n_1(x + y) + n_2(xy + yx) + n_3(xyx + yxy) + n_4(xyxy + yxyx) + \cdots,$$

where $n_1 = n$ and each n_i is an integer. Multiplying both sides of this equation from the left by x gives

$$x = n_1 xy + n_2 xyx + n_3 xyxy + n_4 xyxyx + \cdots$$

Multiplying both sides of this equality from the right by y gives

$$xy = n_2 xyxy + n_4 xyxyxy + \cdots = a(xy)^2 = (xy)^2 a,$$

where $a = n_2 + n_4 xy + \cdots$. It follows that $a(xy)$ is an idempotent. Since $x^2 = 0$, we have $a(xy) \neq 1$. We obtain $a(xy) = 0$, hence $xy = 0$. It follows that $x = 0$, a contradiction. □

PROPOSITION 2.4. *Suppose that R is of characteristic 0 and has the 2-nil-sum property of type (p, q) where $p \leq 3$ and $q \leq 5$. Then R is commutative.*

PROOF. We can assume that $p = 3$ and $q = 5$. Suppose that R is not commutative. As in the proof of Proposition 2.3, there exists a square-zero element x that is not central. So, there exist $y, z \in R$ such that $1 - x = y + z$ and $y^3 = z^5 = 0$. This contradicts [**16**, Proposition 9]. □

We close this section with the following

Open Question. If a ring R has the 2-nil-sum property of type (p, q) with $p \in \mathbb{N}$ and $q \in \mathbb{N} \cup \{\infty\}$, is it commutative?

3. Simple rings

EXAMPLE 3.1. There exists a simple ring with the 2-nil-sum property that is not commutative.

PROOF. Let $\mathbb{F}_2$ be the field of 2 elements. For each integer $n \geq 0$ we consider the diagonal morphism

$$\varphi_n : \mathbb{M}_{2^n}(\mathbb{F}_2) \to \mathbb{M}_{2^{n+1}}(\mathbb{F}_2), A \mapsto \operatorname{diag}(A, A).$$

This is an inductive system, and we denote its colimit by R, while $\varphi_n : \mathbb{M}_{2^n}(\mathbb{F}_2) \to R$ denote the canonical ring morphisms. Then R is a simple ring (see [**8**, Example 8.1]).

Let $0 \neq r \in R$ a non central-unit of R. Then we can view it as an image $r = \varphi_n(A)$ for some $0 \neq A \in \mathbb{M}_{2^n}(\mathbb{F}_2)$. Note that $A \neq I_{2^n}$ since otherwise r is the identity of R. Observe that $r = \varphi_{n+1}(\operatorname{diag}(A, A))$. From $A \neq I_{2^n}$ it follows that $\operatorname{diag}(A, A)$ is not a scalar matrix. Since its trace is zero, there exist two nilpotent matrices N_1 and N_2 in $\mathbb{M}_{2^{n+1}}(\mathbb{F}_2)$ such that $\operatorname{diag}(A, A) = N_1 + N_2$, see [**3**, Proposition 3(ii)]. Then $r = \varphi_{n+1}(N_1) + \varphi_{n+1}(N_2)$ is a sum of two nilpotents. □

Similar examples can be obtained by using other fields (of positive characteristic). We do not know other kinds of examples of non-commutative rings with the 2-nil-sum property. In fact, we will prove that if a simple ring with the 2-nil-sum property satisfies some standard finiteness conditions then it is a field. A basic example is the following.

EXAMPLE 3.2. Let F be a field and $n \geq 2$ an integer. Since the traces of all nilpotent matrices in $\mathbb{M}_n(F)$ are 0, it follows that $\mathbb{M}_n(F)$ does not have the 2-nil-sum property. This can be easily extended to matrices over commutative rings (since in this case the trace of a nilpotent matrix has to be nilpotent).

Below we will study matrix rings over division rings. In this case, the above argument does not work because nilpotent matrices over division rings may have non-zero traces and, moreover, the trace is no longer invariant under similarity. For instance, if $A = \begin{pmatrix} \mathbf{i} & \mathbf{j} \\ -\mathbf{j} & \mathbf{i} \end{pmatrix} \in \mathbb{M}_2(\mathbb{H})$, where $\mathbb{H}$ is the ring of real quaternions, then $A^2 = 0$ and $\begin{pmatrix} 1 & 0 \\ \mathbf{k} & 1 \end{pmatrix} A \begin{pmatrix} 1 & 0 \\ -\mathbf{k} & 1 \end{pmatrix} = \begin{pmatrix} 0 & \mathbf{j} \\ 0 & 0 \end{pmatrix}$. We start with some elementary lemmas. We include the details of the proofs for the reader's convenience.

LEMMA 3.3. *Let D be a division ring. If $A \in \mathbb{M}_n(D)$ is nilpotent, then $A^n = 0$.*

PROOF. We identify A with a linear transformation of the right vector space $V := D^n$ over D. Then the chain

$$Im(A) \supset Im(A^2) \supset Im(A^3) \supset \cdots$$

should be strictly decreasing (otherwise, A is not nilpotent). Since $\dim(V_D) = n$, it follows that $A^n = 0$. □

LEMMA 3.4. *Let R be a ring and $n \geq 2$. If* $A = \begin{pmatrix} a_{11} & \cdots & \cdots & a_{1n} \\ 1 & \ddots & \ddots & \vdots \\ & \ddots & \ddots & \vdots \\ & & 1 & a_{nn} \end{pmatrix} \in$ $\mathbb{M}_n(R)$*, then*

$$A^k = \left(\begin{array}{cccc|c} & & * & & * \\ \hline \sum_{i=1}^{k} a_{ii} & \cdots & \cdots & \cdots & \\ 1 & \ddots & \ddots & \vdots & * \\ & \ddots & \ddots & \vdots & \\ & & 1 & \sum_{i=n-k+1}^{n} a_{ii} & \end{array}\right),$$

where the lower left block has size $(n-k+1) \times (n-k+1)$. So the $(k,1)$-entry of A^k is $a_{11} + \cdots + a_{kk}$ for $k = 1, \ldots, n$; in particular, the $(n,1)$-entry of A^n is $a_{11} + \cdots + a_{nn}$.

PROOF. The claim can be proved by induction on k. □

LEMMA 3.5. *Let D be a division ring and $n \geq 2$ an integer. If $X = (x_{ij}) \in$ $\mathbb{M}_n(D)$ is a matrix such that the $(n-1)$-dimensional column vector* $\beta = \begin{bmatrix} x_{21} \\ \vdots \\ x_{n1} \end{bmatrix}$ *is not zero, then there exists an invertible matrix* $U = \begin{pmatrix} 1 & \mathbf{0} \\ \mathbf{0} & V \end{pmatrix} \in \mathbb{M}_n(D)$ *with* $V \in \mathbb{M}_{n-1}(D)$*, and $1 < k \leq n$ such that*

$$UXU^{-1} = \left(\begin{array}{cccc|c} w_{11} & \cdots & \cdots & w_{1k} & \\ 1 & \ddots & \ddots & \vdots & Y \\ & \ddots & \ddots & \vdots & \\ & & 1 & w_{kk} & \\ \hline & & \mathbf{0} & & Z \end{array}\right).$$

PROOF. Write $X = \begin{pmatrix} x_{11} & Y_1 \\ W_1 & Z_1 \end{pmatrix}$ in block form where Z_1 is an $(n-1) \times (n-1)$ matrix. Since $W_1 = \beta \neq 0$ there exists an invertible matrix V_1 in $\mathbb{M}_{n-1}(D)$ such that $V_1 W_1 = \begin{bmatrix} 1 \\ 0 \\ \vdots \\ 0 \end{bmatrix}$. Then $U_1 = \begin{pmatrix} 1 & \mathbf{0} \\ \mathbf{0} & V_1 \end{pmatrix}$ is invertible in $\mathbb{M}_n(D)$ and $U_1 X U_1^{-1} =$ $\begin{pmatrix} x_{11} & \gamma_1 V_1^{-1} \\ V_1 W_1 & V_1 X_1 V_1^{-1} \end{pmatrix}$, so the first column of $U_1 X U_1^{-1}$ is $\begin{bmatrix} x_{11} \\ 1 \\ 0 \\ \vdots \\ 0 \end{bmatrix}$.

Write $U_1XU_1^{-1} = \left(\begin{array}{cc|c} y_{11} & y_{12} & \multirow{2}{*}{Y_2} \\ 1 & y_{22} & \\ \hline \multicolumn{2}{c|}{W_2} & Z_2 \end{array}\right)$, where Z_2 is an $(n-2)\times(n-2)$ matrix and the first column of W_2 is $\mathbf{0}$. If $W_2 = \mathbf{0}$ we can simply take $U = U_1$. If $W_2 \neq \mathbf{0}$, there exists an invertible matrix V_2 in $\mathbb{M}_{n-2}(D)$ such that $V_2W_2 = \begin{bmatrix} 0 & 1 \\ 0 & 0 \\ \vdots & \vdots \\ 0 & 0 \end{bmatrix}$.

Observe that $U_2 = \begin{pmatrix} I_2 & \mathbf{0} \\ \mathbf{0} & V_2 \end{pmatrix}$ is invertible in $\mathbb{M}_n(D)$ and

$$U_2U_1XU_1^{-1}U_2^{-1} = \left(\begin{array}{ccc|c} z_{11} & z_{12} & z_{13} & \\ 1 & z_{22} & z_{23} & Y_3 \\ 0 & 1 & z_{33} & \\ \hline \multicolumn{3}{c|}{W_3} & Z_3 \end{array}\right).$$

The block consisting of the first two columns of $U_2U_1XU_1^{-1}U_2^{-1}$ is of the form $\begin{bmatrix} z_{11} & z_{12} \\ 1 & z_{22} \\ 0 & 1 \\ 0 & 0 \\ \vdots & \vdots \\ 0 & 0 \end{bmatrix}$, so the first two columns of W_3 are zero. If $W_3 = \mathbf{0}$, we can take $U = U_2U_1$. If $W_3 \neq \mathbf{0}$, we continue in a similar way.

This process has to stop after a finite number of steps. Therefore, there exists an integer $1 < k \leq n$, and a family of invertible matrices $U_l = \begin{pmatrix} I_l & \mathbf{0} \\ \mathbf{0} & V_l \end{pmatrix}$, $1 \leq l \leq k-1$, such that

- $(U_l \cdots U_1)X(U_l \cdots U_1)^{-1} = \left(\begin{array}{ccccc|c} w_{11} & w_{12} & \dots & w_{1l} & w_{1(l+1)} & \\ 1 & w_{22} & \dots & w_{2l} & w_{2(l+1)} & X_{l+1} \\ \vdots & \vdots & & \vdots & \vdots & \\ 0 & 0 & \dots & 1 & w_{(l+1)(l+1)} & \\ \hline \multicolumn{5}{c|}{W_{l+1}} & Y_{l+1} \end{array}\right)$,
- the first l columns of W_{l+1} are zero,
- $k = n$, or $k < n$ and $W_k = \mathbf{0}$.

Now take $U = U_{k-1} \cdots U_1$. By the constructions of $U_1, \ldots, U_{k-1}$, one easily sees that U has the desired property. □

REMARK 3.6. The above lemma will be applied for the case when X is nilpotent. We want to point out that it was proved in [**6**] that every nilpotent matrix is similar to a rational form (as in the commutative case). But we need the weaker property presented above since we have to use the form of the matrix U.

LEMMA 3.7. *Let D be a division ring and $A = (a_{ij}) \in \mathbb{M}_n(D)$ a matrix with the property that the k-th row of A is the only non-zero row of A. Then A is a sum of two nilpotents if and only if $a_{kk} = 0$.*

PROOF. The sufficiency is obvious. For the necessity, let us observe that A is similar to $A' = \begin{pmatrix} a & \alpha \\ \mathbf{0} & \mathbf{0} \end{pmatrix}$, where $a = a_{kk} \in D$ and α is an $(n-1)$-dimensional row vector. Hence $A' = X + Y$, where X, Y are nilpotents. Write $X = \begin{pmatrix} x_{11} & \gamma \\ \beta & X_1 \end{pmatrix}$ in block form where X_1 is an $(n-1)\times(n-1)$ matrix. If $\beta = 0$, then $a = x_{11} + (a - x_{11})$ is a sum of two nilpotents in D; so $a = 0$. Thus, we can assume that $\beta \neq 0$. There exists an invertible matrix $U = \begin{pmatrix} 1 & \mathbf{0} \\ \mathbf{0} & V \end{pmatrix} \in \mathbb{M}_n(D)$ and an integer $k > 1$ as in Lemma 3.5 such that

$$UXU^{-1} = \left(\begin{array}{cccc|c} w_{11} & \cdots & \cdots & w_{1k} & \\ 1 & \ddots & \ddots & \vdots & \\ & \ddots & \ddots & \vdots & T \\ & & 1 & w_{kk} & \\ \hline & & \mathbf{0} & & Z \end{array}\right).$$

Note that $UA'U^{-1} = \begin{pmatrix} a & (a_2 & \cdots & a_n) \\ \mathbf{0} & & \mathbf{0} & \end{pmatrix}$ for some $a_2, \ldots, a_n$ in D, and that UXU^{-1} and UYU^{-1} are nilpotent matrices. It follows that the matrices

$$\begin{pmatrix} w_{11} & \cdots & \cdots & w_{1k} \\ 1 & \ddots & \ddots & \vdots \\ & \ddots & \ddots & \vdots \\ & & 1 & w_{kk} \end{pmatrix} \text{ and } \begin{pmatrix} w_{11} - a & \cdots & \cdots & w_{1k} - a_k \\ 1 & \ddots & \ddots & \vdots \\ & \ddots & \ddots & \vdots \\ & & 1 & w_{kk} \end{pmatrix}$$

are both nilpotent matrices. Using Lemma 3.3 and Lemma 3.4, we obtain the equalities $w_{11} + \cdots + w_{kk} = 0 = w_{11} + \cdots + w_{kk} - a$, and hence $a = 0$. □

THEOREM 3.8. *Let R be a simple right Goldie ring with the 2-nil-sum property. Then R is a field.*

PROOF. Note that we can view R as a right order of $\mathbb{M}_n(D)$ where $n \geq 1$ and D is a division ring. If $n = 1$, then R is a subring of a division ring. Since $\mathrm{nil}(R) = 0$, the only non central-unit of R is 0, so R is a field.

We next show that $n \geq 2$ yields a contradiction. Assume that $n \geq 2$. For all $1 \leq k \leq n$, we denote by $\mathcal{A}_k$ the set of all non-zero matrices $A = (a_{ij}) \in R$ with the property that the k-th row of A is the only non-zero row of A.

Since R is a right order of $\mathbb{M}_n(D)$, for $1 \leq k \leq n$ we can write $E_{kk} = U_k S_k^{-1}$ where $U_k, S_k \in R$ and where $U_k = (u_{ij}^{(k)})$ and $S_k = (s_{ij}^{(k)})$. It follows that $U_k = E_{kk} S_k \in \mathcal{A}_k$, so all sets $\mathcal{A}_k$ are non-empty.

Moreover, if $1 \leq k \leq n$, we denote by $i(k)$ the minimal integer with the property that there exists $(a_{ij}) \in \mathcal{A}_k$ such that $a_{kl} = 0$ for all $l < i(k)$ and $a_{ki(k)} \neq 0$.

We show that if $k < i(k)$ then $i(k) < i(i(k))$. In order to prove this, let $A = (a_{ij}) \in \mathcal{A}_k$ be a matrix such that $a_{kl} = 0$ for all $l < i(k)$ and $a := a_{ki(k)} \neq 0$. For any $B = (b_{ij}) \in \mathcal{A}_{i(k)}$, the k-th row of AB is $(ab_{i(k)1}, \ldots, ab_{i(k)i(k)}, \ldots, ab_{i(k)n})$ and this is the only possible non-zero row of AB. Since $a \neq 0$ it follows, by the minimality of $i(k)$, that $b_{i(k)1} = \cdots = b_{i(k)(i(k)-1)} = 0$. Moreover, by Lemma 3.7 it follows that $b_{i(k)i(k)} = 0$ as $B \in \mathcal{A}_{i(k)}$. Hence, $i(k) < i(i(k))$.

Now we close the proof by observing that $1 < i(1)$, hence we obtain a strictly increasing infinite sequence bounded above by n: $1 < i(1) < i(i(1)) < \cdots$. This is a contradiction. □

REMARK 3.9. By Theorem 3.8, for a division ring D, the ring $\mathbb{M}_n(D)$ satisfies the 2-nil-sum property iff $n = 1$ and D is a field. Note that there exists a division ring D such that all matrices in $\mathbb{M}_3(D)$ are sums of three nilpotent matrices (see [**15**]).

We close the paper, with some applications of Theorem 2.2 and Theorem 3.8. A nonzero right ideal I of a ring R is called a uniform right ideal if the intersection of any two nonzero right ideals contained in I is nonzero. A ring R is semipotent if every nonzero right ideal not contained in $J(R)$ contains a nonzero idempotent. A ring R is said to be of bounded index (of nilpotence) if there is a positive integer n such that $a^n = 0$ for all nilpotents a of R.

COROLLARY 3.10. *A ring R with the 2-nil-sum property is commutative under any of the following additional assumptions:*

(1) *R contains a uniform ideal;*
(2) *R is a semipotent ring of bounded index;*
(3) *R is a semilocal ring.*

PROOF. By Theorem 2.2, suppose that R is a simple ring.

(1) It follows from [**9**] that R contains no infinite direct sums of right ideals (i.e., R_R is of finite uniform dimension). Moreover, from the last remark of [**12**] it follows that the maximal right ring of quotients of R is simple artinian. Thus, R is right Goldie by [**13**, Proposition 13.41].
(2) If R is semipotent of bounded index, then R is simple artinian by [**1**, Corollary 6].
(3) If R is semilocal, then R is simple artinian.

So the claim follows from Theorem 3.8. □

References

[1] K. I. Beidar, *On rings with zero total*, Beiträge Algebra Geom. **38** (1997), no. 2, 233–239. MR1473103

[2] Fabrizio Barroero, Christopher Frei, and Robert F. Tichy, *Additive unit representations in rings over global fields—a survey*, Publ. Math. Debrecen **79** (2011), no. 3-4, 291–307, DOI 10.5486/PMD.2011.5126. MR2907966

[3] Simion Breaz and Grigore Călugăreanu, *Sums of nilpotent matrices*, Linear Multilinear Algebra **65** (2017), no. 1, 67–78, DOI 10.1080/03081087.2016.1167816. MR3575879

[4] G. Călugăreanu and T. Y. Lam, *Fine rings: a new class of simple rings*, J. Algebra Appl. **15** (2016), no. 9, 1650173, 18, DOI 10.1142/S0219498816501735. MR3539597

[5] M. Chebotar, P.-H. Lee, and E. R. Puczyłowski, *On commutators and nilpotent elements in simple rings*, Bull. Lond. Math. Soc. **42** (2010), no. 2, 191–194, DOI 10.1112/blms/bdp089. MR2601545

[6] Paul M. Cohn, *The similarity reduction of matrices over a skew field*, Math. Z. **132** (1973), 151–163, DOI 10.1007/BF01213920. MR325646

[7] M. Ferrero, E. R. Puczyłowski, and S. Sidki, *On the representation of an idempotent as a sum of nilpotent elements*, Canad. Math. Bull. **39** (1996), no. 2, 178–185, DOI 10.4153/CMB-1996-022-0. MR1390353

[8] K. R. Goodearl, *Ring theory: Nonsingular rings and modules*, Pure and Applied Mathematics, No. 33, Marcel Dekker, Inc., New York-Basel, 1976. MR0429962

[9] R. Hart, *Simple rings with uniform right ideals*, J. London Math. Soc. **42** (1967), 614–617, DOI 10.1112/jlms/s1-42.1.614. MR218390

[10] Bruno Harris, *Commutators in division rings*, Proc. Amer. Math. Soc. **9** (1958), 628–630, DOI 10.2307/2033220. MR96697

[11] Dinesh Khurana and Ashish K. Srivastava, *Right self-injective rings in which every element is a sum of two units*, J. Algebra Appl. **6** (2007), no. 2, 281–286, DOI 10.1142/S0219498807002181. MR2316422

[12] Kwangil Koh, *On simple rings with uniform right ideals*, Amer. Math. Monthly **74** (1967), 685–687, DOI 10.2307/2314256. MR220766

[13] T. Y. Lam, *Lectures on modules and rings*, Graduate Texts in Mathematics, vol. 189, Springer-Verlag, New York, 1999, DOI 10.1007/978-1-4612-0525-8. MR1653294

[14] C. Meehan, *Sums of automorphisms of free modules and completely decomposable groups*, J. Algebra **299** (2006), no. 2, 467–479, DOI 10.1016/j.jalgebra.2005.08.006. MR2228323

[15] A. Salwa, *Simple Artinian rings as sums of nilpotent subrings*, Arch. Math. (Basel) **78** (2002), no. 5, 350–357, DOI 10.1007/s00013-002-8257-6. MR1903668

[16] Arkadiusz Salwa, *Representing idempotents as a sum of two nilpotents—an approach via matrices over division rings*, Colloq. Math. **77** (1998), no. 1, 59–83, DOI 10.4064/cm-77-1-59-83. MR1622768

[17] Ashish K. Srivastava, *A survey of rings generated by units* (English, with English and French summaries), Ann. Fac. Sci. Toulouse Math. (6) **19** (2010), no. Fascicule Spécial, 203–213. MR2675727

[18] Peter Vámos, *2-good rings*, Q. J. Math. **56** (2005), no. 3, 417–430, DOI 10.1093/qmath/hah046. MR2161255

[19] Kenneth G. Wolfson, *An ideal-theoretic characterization of the ring of all linear transformations*, Amer. J. Math. **75** (1953), 358–386, DOI 10.2307/2372458. MR53080

[20] Daniel Zelinsky, *Every linear transformation is a sum of nonsingular ones*, Proc. Amer. Math. Soc. **5** (1954), 627–630, DOI 10.2307/2032048. MR62728

Department of Mathematics, Faculty of Mathematics and Computer Science, Babes-Bolyai University, Ro-400084 Cluj-Napoca, Romania

Email address: bodo@math.ubbcluj.ro

Department of Mathematics and Statistics, Memorial University of Newfoundland, St.John's, NL A1C 5S7, Canada

Email address: zhou@mun.ca

Contemporary Mathematics
Volume **785**, 2023
https://doi.org/10.1090/conm/785/15775

Asymptotic Brauer p-dimension

Adam Chapman and Kelly McKinnie

Dedicated to Professor Syed Tariq Rizvi for his contribution to Ring and Module Theory

Abstract. We define and compute the asymptotic Brauer p-dimension of a field F, denoted $\mathrm{ABrd}_p(F)$, in cases where F is a rational function field or Laurent series field. $\mathrm{ABrd}_p(F)$ is defined like the Brauer p-dimension except it considers finite sets of Brauer classes instead of single classes. Our main result shows that for fields $F_0(\alpha_1,\dots,\alpha_n)$ and $F_0(\!(\alpha_1)\!)\dots(\!(\alpha_n)\!)$ where F_0 is a perfect field of characteristic $p>0$ when $n\geq 2$ the asymptotic Brauer p-dimension is n. We also show that it is $n-1$ when $F=F_0(\!(\alpha_1)\!)\dots(\!(\alpha_n)\!)$ and F_0 is algebraically closed of characteristic not p. We conclude the paper with examples of pairs of cyclic algebras of odd prime degree p over a field F for which $\mathrm{Brd}_p(F)=2$ that share no maximal subfields despite their tensor product being non-division.

1. Introduction

The Brauer dimension $\mathrm{Brd}(F)$ of a field F is defined to be the minimal m such that for any central simple algebra A over E, $\mathrm{ind}(A)|\exp(A)^m$, where E ranges over the finite field extensions of F. Since $\mathrm{ind}(A)$ is also the minimal degree of a splitting field of A, an equivalent definition would be that every central simple algebra of exponent e has a splitting field of degree dividing e^m. If we consider only central simple algebras A of p-power exponent, then the minimal m is called the Brauer p-dimension $\mathrm{Brd}_p(F)$. This invariant has been studied by many mathematicians. See [**11**] and [**4**] for recent results on the subject.

Inspired by Gosavi's notion of "generalized Brauer dimension" (see [**13**] and [**14**]), we introduce the following generalization of the Brauer p-dimension: the Brauer (p,ℓ)-dimension of a field F, denoted $\mathrm{Brd}_{p,\ell}(F)$, is defined to be the minimal m such that for any finite sequence of ℓ central simple E-algebras $A_1,\dots,A_\ell$ whose exponents are powers of p there exists a common splitting field of degree dividing t^m, where $t=\max(\exp(A_1),\dots,\exp(A_\ell))$ and E ranges over all the finite field extensions of F. We define the asymptotic Brauer p-dimension of F, denoted $\mathrm{ABrd}_p(F)$, to be $\lim_{\ell\to\infty}\mathrm{Brd}_{p,\ell}(F)$. (Note that this is also $\sup_\ell\{\mathrm{Brd}_{p,\ell}(F)\}$ since the sequence is monotonous non-decreasing.) Clearly $\mathrm{Brd}_p(F)\leq\mathrm{ABrd}_p(F)$.

For fields of characteristic p, there is an easy upper bound on the asymptotic Brauer p-dimension, which is the p-rank of the field, i.e., the number n for which

---

2020 *Mathematics Subject Classification.* Primary 16K20; Secondary 16W60.
Key words and phrases. Division algebras, cyclic algebras, valuation theory, linkage, fields of positive characteristic.

$[F : F^p] = p^n$. Since the p-rank does not change under finite field extensions (see for example [**5**, Chapter V, Section 16.6, Corollary 3]), for any finite field extension E of F, the p-rank of E remains n, and so ${}_t\operatorname{Br}(E)$ is completely trivialized by $E(\sqrt[t]{\beta_1}, \dots, \sqrt[t]{\beta_n})$ where $\beta_1, \dots, \beta_n$ is a p-basis for E and t is any power of p (see for example [**19**, Corollary 1.2]). Since this is a field extension of degree t^n, we have $\operatorname{ABrd}_p(F) \leq n$. Such fields include $F_0(\alpha_1, \dots, \alpha_n)$ and $F_0(\!(\alpha_1)\!) \dots (\!(\alpha_n)\!)$ where F_0 is a perfect field of $\operatorname{char}(F_0) = p$. In particular, if $F = \mathbb{F}_p(\alpha_1, \dots, \alpha_n)$ then there exists a division algebra of exponent p and index p^n over F (see [**4**, Section 5]), hence $\operatorname{Brd}_p(F) \geq n$, but since $\operatorname{Brd}_p(F) \leq \operatorname{ABrd}_p(F) \leq n$, we have $\operatorname{Brd}_p(F) = \operatorname{ABrd}_p(F) = n$.

Our main goal in this paper is to prove the following:

THEOREM 1.1. *For any perfect field F_0 of characteristic $p > 0$ and integer $n \geq 2$, $\operatorname{ABrd}_p(F) = n$ where F is either $F_0(\!(\alpha_1)\!) \dots (\!(\alpha_n)\!)$ or $F_0(\alpha_1, \dots, \alpha_n)$.*

Note that the Brauer p-dimension of $F = F_0(\!(\alpha_1)\!) \dots (\!(\alpha_n)\!)$ is known to be $n-1$ when $\dim_{\mathbb{F}_p}(F_0/\wp(F_0)) < n$ and n when $\dim_{\mathbb{F}_p}(F_0/\wp(F_0)) \geq n$ by [**11**, Proposition 5.3], $\wp(F)$ being $\{\lambda^p - \lambda : \lambda \in F\}$, so when $\dim_{\mathbb{F}_p}(F_0/\wp(F_0)) < n$ (e.g., when F_0 is $\mathbb{F}_p$ or $\mathbb{F}_p^{sep}$) F is a field whose asymptotic Brauer p-dimension is strictly larger than its Brauer p-dimension. In Section 4 we prove this theorem by explicitly constructing a finite set of division algebras of exponent p which have no common splitting field of degree p^{n-1}. The proof uses techniques on valued division algebras (see [**12**] for background on division algebras and [**22**] for their valuation theory).

We conclude the paper with another nice application of the main theorem of [**18**] in the construction of examples of pairs of cyclic algebras of prime degree $p > 2$ over a field F of $\operatorname{char}(F) = p$ that share no maximal subfields despite their tensor product containing zero divisors. Such examples were first constructed in [**21**] for $\operatorname{char}(F) \neq p$, then improved in [**15**] to include also pairs of cyclic division algebras A and B of odd prime degree p with no common maximal subfields, although $\operatorname{ind}(A^{\otimes i} \otimes B) = p$ for all i. More examples were constructed in [**16**] where over any field F (without assumptions on the characteristic) any pair of algebras A and B with $A \otimes B$ division can be scalar extended to some field L over which $\operatorname{ind}(A_L^{\otimes i} \otimes B_L) = p\ \forall i$ but A_L and B_L still share no common subfields. The example we give here, we believe, is neat enough to merit its inclusion, since it is over the field F of iterated Laurent series in three variables over any field F_0 of characteristic p, as opposed to the generically constructed fields of [**16**], and in our example, F can be chosen to have Brauer p-dimension 2 (e.g., when F_0 is algebraically closed), which is lower than in Karpenko's example, and leaves the question of whether such examples exist when $\operatorname{Brd}_p(F) = 1$. Our example is analogous to [**21**] and not to [**15**] or [**16**] because $A \otimes B$ is not division but $A^{\otimes i} \otimes B$ is for some $i \in \{2, \dots, p-1\}$.

2. Preliminaries

Recall that when $\operatorname{char}(F) = p$, the p-torsion ${}_p\operatorname{Br}(F)$ of $\operatorname{Br}(F)$ is generated by algebras of the form $F\langle x, y : x^p - x = \alpha, y^p = \beta, yxy^{-1} = x + 1\rangle$, for some $\alpha \in F$ and $\beta \in F^\times$, which we denote by the symbol $[\alpha, \beta)_{p,F}$, and thus these algebras are called "symbol algebras" (see [**12**, Theorem 9.1.4]). They are also cyclic algebras, for $F[x]/F$ is a cyclic extension.

Note that as a vector space, the algebra $[\alpha, \beta)_{p,F}$ decomposes as $\bigoplus_{i,j=0}^{p-1} Fx^iy^j$, and for any $v = \sum_{i,j=0}^{p-1} c_{i,j}x^iy^j$ the (reduced) trace form maps v to $\operatorname{Tr}(v) = -c_{p-1,0}$

(see [**8**, Lemma 2.1]). Now, if A and B are central simple algebras over F and $r \in A$ and $t \in B$, then the trace of $r \otimes t$ in $A \otimes B$ is the product of the trace of r in A and the trace of t in B. Consequently, the subspace of trace zero elements in $[\alpha_1, \beta_1)_{p,F} \otimes \cdots \otimes [\alpha_n, \beta_n)_{p,F}$ generated by $x_1, y_1, \dots, x_n, y_n$ is spanned by all $x_1^{d_1} y_1^{e_1} \dots x_n^{d_n} y_n^{e_n}$ except for the case of $d_1 = \cdots = d_n = p-1$ and $e_1 = \cdots = e_n = 0$. In particular, it is of co-dimension 1, i.e., of dimension $p^{2n} - 1$.

For a given valued division algebra D we follow the notation of [**22**] for the residue division algebra $\overline{D}$ and the value group Γ_D and recall freely several facts from that book. In particular, if a field F is equipped with a Henselian valuation (e.g., the field $F = F_0((\alpha_1)) \dots ((\alpha_n))$ with the right-to-left $(\alpha_1, \dots, \alpha_n)$-adic valuation), then the valuation extends to any central division algebra D over F. Another fact we will use is the fundamental inequality $[D : F] \geq |\Gamma_D / \Gamma_F| \cdot [\overline{D} : \overline{F}]$ (see [**22**, Section 1.2.1]).

3. Equivalent conditions

THEOREM 3.1 (cf. [**19**, Lemma 1.1]). *Let p be a prime integer and F be a field. Then* $\mathrm{Brd}_{p,\ell}(F) \leq n$ *if and only if for any finite extension E/F, every ℓ algebras of exponent p over E have a common degree p^n splitting field.*

PROOF. The forward implication is clear. To prove the backwards implication we need to prove that if for any finite extension E/F, every sequence of ℓ algebras of exponent p over E have a common degree p^n splitting field, then for any $m \in \mathbb{N}$ and any finite extension E/F, every sequence of ℓ algebras of exponent dividing p^m over E have a common degree $(p^m)^n$ splitting field.

The proof is by induction on m. The case of $m = 1$ coincides with the assumption. Suppose the statement holds for $1, \dots, m-1$. Consider a sequence of ℓ algebras $A_1, \dots, A_\ell$ of exponents dividing p^m over a finite extension E of F. Then $A_1^{\otimes p}, \dots, A_\ell^{\otimes p}$ are of exponents dividing p^{m-1}, which share a splitting field K of degree $(p^{m-1})^n$ over E by the induction hypothesis. Now, the algebras $A_1 \otimes K, \dots, A_\ell \otimes K$ are of exponent dividing p, and thus by the assumption, have a degree p^n splitting field L over K. Hence, L is a splitting field for $A_1, \dots, A_\ell$, and it is of degree $(p^{m-1})^n \cdot p^n = (p^m)^n$ over E. □

EXAMPLE 3.2 (Transcendence degree 2). Let F_0 be an algebraically closed field, let F be either the function field $F_0(\alpha, \beta)$ in two variables over F_0 or the field $F_0((\alpha))((\beta))$. Let E be a finite extension of F. For $p = 2$, every three quaternion algebras over E share a quadratic splitting field (see [**7**, Example 2.4] for $\mathrm{char}(F_0) = 2$ and [**9**, Corollary 5.3] for $\mathrm{char}(F_0) \neq 2$), and so $\mathrm{Brd}_{2,3}(F) = 1$ by Theorem 3.1. For $p = 3$, every two algebras of exponent 3 over E share a cubic splitting field (see [**2**, Appendix]), and so $\mathrm{Brd}_{3,2}(F) = 1$ by Theorem 3.1. For any prime p, it follows from [**8**] and [**10**] that $\mathrm{Brd}_{p,p^2-1}(F) > 1$ when $\mathrm{char}(F) = p > 2$, and $\mathrm{Brd}_{p,p^2}(F) > 1$ when $\mathrm{char}(F) = 0$. Therefore, given the p-rank upper bound $\mathrm{ABrd}_p(F) \leq 2$, when $\mathrm{char}(F) = p > 2$, $\mathrm{Brd}_{p,p^2-1}(F) = \mathrm{ABrd}_p(F) = 2$.

COROLLARY 3.3. *For any field F,* $\mathrm{Brd}_2(F) = 1$ *if and only if* $\mathrm{Brd}_{2,2}(F) = 1$.

PROOF. Suppose $\mathrm{Brd}_2(F) = 1$. Then for any finite extension E of F, any biquaternion algebra $A \otimes B$ over E contains zero divisors, and therefore A and B share a quadratic splitting field by Albert (see [**1**]). Therefore $\mathrm{Brd}_{2,2}(F) = 1$. The opposite direction is clear. □

EXAMPLE 3.4 (Transcendence degree 3). When F_0 is algebraically closed, the field of iterated Laurent series $F = F_0((\alpha))((\beta))((\gamma))$ in three variables over F_0 has $\mathrm{Brd}_2(F) = 1$ by [**11**], and therefore also $\mathrm{Brd}_{2,2}(F) = 1$. However, $\mathrm{Brd}_{2,3}(F) > 1$ because if it were 1, by [**9**] and [**3**] we would have $I_q^3 F = 0$, but $I_q^3 F \neq 0$. Since for every three quaternion algebras over a finite extension E of F, the first two have a common quadratic splitting field K and the last has a quadratic splitting field L, the three algebras have the compositum $K.L$ as the common bi-quadratic splitting field, and thus $\mathrm{Brd}_{2,3}(F) = 2$.

REMARK 3.5. The last argument, that given a common splitting field K of a list of ℓ cyclic algebras of degree p and a maximal subfield L of another cyclic algebra of degree p, the compositum $K.L$ is a common splitting field of all $\ell + 1$ algebras, explains why the more general formula $\mathrm{Brd}_{p,\ell+1}(F) \leq \mathrm{Brd}_{p,\ell}(F) + 1$ holds true for any field F, prime number p and natural number ℓ.

QUESTION 3.6. Does $\mathrm{Brd}_p(F) = 1$ imply $\mathrm{Brd}_{p,2}(F) = 1$ for every prime p?

It is known that Albert's theorem [**1**] does not extend to algebras of exponent $p > 2$. However, the existing examples in the literature (see [**21**], [**15**], [**16**] and Section 7) are constructed over fields F with $\mathrm{Brd}_p(F) \neq 1$, and therefore do not resolve the question.

4. Iterated Laurent series over perfect fields

In this section we prove Theorem 1.1. We focus on $F = F_0((\alpha_1))\dots((\alpha_n))$ for convenience, but all the algebras under discussion are defined over $F_0(\alpha_1, \dots, \alpha_n)$, and one obtains the inexistence of a common degree p^{n-1} splitting field from the inexistence of such a splitting field under restriction of these algebras to $F_0((\alpha_1))\dots((\alpha_n))$. In what follows, $F = F_0((\alpha_1))\dots((\alpha_n))$ where F_0 is a field of characteristic $p > 0$ and $n \geq 2$. Note that for $n = 1$, all the central simple algebras over F of exponent dividing p^ℓ are split by $F[\sqrt[p^\ell]{\alpha_1}]$ and the Asymptotic Brauer p-dimension is 1 (unless $F_0/\wp(F_0) = 0$, in which case it is 0).

We cite the following useful result from [**18**, Theorem 1]: If F is a Henselian valued field, D and E are division algebras over F such that D is defectless, $\overline{D} \otimes \overline{E}$ is a division algebra and $\Gamma_D \cap \Gamma_E = \Gamma_F$, then $D \otimes E$ is a division algebra as well.

LEMMA 4.1. *For $n \geq 3$, pick $i \in \{1, \dots, n-1\}$ and write $\beta_1, \dots, \beta_{n-2}$ for $\alpha_1, \dots, \widehat{\alpha_i}, \dots, \alpha_{n-1}$. Then the algebra*

$$A_i = [\alpha_n^{-1}, \beta_{n-2})_{p,F} \otimes [\beta_{n-2}^{-1}, \beta_{n-3})_{p,F} \otimes \dots \otimes [\beta_2^{-1}, \beta_1)_{p,F} \otimes [\beta_1^{-1}, \alpha_i)_{p,F}$$

is a division algebra.

PROOF. The case of $i = 1$ was covered in [**6**]. Suppose $i > 1$. Write $D = [\alpha_n^{-1}, \beta_{n-2})_{p,F} \otimes \dots \otimes [\beta_2^{-1}, \beta_1)_{p,F}$ and $E = [\beta_1^{-1}, \alpha_i)_{p,F}$. The algebra D has degree p^{n-2} and is division by [**6**]. The algebra E is division because the values of β_1^{-1} and α_i are $\mathbb{F}_p$-independent in $\Gamma_F/p\Gamma_F$ w.r.t the standard valuation on F. Consider the $(\alpha_2, \dots, \alpha_n)$-adic valuation on F. The algebra D is defectless, because $|\Gamma_D/\Gamma_F| = p^{2n-5}$ and $\overline{D} = \overline{F}[y : y^p = \alpha_1]$. Moreover, $\Gamma_D \cap \Gamma_E = \Gamma_F$, because the value group of D is

$$\frac{1}{p}\mathbb{Z} \times \frac{1}{p^2}\mathbb{Z} \times \dots \times \frac{1}{p^2}\mathbb{Z} \times \underbrace{\mathbb{Z}}_{\alpha_i} \times \frac{1}{p^2}\mathbb{Z} \times \dots \times \frac{1}{p^2}\mathbb{Z} \times \frac{1}{p}\mathbb{Z},$$

whereas the value group of E is

$$\mathbb{Z} \times \cdots \times \mathbb{Z} \times \underbrace{\frac{1}{p}\mathbb{Z}}_{\alpha_i} \times \mathbb{Z} \times \cdots \times \mathbb{Z}.$$

(See the proof of Proposition 4.2 for a similar computation of the value groups.) $\overline{D} \otimes \overline{E} = \overline{F}[x, y : x^p - x = \alpha_1^{-1} = y^p]$ is the field generated over $\overline{F} = F_0((\alpha_1))$ by $z = x - y$ satisfying $z^{p^2} - z^p = \alpha_1^{-1}$. By [**18**] $D \otimes E$ is division. $\square$

Since this lemma referred to the cases with $n \geq 3$, we can add that for $n = 2$, we can define the algebra $A_1 = [\alpha_2^{-1}, \alpha_1)_{p,F}$, which is division because the values of α_2^{-1} and α_1 are $\mathbb{F}_p$-independent in $\Gamma_F/p\Gamma_F$ w.r.t the standard valuation on F (see [**6**] for more details).

PROPOSITION 4.2. *Consider from now on the $(\alpha_1, \ldots, \alpha_n)$-adic valuation on F, denoted by $\mathfrak{v}$. For each $i \in \{1, \ldots, n-1\}$, the algebra A_i is totally ramified and its value group Γ_{A_i} is*

$$\frac{1}{p^2}\mathbb{Z} \times \cdots \times \frac{1}{p^2}\mathbb{Z} \times \underbrace{\frac{1}{p}\mathbb{Z}}_{ith\ place} \times \frac{1}{p^2}\mathbb{Z} \times \cdots \times \frac{1}{p^2}\mathbb{Z} \times \frac{1}{p}\mathbb{Z}.$$

As a result, $\bigcap_{i=1}^{n-1} \Gamma_{A_i} = \frac{1}{p}\mathbb{Z} \times \cdots \times \frac{1}{p}\mathbb{Z}$.

PROOF. Since A_i is a division algebra and F is Henselian, the valuation extends to A_i. By the fundamental inequality, $[A_i : F] \geq |\Gamma_{A_i}/\Gamma_F| \cdot [\overline{A_i} : \overline{F}]$, it is enough to show that $\Gamma_{A_i} \supseteq \frac{1}{p^2}\mathbb{Z} \times \cdots \times \frac{1}{p^2}\mathbb{Z} \times \underbrace{\frac{1}{p}\mathbb{Z}}_{ith\ place} \times \frac{1}{p^2}\mathbb{Z} \times \cdots \times \frac{1}{p^2}\mathbb{Z} \times \frac{1}{p}\mathbb{Z}$, and then all the inequalities would become equalities, and $\overline{A_i} = \overline{F}$. For each $k \in \{1, \ldots, n-1\} \setminus \{i\}$, A_i contains the subfield $F[x, y : x^p - x = \alpha_k^{-1}, y^p = \alpha_k]$, which is in fact a simple extension of F generated by $t = x - y^{-1}$ satisfying $t^p - t = y^{-1}$, and thus $\mathfrak{v}(t) = \frac{1}{p}\mathfrak{v}(y^{-1}) = \frac{1}{p^2}\mathfrak{v}(\alpha_k^{-1})$. In addition, A_i contains the field $F[x : x^p - x = \alpha_n^{-1}]$, and here $\mathfrak{v}(x) = \frac{1}{p}\mathfrak{v}(\alpha_n^{-1})$, and the field $F[y : y^p = \alpha_i]$, and here $\mathfrak{v}(y) = \frac{1}{p}\mathfrak{v}(\alpha_i)$. $\square$

REMARK 4.3. In addition to the algebras constructed in Lemma 4.1, the following algebras are also division algebras, by the same reasoning:

$$B_{d_1,\ldots,d_n} = [\prod_{i=1}^{n} \alpha_i^{-d_i}, \alpha_{n-1})_{p,F} \otimes [\alpha_{n-1}^{-1}, \alpha_{n-2})_{p,F} \otimes \cdots \otimes [\alpha_2^{-1}, \alpha_1)_{p,F},$$

where $d_1, \ldots, d_{n-1} \in \{0, \ldots, p-1\}$ and $d_n \in \{1, \ldots, p-1\}$, and

$$B_{d_1,\ldots,d_{n-1},0} = [\prod_{i=1}^{n-1} \alpha_i^{-d_i}, \alpha_{n-2})_{p,F} \otimes [\alpha_{n-2}^{-1}, \alpha_{n-3})_{p,F} \otimes \cdots \otimes [\alpha_2^{-1}, \alpha_1)_{p,F} \otimes [\alpha_1^{-1}, \alpha_n)_{p,F},$$

where $d_1, \ldots, d_{n-2} \in \{0, \ldots, p-1\}$ and $d_{n-1} \in \{1, \ldots, p-1\}$. Note that $B_{0,\ldots,0,1} = A_1$. Each $B_{d_1,\ldots,d_n}$ decomposes as $W \oplus F \prod_{i=1}^{n} x_i^{p-1}$ where W is the subspace of trace zero elements and $x_1, \ldots, x_n$ are the generators of the cyclic subfields of the cyclic algebras in the tensor product (see Section 2). By examining the basis elements, $\Gamma_{B_{d_1,\ldots,d_n}}$ is the disjoint union of the value set of W and the value set of $F\prod_{i=1}^n x_i^{p-1}$, the latter being $\Gamma_F+(\frac{d_1}{p},\frac{d_2+1}{p},\dots,\frac{d_{n-1}+1}{p},\frac{d_n}{p})$ when $d_n\neq 1$ and $\Gamma_F+(\frac{d_1+1}{p},\frac{d_2+1}{p},\dots,\frac{d_{n-2}+1}{p},\frac{d_{n-1}}{p},0)$ when $d_n=0$.

THEOREM 4.4. *Assuming p is a prime integer, $n\geq 2$ and $(n,p)\neq(2,2)$, the $n-2+p^n-p^{n-2}$ algebras $\{A_i,B_{d_1,\dots,d_n}: i\in\{2,\dots,n-1\},(d_1,\dots,d_n)\in\{0,\dots,p-1\}^{\times n},(d_{n-1},d_n)\neq(0,0)\}$ share no splitting field of degree p^{n-1}.*

PROOF. Suppose K is a common degree p^{n-1} splitting field. Since all these division algebras are of degree p^{n-1}, K embeds into all of them. Since K embeds into each A_i, we have $\Gamma_K\subseteq\bigcap_{i=1}^{n-1}\Gamma_{A_i}=\frac{1}{p}\mathbb{Z}\times\dots\times\frac{1}{p}\mathbb{Z}$, considering the right-to-left $(\alpha_1,\dots,\alpha_n)$-adic valuation $\mathfrak{v}$ on F extended to each A_i. The subspace V of K of elements of trace 0 is of dimension at least $p^{n-1}-1$, and since K is totally ramified over F, the set of classes of $\mathfrak{v}(V)$ modulo Γ_F is of cardinality at least $p^{n-1}-1$. As we noted, $\mathfrak{v}(V)$ modulo Γ_F is a subset of $\frac{1}{p}\mathbb{Z}\times\dots\times\frac{1}{p}\mathbb{Z}/\Gamma_F$. Now, $\mathfrak{v}(V)$ must be a subset of the intersection of the sets of values of trace zero elements in all $B_{d_1,\dots,d_n}$ (see Remark 4.3), which means that $\mathfrak{v}(V)$ mod Γ_F is a subset of

$$(\frac{1}{p}\mathbb{Z}\times\dots\times\frac{1}{p}\mathbb{Z}/\Gamma_F)\setminus\{(\frac{d_1}{p},\dots,\frac{d_n}{p}):(d_1,\dots,d_n)\in\{0,\dots,p-1\}^{\times n},(d_{n-1},d_n)\neq(0,0)\}.$$

However, the latter is of cardinality $p^n-(p-1)(p^{n-1}+p^{n-2})=p^{n-2}$, and $p^{n-2}<p^{n-1}-1$, contradiction. □

This theorem shows that $\mathrm{ABrd}_p(F)>n-1$ for these fields in all cases but $n=p=2$, a case that was covered in [**7**]. When F_0 is perfect the field F has $\mathrm{ABrd}_p(F)\leq n$ by the remarks above Theorem 1.1 and the main theorem (Theorem 1.1) follows.

REMARK 4.5. Theorem 3.4 actually proves that $\mathrm{Brd}_{p,n-2+p^n-p^{n-2}}(F)=n$ when F_0 is perfect of $\mathrm{char}(F_0)=p>0$, F is either $F_0(\alpha_1,\dots,\alpha_n)$ or $F_0((\alpha_1))\dots((\alpha_n))$ and $n\geq 2$ with $(n,p)\neq(2,2)$. For $n=p=2$, $n-2+p^n-p^{n-2}=3$, and it is not always true that $\mathrm{Brd}_{2,3}(F)=2$. For $n=p=2$, when $F_0=\mathbb{F}_2^{sep}$, we have $\mathrm{Brd}_{2,3}(F)=1$ (see Example 3.2). However, for any perfect F_0, there exist four quaternion algebras over F with no common quadratic splitting field (see [**7**, Proposition 2.3]), and therefore $\mathrm{Brd}_{2,4}(F)=2$.

5. Iterated Laurent Series over Fields of Characteristic not p

Let p be a prime integer. Recall that over fields F of $\mathrm{char}(F)\neq p$ containing primitive pth roots of unity ρ, ${}_p\mathrm{Br}(F)$ is generated by symbol algebras of degree p ([**17**]), which take the form $(\alpha,\beta)_{p,F}=F\langle i,j:i^p=\alpha,j^p=\beta,ji=\rho ij\rangle$. Let F_0 be an algebraically closed field of characteristic not p. Let n be a natural number ≥ 2, and $F=F_0((\alpha_1))\dots((\alpha_n))$ the field of iterated Laurent series in n variables over F_0.

The Brauer p-dimension of F in this case is known to be $\lfloor\frac{n}{2}\rfloor$ (see [**11**]). When $n=2$, the group ${}_p\mathrm{Br}(F)$ is the cyclic group $\langle(\alpha_1,\alpha_2)_{p,F}\rangle$ of order p, and thus $\mathrm{ABrd}_p(F)=\mathrm{Brd}_p(F)$. For $n\geq 3$, the asymptotic Brauer p-dimension may be greater. Since the set S of classes of index $p^{\lfloor\frac{n}{2}\rfloor}$ in ${}_p\mathrm{Br}(F)$ is finite, it is enough to consider the common splitting fields of the classes in S.

THEOREM 5.1. *Suppose $n\geq 3$. Then* $\mathrm{ABrd}_p(F)=n-1$.

PROOF. Let K/F be a field extension of degree dividing p^{n-2}. The natural valuation on F extends to K for F is henselian. We claim that there exist γ and δ in $F^\times$ whose values are $\mathbb{F}_p$-independent in $\Gamma_K/p\Gamma_K$: consider the map $\phi : \Gamma_F/p\Gamma_F \to \Gamma_K/p\Gamma_K$ induced by the inclusion $F \subset K$. Its kernel is $(\Gamma_F \cap p\Gamma_K)/p\Gamma_F$, which is contained in $p\Gamma_K/p\Gamma_F$. Since the latter is isomorphic to Γ_K/Γ_F whose dimension is $\leq n-2$, we conclude that $\dim \ker \phi \leq n-2$, and thus the image of ϕ is of dimension at least 2, so the existence of γ and δ follows. Therefore, $(\gamma,\delta)_{p,K}$ is a division algebra. In particular, K does not split $(\gamma,\delta)_{p,F}$, which means there is no common splitting field of ${}_p\operatorname{Br}(F)$ of degree dividing p^{n-2}. Hence, $\operatorname{ABrd}_p(F) \geq n-1$.

Now, for any finite field extension E of F, take $\beta_1,\dots,\beta_n$ to be elements whose values generate Γ_E, and then the field extension $E[\sqrt[p]{\beta_1},\dots,\sqrt[p]{\beta_{n-1}}]$ splits the entire group ${}_p\operatorname{Br}(E)$. Consequently, $\operatorname{ABrd}_p(F) = n-1$. $\square$

Since for $n \geq 3$, we have $n-1 > \lfloor \frac{n}{2} \rfloor$, this provides a family of fields where $\operatorname{ABrd}_p(F) > \operatorname{Brd}_p(F)$ with a gap as large as we want.

6. Further open questions

Clearly fields of infinite Brauer p-dimension exits, such as the field of functions with $\aleph_0$ variables $F_0(\alpha_1,\dots)$. For this field, the asymptotic Brauer p-dimension is also infinite.

QUESTION 6.1. Are there fields F with $\operatorname{ABrd}_p(F) = \infty > \operatorname{Brd}_p(F)$?

A natural candidate would be the field $F = F_0(\alpha,\beta)$ with F_0 algebraically closed of characteristic 0.

QUESTION 6.2. What is $\operatorname{ABrd}_p(\mathbb{C}(\alpha,\beta))$?

7. Cyclic algebras without common maximal subfields

PROPOSITION 7.1 (cf. [**15**, Theorem 1.1 (i)]). *Let F_0 be a field of odd characteristic p, let a,c,d be elements in F_0, and let $F = F_0((t))$ the field of Laurent series ovcr F.*

(1) *Write $A = [a,t)_{p,F}$ and $B = [c,dt)_{p,F}$. Then $A \otimes B$ is division if and only if $[c,d)_{p,F_0[\wp^{-1}(a+c)]}$ is division.*
(2) *Write $A = [t^{-1},a)_{p,F}$ and $B = [d+t^{-1},c)_{p,F}$. Then $A \otimes B$ is division if and only if $[d,c)_{p,F_0[\sqrt[p]{ac}]}$ is division.*

PROOF. Write $A = [a,t)_{p,F}$ and $B = [c,dt)_{p,F}$. The tensor product $A \otimes B$ is isomorphic to $D \otimes E$ where $D = [a+c,t)_{p,F}$ and $E = [c,d)_{p,F}$. Consider the t-adic valuation on F. Since $\overline{D} \otimes \overline{E} = [c,d)_{p,F_0[\wp^{-1}(a+c)]}$, if it is division, then by [**18**], $D \otimes E$ is division. In the opposite direction, if $[c,d)_{p,F_0[\wp^{-1}(a+c)]}$ is not division, then since it is isomorphic to a subalgebra of $D \otimes E$, the latter is not division either.

Now, write $A = [t^{-1},a)_{p,F}$ and $B = [d+t^{-1},c)_{p,F}$. The tensor product $A \otimes B$ is isomorphic to $D \otimes E$ where $D = [t^{-1},ac)_{p,F}$ and $E = [d,c)_{p,F}$. Consider the t-adic valuation on F. Since $\overline{D} \otimes \overline{E} = [d,c)_{p,F_0[\sqrt[p]{ac}]}$, if it is division, then by [**18**], $D \otimes E$ is division. In the opposite direction, if $[d,c)_{p,F_0[\sqrt[p]{ac}]}$ is not division, then since it is isomorphic to a subalgebra of $D \otimes E$, the latter is not division either. $\square$

LEMMA 7.2. *Given a prime integer p and a field k of* $\mathrm{char}(k) = p$, *let* $F_0 = k((d))((c))$.

(1) *The field $K = F_0[x : x^p - x = d^{-1}]$ is not a subfield of the symbol division algebra $D = [c^{-1}, d^{-1})_{p,F_0}$.*
(2) *The field $K = F_0[y : y^p = d]$ is not a subfield of the symbol division algebra $D = [d^{-1}, c)_{p,F_0}$.*

PROOF. Write $K = F_0[x : x^p - x = d^{-1}]$ and $D = [c^{-1}, d^{-1})_{p,F_0}$. Consider the right-to-left (d, c)-adic valuation $\mathfrak{v}$. Recall that for any division algebra A over F, $w(A)$ denotes the minimum of $\mathfrak{v}(\mathrm{Tr}(z)) - \mathfrak{v}(z)$ where z ranges over all elements of A. By [**20**, Theorem 2.6] $w(\mathrm{Trd}_{D/F}) = (0, \frac{p-1}{p})$ for the division algebra D, and by [**20**, Theorem 1.8] $w(\mathrm{Tr}_{K/F_0}) = (\frac{p-1}{p}, 0)$ for the field K. But by definition, $w(\mathrm{Trd}_{D/F}) \leq w(\mathrm{Tr}_{K/F_0})$ if K is a subfield of D, whereas $(0, \frac{p-1}{p}) > (\frac{p-1}{p}, 0)$.

Now, write $K = F_0[y : y^p = d]$ and $D = [d^{-1}, c)_{p,F_0}$. Since $K = k((y))((c))$ and $D = [d^{-1}, c)_{p,F_0} = [y^{-p}, c)_{p,F_0}$, we have $D \otimes K = [y^{-1}, c)_{p,k((y))((c))}$, which is a division algebra because the values of y^{-1} and c are $\mathbb{F}_p$-independent in $\Gamma_K / p\Gamma_K$. □

EXAMPLE 7.3. Suppose p is an odd prime. Let k be a field of $\mathrm{char}(k) = p$, $F_0 = k((d))((c))$, $F = F_0((t))$, $A = [d^{-1} - c^{-1}, t)_{p,F}$, $B = [2d^{-1} - 2c^{-1}, t)_{p,F}$ and $C = [c^{-1}, d^{-1}t)_{p,F}$. Since $[c^{-1}, d^{-1})_{p,K}$ is division by Lemma 7.2 where $K = F[\wp^{-1}(d^{-1})] = F[\wp^{-1}(d^{-1} - c^{-1} + c^{-1})]$, the algebra $A \otimes C$ is division by Proposition 7.1, and thus A and C do not share any maximal subfield. The algebras B and C do not share any maximal subfield either, because B is Brauer equivalent to $A \otimes A$. However, $[c^{-1}, d^{-1})_{p,L}$ is split when $L = F[\wp^{-1}(2d^{-1} - c^{-1})] = F[\wp^{-1}(2d^{-1} - 2c^{-1} + c^{-1})]$ because $[c^{-1}, d^{-1})_{p,L} = [c^{-1} - 2d^{-1}, d^{-1})_{p,L} = [2d^{-1} - c^{-1}, d)_{p,L}$. Therefore, B and C form a pair of cyclic degree p algebras that share no maximal subfield despite their tensor product being non-division by Proposition 7.1.

Now, write $A = [t^{-1}, dc^{-1})_{p,F}$, $B = [t^{-1}, d^2c^{-2})_{p,F}$ and $C = [d^{-1} + t^{-1}, c)_{p,F}$. Since $[d^{-1}, c)_{p,K}$ is division where $K = F[\sqrt[p]{d}] = F[\sqrt[p]{dc^{-1} \cdot c}]$ by Lemma 7.2, the algebra $A \otimes C$ is division by Proposition 7.1, and thus A and C do not share any maximal subfield. The algebras B and C do not share any maximal subfield either, because B is Brauer equivalent to $A \otimes A$. However, $[d^{-1}, c)_{p,L}$ is split when $L = F[\sqrt[p]{d^2c^{-1}}] = F[\sqrt[p]{d^2c^{-2} \cdot c}])]$ because $[d^{-1}, c)_{p,L} = [d^{-1}, cd^{-2})_{p,L} = [-d^{-1}, d^2c^{-1})_{p,L}$. Therefore, B and C form a pair of cyclic degree p algebras that share no maximal subfield despite their tensor product being non-division by Proposition 7.1.

When F_0 is algebraically closed, the field F from the previous example has Brauer p-dimension 2, like in the analogous examples from [**21**].

Acknowledgments

We thank anonymous referee for his careful reading and clever observations that improved the quality of this paper.

References

[1] A. A. Albert, *Tensor products of quaternion algebras*, Proc. Amer. Math. Soc. **35** (1972), 65–66, DOI 10.2307/2038442. MR297803

[2] M. Artin, *Brauer-Severi varieties*, Brauer groups in ring theory and algebraic geometry (Wilrijk, 1981), Lecture Notes in Math., vol. 917, Springer, Berlin-New York, 1982, pp. 194–210. MR657430

[3] Karim Johannes Becher, *Triple linkage*, Ann. K-Theory **3** (2018), no. 3, 369–378, DOI 10.2140/akt.2018.3.369. MR3830196

[4] Nivedita Bhaskhar and Bastian Haase, *Brauer p-dimension of complete discretely valued fields*, Trans. Amer. Math. Soc. **373** (2020), no. 5, 3709–3732, DOI 10.1090/tran/8038. MR4082254

[5] N. Bourbaki, *Algebra. II. Chapters 4–7*, Elements of Mathematics (Berlin), Springer-Verlag, Berlin, 1990. Translated from the French by P. M. Cohn and J. Howie. MR1080964

[6] Adam Chapman, *Field of iterated Laurent series and its Brauer group*, Bull. Belg. Math. Soc. Simon Stevin **27** (2020), no. 1, 1–6, DOI 10.36045/bbms/1590199296. MR4102695

[7] Adam Chapman, *Linkage of sets of quaternion algebras in characteristic 2*, Comm. Algebra **49** (2021), no. 9, 3680–3684, DOI 10.1080/00927872.2021.1902532. MR4290101

[8] Adam Chapman, *Linkage of sets of cyclic algebras*, J. Pure Appl. Algebra **226** (2022), no. 3, Paper No. 106855, 6, DOI 10.1016/j.jpaa.2021.106855. MR4291532

[9] Adam Chapman, Andrew Dolphin, and David B. Leep, *Triple linkage of quadratic Pfister forms*, Manuscripta Math. **157** (2018), no. 3-4, 435–443, DOI 10.1007/s00229-017-0996-6. MR3858411

[10] Adam Chapman and Jean-Pierre Tignol, *Linkage of Pfister forms over* $\mathbb{C}(x_1, \dots, x_n)$, Ann. K-Theory **4** (2019), no. 3, 521–524, DOI 10.2140/akt.2019.4.521. MR4043468

[11] Ivan D. Chipchakov, *On Brauer p-dimensions and absolute Brauer p-dimensions of Henselian fields*, J. Pure Appl. Algebra **223** (2019), no. 1, 10–29, DOI 10.1016/j.jpaa.2018.02.032. MR3833449

[12] Philippe Gille and Tamás Szamuely, *Central simple algebras and Galois cohomology*, Cambridge Studies in Advanced Mathematics, vol. 101, Cambridge University Press, Cambridge, 2006, DOI 10.1017/CBO9780511607219. MR2266528

[13] Saurabh Gosavi, *Generalized period-index problem with an application to quadratic forms*, (2019), arXiv:1910.02473.

[14] Saurabh Gosavi, *Generalized Brauer Dimension of Semi-Global Fields*, ProQuest LLC, Ann Arbor, MI, 2020. Thesis (Ph.D.)–Rutgers The State University of New Jersey, School of Graduate Studies. MR4225056

[15] Bill Jacob and Adrian R. Wadsworth, *Division algebras with no common subfields*, Israel J. Math. **83** (1993), no. 3, 353–360, DOI 10.1007/BF02784062. MR1239068

[16] Nikita A. Karpenko, *Three theorems on common splitting fields of central simple algebras*, Israel J. Math. **111** (1999), 125–141, DOI 10.1007/BF02810681. MR1710735

[17] A. S. Merkur′ev and A. A. Suslin, *K-cohomology of Severi-Brauer varieties and the norm residue homomorphism* (Russian), Izv. Akad. Nauk SSSR Ser. Mat. **46** (1982), no. 5, 1011–1046, 1135–1136. MR675529

[18] Pat Morandi, *The Henselization of a valued division algebra*, J. Algebra **122** (1989), no. 1, 232–243, DOI 10.1016/0021-8693(89)90247-0. MR994945

[19] R. Parimala and V. Suresh, *Period-index and u-invariant questions for function fields over complete discretely valued fields*, Invent. Math. **197** (2014), no. 1, 215–235, DOI 10.1007/s00222-013-0483-y. MR3219517

[20] J.-P. Tignol, *Classification of wild cyclic field extensions and division algebras of prime degree over a Henselian field*, Proceedings of the International Conference on Algebra, Part 2 (Novosibirsk, 1989), Contemp. Math., vol. 131, Amer. Math. Soc., Providence, RI, 1992, pp. 491–508. MR1175852

[21] J.-P. Tignol and A. R. Wadsworth, *Totally ramified valuations on finite-dimensional division algebras*, Trans. Amer. Math. Soc. **302** (1987), no. 1, 223–250, DOI 10.2307/2000907. MR887507

[22] Jean-Pierre Tignol and Adrian R. Wadsworth, *Value functions on simple algebras, and associated graded rings*, Springer Monographs in Mathematics, Springer, Cham, 2015. MR3328410

School of Computer Science, Academic College of Tel-Aviv-Yaffo, Rabenu Yeruham St., P.O.B 8401 Yaffo, 6818211, Israel

Email address: adam1chapman@yahoo.com

Department of Mathematical Sciences, University of Montana, Missoula, MT 59812, USA

Email address: kelly.mckinnie@umontana.edu

Contemporary Mathematics
Volume **785**, 2023
https://doi.org/10.1090/conm/785/15776

Five theorems on Gorenstein global dimensions

Lars Winther Christensen, Sergio Estrada, and Peder Thompson

Dedicated to Professor Syed Tariq Rizvi for his contribution to Ring and Module Theory

ABSTRACT. We expand on two existing characterizations of rings of Gorenstein (weak) global dimension zero and give two new characterizations of rings of finite Gorenstein (weak) global dimension. We also include the answer to a question of Y. Xiang on Gorenstein weak global dimension of group rings.

Introduction

In this paper, A denotes a unital associative ring. By an A-module we mean a left A-module; right A-modules are considered as modules over the opposite ring A°.

Two classic facts in algebra are that A has finite weak global dimension if and only if so does A°, and that A has global dimension zero if and only if so does A°. Corresponding results are known to hold in the Gorenstein setting: In [**7**] we observe that the Gorenstein weak global dimension is symmetric, and in [**4**] Bennis and Mahdou show a ring has Gorenstein global dimension zero if and only if it is Quasi-Frobenius, which is a left–right symmetric property.

Gorenstein flat-cotorsion modules, defined in [**6**], and the associated dimension developed in [**5**] turn out to be useful for refining our understanding of Gorenstein global dimensions. Two of our main results, Theorems 2.2 and 3.2, characterize rings of Gorenstein (weak) global dimension zero. In the first, we show that IF rings are precisely those rings where all modules have Gorenstein flat-cotorsion dimension at most zero. In the second, we show that Quasi-Frobenius rings are those rings where all modules are Gorenstein flat-cotorsion. Further, we show in Theorems 4.5 and 4.9 that under assumptions of finite finitistic dimensions, rings of finite Gorenstein (weak) global dimension are precisely the rings over which strongly cotorsion modules and Gorenstein injective modules coincide.

An important ingredient in the proofs of these results are generalizations of two results due to Holm [**18**]: We prove in [**7**, Theorem 2.1] that the flat, Gorenstein flat, and Gorenstein flat-cotorsion dimensions agree for modules of finite injective

2020 *Mathematics Subject Classification.* Primary 16E10; Secondary 16E65; 16S34.

Key words and phrases. Ding-Chen ring; Gorenstein global dimension; Gorenstein weak global dimension; Gorenstein flat-cotorsion dimension; IF ring; Quasi-Frobenius ring; strongly cotorsion module.

The first author was partly supported by Simons Foundation collaboration grant 428308. The second author was partly supported by grant PID2020-113206GB-I00 funded by MCIN/AEI/10.13039/ 501100011033.

dimension, and in Theorem 1.2 we show that the injective and Gorenstein injective dimensions agree for modules of finite flat dimension. In the final section, Theorem 5.2 provides an answer to a question of Y. Xiang [**27**] on the Gorenstein global dimension of group rings.

The notation employed in the paper is standard: we write pd, id, and fd for the projective, injective, and flat dimensions of modules and complexes. The corresponding Gorenstein dimensions are denoted Gpd, Gid, and Gfd. The Gorenstein flat-cotorsion dimension from [**5**] gets abbreviated Gfcd. We write $\mathrm{FPD}(A)$ and $\mathrm{FFD}(A)$ for the finitistic projective and finitistic flat dimension of A. The invariant $\mathrm{sfli}(A)$ is defined as $\sup\{\mathrm{fd}_A M \mid M \text{ is an injective } A\text{-module}\}$. The Gorenstein global dimension is defined as

$$\mathrm{Ggldim}(A) = \sup\{\mathrm{Gpd}_A M \mid M \text{ is an } A\text{-module}\},$$

and it equals $\sup\{\mathrm{Gid}_A M \mid M \text{ is an } A\text{-module}\}$. The Gorenstein weak global dimension is defined similarly,

$$\mathrm{Gwgldim}(A) = \sup\{\mathrm{Gfd}_A M \mid M \text{ is an } A\text{-module}\}.$$

In dealing with A-complexes we use homological notation. For an A-complex M and $n \in \mathbb{Z}$ set $\mathrm{Z}_n(M) = \mathrm{Ker}\,\partial_n^M$ and $\mathrm{C}_n(M) = \mathrm{Coker}\,\partial_{n+1}^M$ so that they are, respectively, submodules and quotient modules of the module in degree n.

1. Gorenstein injective dimension of flat modules

Recall that an A-module M is *cotorsion* if $\mathrm{Ext}_A^1(F, M) = 0$ holds for every flat A-module F, and M is *strongly cotorsion* if $\mathrm{Ext}_A^1(L, M) = 0$ holds for every A-module L of finite flat dimension; see Xu [**28**, Definition 5.4.1].

By convention the homological supremum of an acyclic complex is $-\infty$, so the next result says, in particular, that every cycle module in an acyclic complex of strongly cotorsion modules is strongly cotorsion.

PROPOSITION 1.1. *Let M be a complex of strongly cotorsion A-modules. For every $n \geqslant \sup\{v \in \mathbb{Z} \mid \mathrm{H}_v(M) \neq 0\}$ the module $\mathrm{C}_n(M)$ is strongly cotorsion.*

PROOF. If M is acyclic, then $\mathrm{C}_n(M)$ for every $n \in \mathbb{Z}$ is a cokernel in an acyclic complex of strongly cotorsion modules; we first reduce the general case to this special case. Set $s = \sup\{v \in \mathbb{Z} \mid \mathrm{H}_v(M) \neq 0\}$. Splicing a shifted injective resolution of the module $\mathrm{C}_s(M)$ with the acyclic complex $\cdots \to M_{s+1} \to M_s \to \mathrm{C}_s(M) \to 0$ one gets an acyclic complex I with $\mathrm{C}_n(I) = \mathrm{C}_n(M)$ for $n \geqslant s$. It now suffices to show that the modules $\mathrm{C}_i(I)$ are strongly cotorsion.

Let L be a module with $\mathrm{fd}_A L \leqslant f$. There is an exact sequence,

$$0 \longrightarrow F_f \longrightarrow \cdots \longrightarrow F_0 \longrightarrow L \longrightarrow 0,$$

with each F_i a flat A-module. Given a cotorsion module C, dimension shifting along this sequence yields $\mathrm{Ext}_A^{f+1}(L, C) \cong \mathrm{Ext}_A^1(F_f, C) = 0$. By a result of Bazzoni, Cortés-Izurdiaga, and Estrada [**2**, Theorem 1.3] each module $\mathrm{C}_i(I)$ is cotorsion, so for every $i \in \mathbb{Z}$ one has $\mathrm{Ext}_A^{f+1}(L, \mathrm{C}_i(I)) = 0$. Now, for every $i \in \mathbb{Z}$ there is an exact sequence,

$$0 \longrightarrow \mathrm{C}_{i+f}(I) \longrightarrow I_{i+f-1} \longrightarrow \cdots \longrightarrow I_i \longrightarrow \mathrm{C}_i(I) \longrightarrow 0.$$

Dimension shifting yields $\mathrm{Ext}_A^1(L, \mathrm{C}_i(I)) \cong \mathrm{Ext}_A^{f+1}(L, \mathrm{C}_{i+f}(I)) = 0$, whence $\mathrm{C}_i(I)$ is strongly cotorsion. $\square$

Recall that an acyclic complex I of injective A-modules is called *totally acyclic* if $\mathrm{Hom}_A(E, I)$ is acyclic for every injective A-module E. Recall further that an A-module G is called *Gorenstein injective* if there exists a totally acyclic complex I of injective A-modules with $\mathrm{Z}_0(I) \cong G$. It follows from Proposition 1.1 that a Gorenstein injective A-module is strongly cotorsion.

Now, if a flat A-module F is Gorenstein injective, then one gets from the defining totally acyclic complex an exact sequence $0 \to G \to I \to F \to 0$, with I injective and G Gorenstein injective. Since $\mathrm{Ext}^1_A(F, G) = 0$ holds this sequence splits, which means that F is a summand of I and hence injective. This argument can be souped up to yield the next theorem, which is dual to [**7**, Theorem 1.1] and subsumes [**18**, Theorem 2.1]

Recall, for example from [**8**, Proposition 3.6], that the Gorenstein injective dimension of an A-complex M can be defined as the least integer n such that (1) $\mathrm{H}_v(M) = 0$ holds for all $v < -n$ and (2) there exists a semi-injective A-complex I, isomorphic to M in the derived category, such that the cycle submodule $\mathrm{Z}_{-n}(I)$ is Gorenstein injective.

THEOREM 1.2. *Let M be an A-complex with bounded homology. If* $\mathrm{fd}_A M$ *is finite, then* $\mathrm{Gid}_A M = \mathrm{id}_A M$ *holds.*

PROOF. The equality $\mathrm{Gid}_A M = \mathrm{id}_A M$ holds trivially if M is acyclic, so assume that M is not acyclic and, without loss of generality, assume further that $\mathrm{fd}_A M = 0$ holds. Set $u = \min\{v \in \mathbb{Z} \mid \mathrm{H}_v(M) \neq 0\}$ and let F be a bounded complex of flat A-modules with $F_v = 0$ for $v > 0$ and $v < u$, such that F and M are isomorphic in the derived category of A. As one has $\mathrm{Gid}_A F \leqslant \mathrm{id}_A F$ it suffices to show that $\mathrm{Gid}_A F = n$ implies $\mathrm{id}_A F \leqslant n$. Let $F \to I$ be a semi-injective resolution with $I_v = 0$ for $v > 0$ and C be its mapping cone. As $I_v = 0$ holds for $v > 0$ one has $C_1 = F_0$ and $C_v = 0$ for $v > 1$. As $F_v = 0$ holds for $v < u$ one has

$$\mathrm{Z}_v(C) = \mathrm{Z}_v(I) \quad \text{for} \quad v \leqslant u. \tag{1}$$

Assume that $\mathrm{Gid}_A F = n$ holds; one then has $-n \leqslant u$. Let G be a Gorenstein injective A-module. From the defining totally acyclic complex of injective A-modules one gets an exact sequence,

$$0 \longrightarrow H \longrightarrow I_0 \longrightarrow \cdots \longrightarrow I_{-n} \longrightarrow G \longrightarrow 0,$$

with H a Gorenstein injective A-module. Dimension shifting along this exact sequence yields

$$\mathrm{Ext}^1_A(\mathrm{Z}_{-(n+1)}(C), G) \cong \mathrm{Ext}^{n+2}_A(\mathrm{Z}_{-(n+1)}(C), H). \tag{2}$$

Since H is Gorenstein injective, and hence cotorsion by Proposition 1.1, and the modules C_i are direct sums of flat modules and injective modules, dimension shifting along the exact sequence

$$0 \longrightarrow F_0 \longrightarrow C_0 \longrightarrow \cdots \longrightarrow C_{-n} \longrightarrow \mathrm{Z}_{-(n+1)}(C) \longrightarrow 0$$

yields

$$\mathrm{Ext}^{n+2}_A(\mathrm{Z}_{-(n+1)}(C), H) \cong \mathrm{Ext}^1_A(F_0, H) = 0. \tag{3}$$

Combining (1)–(3) one gets $\mathrm{Ext}^1_A(\mathrm{Z}_{-(n+1)}(I), G) = 0$ for every Gorenstein injective A-module G. In particular $\mathrm{Ext}^1_A(\mathrm{Z}_{-(n+1)}(I), \mathrm{Z}_{-n}(I)) = 0$ holds, so the exact sequence $0 \to \mathrm{Z}_{-n}(I) \to I_{-n} \to \mathrm{Z}_{-(n+1)}(I) \to 0$ splits, which means that $\mathrm{Z}_{-n}(I)$ is injective and $\mathrm{id}_A F \leqslant n$ holds as desired. □

2. IF rings

Following Colby [**9**] a ring over which every injective module is flat is called a left IF ring. If a ring A and its opposite ring A° are both left IF, then A is called IF. Bennis [**3**] characterized IF rings in terms of the Gorenstein weak global dimension, which at that point was not known to be symmetric. In the commutative case, where this distinction is irrelevant, the characterization was also obtained by Mahdou, Tamekkante, and Yassemi [**22**]. Here we use the Gorenstein flat-cotorsion dimension to characterize left-IF rings; Bennis' characterization [**3**, Proposition 2.14] is recovered as the equivalence of (*i*) and (*iii*) in Theorem 2.2.

Recall that an acyclic complex F of flat-cotorsion A-modules is called *totally acyclic* if $\operatorname{Hom}_A(F,C)$ is acyclic for every flat-cotorsion A-module C. Recall further that an A-module G is called *Gorenstein flat-cotorsion* if there exists a totally acyclic complex F of flat-cotorsion A-modules with $\mathrm{C}_0(F) \cong G$. It follows from [**2**, Theorem 1.3] that a Gorenstein flat-cotorsion A-module is cotorsion.

Recall from [**5**, Definition 4.1], that the Gorenstein flat-cotorsion dimension of an A-complex M can be defined as the least integer n such that (1) $\mathrm{H}_v(M) = 0$ holds for all $v > n$ and (2) there exists a semi-flat-cotorsion A-complex F—i.e. a semi-flat complex consisting of flat–cotorsion modules—isomorphic to M in the derived category, such that the cokernel $\mathrm{C}_n(F)$ is Gorenstein flat-cotorsion.

PROPOSITION 2.1. *The following conditions are equivalent.*

(*i*) $\operatorname{Gfcd}_A M \leqslant 0$ *holds for every A-module M.*
(*ii*) *An A-module is flat-cotorsion if and only if it is injective.*
(*iii*) *An A-module is cotorsion if and only if it is Gorenstein flat-cotorsion.*
(*iv*) *An A-module is cotorsion if and only if it is Gorenstein injective.*
(*v*) *A is a left IF ring and an A-module is Gorenstein flat-cotorsion if and only if it is Gorenstein injective.*

PROOF. (*i*)$\Longrightarrow$(*ii*): If F is flat-cotorsion, then $\operatorname{Ext}^1_A(M,F) = 0$ holds for every A-module M by [**5**, Theorem 4.5], which implies that F is injective. Let I be injective; [**7**, Theorem 1.1] yields $\operatorname{fd}_A I = \operatorname{Gfcd}_A I \leqslant 0$, so I is flat-cotorsion.

(*ii*)$\Longrightarrow$(*iii*): Every Gorenstein flat-cotorsion module is cotorsion. To prove the converse, let C be a cotorsion A-module. It is a cycle submodule in an acyclic complex of flat-cotorsion A-modules: Indeed, take an injective resolution to the right and a flat-cotorsion resolution to the left. By assumption, every injective module is flat-cotorsion, so this is an acyclic complex of flat-cotorsion modules. Moreover, every flat-cotorsion module is injective, so it is in fact a totally acyclic complex of flat-cotorsion modules; in particular, C is Gorenstein flat-cotorsion.

(*iii*)$\Longrightarrow$(*i*): Let M be an A-module. There exists a semi-flat-cotorsion complex F isomorphic to M in the derived category; see Nakamura and Thompson [**23**, Theorem A.6]. Since the cokernel $\mathrm{C}_0(F)$ has a left resolution by cotorsion modules, it is itself cotorsion by [**5**, Lemma 5.6]. Thus $\mathrm{C}_0(F)$ is Gorenstein flat-cotorsion by assumption. By the definition of Gorenstein flat-cotorsion dimension it follows that $\operatorname{Gfcd}_A M \leqslant 0$ holds.

(*ii*)$\Longrightarrow$(*iv*): By Proposition 1.1 every Gorenstein injective A-module is cotorsion. Now, let C be cotorsion. A left resolution of C constructed by taking flat covers consists of flat-cotorsion modules. Splice this resolution together with a right injective resolution of C. As flat-cotorsion modules are injective, this produces an acyclic complex of injective modules. In particular, the complex has cotorsion cycles, see [**2**, Theorem 1.3]. As injective modules are flat, this complex is $\operatorname{Hom}_A(I,-)$-exact for every injective A-module I. Thus C is Gorenstein injective.

$(iv) \Longrightarrow (ii)$: Let I be an injective A-module. For every cotorsion module C, one has $\operatorname{Ext}^1_A(I,C) = 0$ because C is also Gorenstein injective. Thus I is flat-cotorsion. Conversely, let F be flat-cotorsion. By assumption, F must then be Gorenstein injective, so F is injective by Theorem 1.2.

$(ii) \Longrightarrow (v)$: As injective A-modules are flat, A is a left IF ring. Further, as flat-cotorsion modules are precisely the injective modules, Gorenstein flat-cotorsion modules and Gorenstein injective modules coincide.

$(v) \Longrightarrow (ii)$: Every injective module is flat and hence flat-cotorsion. A flat-cotorsion module F is Gorenstein flat-cotorsion and hence Gorenstein injective, so Theorem 1.2 yields $\operatorname{id}_A F = \operatorname{Gid}_A F = 0$. □

THEOREM 2.2. *The following conditions are equivalent.*

(*i*) $\operatorname{Gwgldim}(A) = 0$.
(*ii*) $\operatorname{Gfcd}_A M \leqslant 0$ *holds for every A-module M and* $\operatorname{Gfcd}_{A^\circ} M \leqslant 0$ *holds for every A°-module M.*
(*iii*) *A is an IF ring.*

PROOF. The Gorenstein weak global dimension is symmetric, see [**7**, Corollary 2.5], so (*i*) implies (*ii*) in view of [**5**, Theorem 5.7]. Further (*ii*) implies (*iii*) by Proposition 2.1. Finally, to see that (*iii*) implies (*i*) note that since every injective A°-module is flat, every acyclic complex of flat A-modules is $\mathbf{F}$-totally acyclic, i.e. it remains exact when tensored by an injective A°-module. Let M be an A-module. It suffices to build an acyclic complex F of flat A-modules such that $\mathrm{Z}_0(F) = M$. The left half is built by taking successive flat covers, whereas the right half is obtained by taking an injective resolution of M and observing that it is a complex of flat modules by assumption. □

REMARK 2.3. While condition (*i*) in Theorem 2.2 can be interpreted as saying that all A- and A°-modules are Gorenstein flat, (*ii*) does not say that all A- and A°-modules are Gorenstein flat-cotorsion. In fact, an A-module M with $\operatorname{Gfcd}_A M = 0$ is Gorenstein flat-cotorsion if and only if it is cotorsion, see [**5**, Remark 4.6].

Colby [**9**, Proposition 5] shows that a ring A is von Neumann regular if and only if it is left IF with $\operatorname{wgldim}(A) < \infty$. Here is the Gorenstein analog of this result:

COROLLARY 2.4. *The following conditions are equivalent.*

(*i*) *A is an IF ring.*
(*ii*) *A is a left or right IF ring with* $\operatorname{Gwgldim}(A) < \infty$.

PROOF. By Theorem 2.2 and [**7**, Corollary 2.5] it suffices to show that a left IF ring A of finite Gorenstein weak global dimension has $\operatorname{Gwgldim}(A) = 0$. Let A be such a ring and M an A-module. An injective resolution of M is a right resolution by flat A-modules, so for every n the module M is a flat syzygy of its nth injective cosyzygy. Thus M is Gorenstein flat. □

REMARK 2.5. There exist right IF rings which are not IF, see [**9**, Example 2]. Per Corollary 2.4 such rings must be of infinite Gorenstein weak global dimension.

3. Quasi-Frobenius rings

The difference between a module being Gorenstein flat-cotorsion and having Gorenstein flat-cotorsion dimension zero was already commented on in Remark 2.3. The former quality is the stronger one, and the main result of this section is that Gorenstein flat-cotorsionness of all modules characterizes rings of Gorenstein global dimension zero. First we characterize left perfect rings in terms of Gorenstein flat-cotorsion modules.

PROPOSITION 3.1. *The following conditions are equivalent.*

(*i*) *Every Gorenstein flat A-module is Gorenstein projective.*
(*ii*) *Every flat A-module is Gorenstein projective.*
(*iii*) *Every Gorenstein projective A-module is Gorenstein flat-cotorsion.*
(*iv*) *Every projective A-module is cotorsion.*
(*v*) *A is left perfect.*

Moreover, when these conditions are satisfied, an A-module is Gorenstein projective if and only if it is Gorenstein flat-cotorsion.

PROOF. Notice first that if A is left perfect, then every A-module is cotorsion and the definitions of Gorenstein projective A-modules and Gorenstein flat-cotorsion A-modules coincide. This justifies the last assertion as well as the implication $(v) \implies (iii)$. The implications $(i) \implies (ii)$ and $(iii) \implies (iv)$ are trivial.

$(ii) \Longrightarrow (iv)$: As projective A-modules are right Ext-orthogonal to Gorenstein projective modules, it follows that each projective A-module is right Ext-orthogonal to flat A-modules and hence cotorsion.

$(iv) \Longrightarrow (v)$: The free A-module $A^{(\mathbb{N})}$ is in particular cotorsion, whence A is left perfect by Guil Asensio and Herzog [**16**, Corollary 20].

$(v) \Longrightarrow (i)$: Every A-module is cotorsion, so per [**5**, Theorem 5.2] every Gorenstein flat A-module is Gorenstein flat-cotorsion, which over a left perfect ring means Gorenstein projective. □

The equivalence of (i), (i'), and (iii) in the next result was proved by Bennis and Mahdou in [**4**, Proposition 2.6]. Recall that a Quasi-Frobenius ring is one where the projective modules are injective. Over such a ring, the projective and injective modules coincide, and the same is true for the opposite ring.

THEOREM 3.2. *The following conditions are equivalent.*

(*i*) $\operatorname{Ggldim}(A) = 0$.
(*i'*) $\operatorname{Ggldim}(A^\circ) = 0$.
(*ii*) *Every A-module is Gorenstein flat-cotorsion.*
(*ii'*) *Every A°-module is Gorenstein flat-cotorsion.*
(*iii*) *A is Quasi-Frobenius.*

PROOF. As the Quasi-Frobenius property is left–right symmetric, it suffices to show the equivalence of the unprimed conditions.

$(i) \Longrightarrow (ii)$: By assumption every A-module is Gorenstein projective, so it follows from Proposition 3.1 that every A-module is Gorenstein flat-cotorsion.

$(ii) \Longrightarrow (iii)$: Let P be a projective A-module. As every module is Gorenstein flat-cotorsion, P is flat-cotorsion and $\operatorname{Ext}^1_A(M, P) = 0$ holds for every A-module M, so P is injective.

$(iii) \Longrightarrow (i)$: A Quasi-Frobenius ring is noetherian and self-injective on both sides, so it is in particular an Iwanaga-Gorenstein ring. Thus for every A-module M one has $\operatorname{Gpd}_A M \leqslant \operatorname{id}_A A = 0$, see for example Holm [**17**, Theorem 2.28] and Bass [**1**, Proposition 4.3]. □

4. Strongly cotorsion modules

As noticed it follows from Proposition 1.1 that every Gorenstein injective module is strongly cotorsion. A question considered in the literature is: For which rings do Gorenstein injective modules and strongly cotorsion modules coincide?

Yoshizawa [**29**, Theorem 2.7] shows that over (commutative) Gorenstein complete local rings, Gorenstein injective modules and strongly cotorsion modules are the same. Huang [**19**, Theorem 3.10] proves that the same statement holds for every Iwanaga-Gorenstein ring A such that the injective envelope of A is flat, and Iacob [**20**, Theorem 10] proves it for all Iwanaga-Gorenstein rings; cf. [**12**, Proposition 9.1.2]. The best result to date can be pieced together from work of Gillespie:

REMARK 4.1. A ring is *Ding-Chen* if it is coherent and has finite self FP-injective dimension on both sides. Over a Ding-Chen ring, the classes of Gorenstein injective and strongly cotorsion modules coincide: Indeed, let A be a Ding-Chen ring. Gillespie shows in [**14**, Theorem 4.2 and Corollary 4.5] that the modules that are right Ext^1_A-orthogonal to the class of A-modules with finite flat dimension—that is, the strongly cotorsion A-modules—coincide with the so-called Ding injective A-modules. Further, Gillespie shows [**15**, Theorem 1.1] that Ding injective A-modules are the same as Gorenstein injective A-modules.

PROPOSITION 4.2. *If the quantity*

$$\sup\{\operatorname{Gfcd}_A M \mid M \textit{ is an } A\textit{-module}\}$$

is finite, then $\operatorname{FFD}(A)$ *is finite and an A-module is strongly cotorsion if and only if it is Gorenstein injective.*

PROOF. Set $n = \sup\{\operatorname{Gfcd}_A M \mid M \text{ is an } A\text{-module}\}$ and assume that it is finite. This assumption implies that both $\operatorname{FFD}(A)$ and $\operatorname{sfli}(A)$ are finite: Indeed [**5**, Theorem 5.12] yields $\operatorname{FFD}(A) \leqslant n$, and [**7**, Theorem 2.1] yields $\operatorname{sfli}(A) \leqslant n$.

As already noticed, every Gorenstein injective A-module is strongly cotorsion. Now let M be a strongly cotorsion A-module. Since $\operatorname{sfli}(A)$ is finite, it follows from Proposition 1.1 that every acyclic complex of injectives is totally acyclic. Therefore, it suffices to show that M is the homomorphic image of an injective A-module with strongly cotorsion kernel. Since $\operatorname{FFD}(A)$ is finite, results of Trlifaj [**25**, Lemma 1.5(3) and Theorem 1.14] apply to yield a short exact sequence,

$$0 \longrightarrow K \longrightarrow E \longrightarrow M \longrightarrow 0,$$

with K a strongly cotorsion A-module and E an A-module of finite flat dimension. By the assumption on M it follows that also E is strongly cotorsion. To complete the proof we show that E is an injective A-module. Consider an exact sequence,

$$0 \longrightarrow E \longrightarrow I \longrightarrow C \longrightarrow 0,$$

with I an injective A-module. Since E and I have finite flat dimension, the A-module C has finite flat dimension. But then, since E is a strongly cotorsion A-module, the sequence splits and thus E is injective. □

A noetherian ring is Ding-Chen if and only if it is Iwanaga-Gorenstein. Thus for noetherian rings, the next result is equivalent to the statement in Remark 4.1.

COROLLARY 4.3. *If* Ggldim(A) *is finite, then an* A*-module is strongly cotorsion if and only if it is Gorenstein injective.*

PROOF. The inequality $\sup\{\mathrm{Gfcd}_A M \mid M \text{ is an } A\text{-module}\} \leqslant \mathrm{Ggldim}(A)$ holds by [**7**, Theorem 3.3], so the assertion is a special case of Proposition 4.2. □

This gives new examples of rings, including non-coherent rings, with the property that strongly cotorsion modules and Gorenstein injective modules coincide.

EXAMPLE 4.4. Over any ring of finite global dimension the strongly cotorsion modules, injective modules, and Gorenstein injective modules coincide. A result of Enochs, Estrada, and Iacob [**11**, Theorem 3.2] yields less trivial examples: the ring of dual numbers over any ring of finite Gorenstein global dimension. A concrete example of a non-coherent ring of finite global dimension is provided by Estrada, Iacob, and Yeomans [**13**, Section 4(1)].

Yoshizawa [**29**] in fact shows that among (commutative) Cohen-Macaulay complete local rings, those that are Gorenstein are characterized by the property that strongly cotorsion modules and Gorenstein injective modules are the same. More generally, Iacob [**20**] shows that this property characterizes Iwanaga-Gorenstein rings among noetherian rings A with $\mathrm{FPD}(A) < \infty$ and $\mathrm{FPD}(A^{\circ}) < \infty$. In view of Proposition 1.1, the characterizing property is really that strongly cotorsion modules are Gorenstein injective. Further, a noetherian ring A is Iwanaga-Gorenstein if and only if Ggldim(A) is finite, so the next result can be interpreted as removing the noetherian assumption from [**20**, Theorem 10]:

THEOREM 4.5. *The following conditions are equivalent.*

(*i*) $\mathrm{Ggldim}(A) < \infty$.
(*ii*) $\mathrm{FPD}(A) < \infty$ *and every strongly cotorsion* A*-module is Gorenstein injective.*

PROOF. The implication $(i) \Longrightarrow (ii)$ holds by Corollary 4.3 and [**17**, Theorem 2.28] which shows that the assumption on A implies that FPD(A) is finite. For $(ii) \Longrightarrow (i)$, set $n = \mathrm{FPD}(A)$. By a result of Jensen [**21**, Proposition 6], every flat A-module, and hence every module of finite flat dimension, has projective dimension at most n. It suffices to show that $\mathrm{Gid}_A N \leqslant n$ holds for every A-module N. Let N be an A-module and consider an exact sequence,

$$0 \longrightarrow N \longrightarrow E_0 \longrightarrow \cdots \longrightarrow E_{-n+1} \longrightarrow D \longrightarrow 0,$$

with each E_i injective. For every A-module M with $\mathrm{fd}_A M \leqslant n$, dimension shifting along this sequence yields $\mathrm{Ext}^1_A(M, D) \cong \mathrm{Ext}^{n+1}_A(M, N) = 0$. Therefore, D is a strongly cotorsion A-module. By assumption D is thus Gorenstein injective, whence Ggldim(A) is finite. □

For ease of comparison to the literature, we include:

COROLLARY 4.6. *If* FPD(A) *is finite, then the following conditions are equivalent.*

(*i*) Ggldim$(A) < \infty$.
(*ii*) *An* A*-module is strongly cotorsion if and only if it is Gorenstein injective.*

REMARK 4.7. In the literature it is often remarked that Ding-Chen rings have properties that generalize those of Iwanaga-Gorenstein rings to a non-noetherian setting; see for example the abstract of [**14**]. Ding and Chen [**10**, Theorem 7] show that a Ding-Chen ring has finite Gorenstein weak global dimension. In light of Remark 4.1, however, Theorem 4.5 says that a Ding-Chen ring A has finite Gorenstein global dimension if and only if FPD$(A) < \infty$ holds. It follows from [**21**, Proposition 6] that a von Neumann regular ring A of infinite global dimension is a a Ding-Chen ring with FPD$(A) = \infty$. By a result of Pierce [**24**, Corollary 5.2], a free boolean ring with $\aleph_\omega$ generators, for an infinite cardinal ω, is a von Neumann regular ring of infinite global dimension. We remark that this observation was essentially already made by Wang [**26**, Example 3.3].

Here is another consequence of Proposition 4.2; without assumptions on the ring it does not readily compare to Corollary 4.3.

COROLLARY 4.8. *If* Gwgldim(A) *is finite, then an* A*-module is strongly cotorsion if and only if it is Gorenstein injective.*

PROOF. The inequality $\sup\{\mathrm{Gfcd}_A M \mid M \text{ is an } A\text{-module}\} \leqslant \mathrm{Gwgldim}(A)$ holds by [**5**, Theorem 5.7], so the assertion is a special case of Proposition 4.2. □

THEOREM 4.9. *If* A *is left or right coherent, then the next conditions are equivalent.*

(*i*) Gwgldim$(A) < \infty$.
(*ii*) FFD$(A) < \infty$ *and every strongly cotorsion* A*-module is Gorenstein injective.*

PROOF. Condition (*i*) implies (*ii*) by Corollary 4.8 and [**17**, Theorem 3.24] as the Gorenstein weak global dimension is symmetric by [**7**, Corollary 2.5]. For the converse, assume that A is left coherent and set $n = \mathrm{FFD}(A)$. Let M be an A°-module with a flat resolution F, and let $N = \mathrm{C}_n(F)$ be the nth syzygy in this resolution. By [**17**, Theorem 3.6] the module N is Gorenstein flat if and only if the dual $N^+ = \mathrm{Hom}_{\mathbb{Z}}(M, \mathbb{Q}/\mathbb{Z})$ is a Gorenstein injective A-module. Let L be an A-module of finite flat dimension. The modules F_i^+ are injective, so dimension shifting along

$$0 \longrightarrow M^+ \longrightarrow (F_0)^+ \longrightarrow \cdots \longrightarrow (F_{n-1})^+ \longrightarrow N^+ \longrightarrow 0$$

yields $\mathrm{Ext}^1_A(L, N^+) \cong \mathrm{Ext}^{n+1}_A(L, M^+)$. As $\mathrm{fd}_A L \leqslant n$ and M^+ is cotorsion, dimension shifting along a flat resolution of L yields $\mathrm{Ext}^{n+1}_A(L, M^+) = 0$. Thus N^+ is strongly cotorsion and hence Gorenstein injective. This shows that every A°-module has Gorenstein flat dimension at most n, and $\mathrm{Gwgldim}(A) = \mathrm{Gwgldim}(A^\circ)$ holds by [**7**, Corollary 2.5]. If A is instead right coherent, then the same argument applies with A and A° interchanged. □

Again for ease of comparison to the literature, we include:

COROLLARY 4.10. *If A is left or right coherent and* $\mathrm{FFD}(A)$ *is finite, then the following conditions are equivalent.*

(*i*) $\mathrm{Gwgldim}(A) < \infty$.

(*ii*) *An A-module is strongly cotorsion if and only if it is Gorenstein injective.*

For modules over a right coherent ring, the Gorenstein flat dimension agrees with the Gorenstein flat-cotorsion dimension, see [**5**, Corollary 5.8]. In view of Theorem 4.9 it is thus natural to ask if Proposition 4.2 has a converse:

QUESTION 4.11. If $\mathrm{FFD}(A)$ is finite and an A-module is strongly cotorsion if and only if it is Gorenstein injective, is $\sup\{\mathrm{Gfcd}_A M \mid M \text{ is an } A\text{-module}\}$ finite?

REMARK 4.12. We remark that finiteness of $\sup\{\mathrm{Gfcd}_A M \mid M \text{ is an } A\text{-module}\}$ implies that another two quantities are finite, namely $\mathrm{sfli}(A)$, as already noticed in the proof of Proposition 4.2, and $\sup\{\mathrm{id}_A M \mid M \text{ is a flat-cotorsion } A\text{-module}\}$. If one adds to the assumption $\mathrm{FFD}(A) < \infty$ the assumption that these two quantities are finite, then it seems feasible to prove that $\sup\{\mathrm{Gfcd}_A M \mid M \text{ is an } A\text{-module}\}$ is finite, but at this time we are not convinced that this is the best one can do.

5. Group rings

We take the opportunity to include a note on how the Gorenstein flat-cotorsion dimension behaves along certain ring homomorphisms. This leads to an answer to a question of Y. Xiang [**27**] on the Gorenstein weak global dimension of group rings.

First make the following observation: Let $A \to B$ be a ring homomorphism. If C is a cotorsion B-module, then it is also cotorsion as an A-module. To see this, let F be a flat A-module and note that standard Hom-tensor adjunction yields $\mathrm{Ext}^1_A(F, C) \cong \mathrm{Ext}^1_A(F, \mathrm{Hom}_B(B, C)) \cong \mathrm{Ext}^1_B(B \otimes_A F, C) = 0$.

PROPOSITION 5.1. *Let $A \to B$ be a homomorphism of rings and M a B-module. If B is free as an A-module and* $\mathrm{Gfcd}_B M < \infty$, *then* $\mathrm{Gfcd}_A M \geqslant \mathrm{Gfcd}_B M$.

PROOF. Set $n = \mathrm{Gfcd}_B M$. By [**5**, Theorem 4.5], one thus has $\mathrm{Ext}^n_B(M, C) \neq 0$ for some flat-cotorsion B-module C. By the earlier remark, C is also flat-cotorsion viewed as an A-module. It now follows that

$$\mathrm{Ext}^n_A(M, C) \cong \mathrm{Ext}^n_A(B \otimes_B M, C) \cong \mathrm{Ext}^n_B(M, \mathrm{Hom}_A(B, C)),$$

which contains $\mathrm{Ext}^n_B(M, C)$ as a direct summand and hence is non-zero. Thus $n \leqslant \mathrm{Gfcd}_A M$. □

As an application, the following result removes the right coherence assumption in [**27**, Proposition 4.9], thus affirmatively answering [**27**, Question 4.11].

THEOREM 5.2. *Let k be a field, G a group, and H a subgroup of G of finite index. If* $\mathrm{Gwgldim}(k[G])$ *is finite, then it is equal to* $\mathrm{Gwgldim}(k[H])$.

PROOF. The inequality $\mathrm{Gwgldim}(k[H]) \leqslant \mathrm{Gwgldim}(k[G])$ holds by [**27**, Proposition 4.3(2)]. Assume that $\mathrm{Gwgldim}(k[G]) = n < \infty$. By [**27**, Proposition 4.2], one has $\mathrm{Gwgldim}(k[G]) = \mathrm{Gfd}_{k[G]} k$, which in turn equals $\mathrm{Gfcd}_{k[G]} k$ since Gorenstein flat-cotorsion and Gorenstein flat dimensions agree when the latter is finite by [**5**, Theorem 5.7]. Since $k[G]$ is free as a $k[H]$-module, Proposition 5.1 yields

$\mathrm{Gfcd}_{k[G]}\, k \leqslant \mathrm{Gfcd}_{k[H]}\, k$, and so we obtain the other inequality: $\mathrm{Gwgldim}(k[H]) \geqslant \mathrm{Gwgldim}(k[G])$. □

References

[1] Hyman Bass, *Injective dimension in Noetherian rings*, Trans. Amer. Math. Soc. **102** (1962), 18–29, DOI 10.2307/1993878. MR138644

[2] Silvana Bazzoni, Manuel Cortés-Izurdiaga, and Sergio Estrada, *Periodic modules and acyclic complexes*, Algebr. Represent. Theory **23** (2020), no. 5, 1861–1883, DOI 10.1007/s10468-019-09918-z. MR4140057

[3] Driss Bennis, *Weak Gorenstein global dimension*, Int. Electron. J. Algebra **8** (2010), 140–152. MR2660545

[4] Driss Bennis and Najib Mahdou, *Global Gorenstein dimensions*, Proc. Amer. Math. Soc. **138** (2010), no. 2, 461–465, DOI 10.1090/S0002-9939-09-10099-0. MR2557164

[5] Lars Winther Christensen, Sergio Estrada, Li Liang, Peder Thompson, Dejun Wu, and Gang Yang, *A refinement of Gorenstein flat dimension via the flat-cotorsion theory*, J. Algebra **567** (2021), 346–370, DOI 10.1016/j.jalgebra.2020.09.024. MR4159258

[6] Lars Winther Christensen, Sergio Estrada, and Peder Thompson, *Homotopy categories of totally acyclic complexes with applications to the flat-cotorsion theory*, Categorical, homological and combinatorial methods in algebra, Contemp. Math., vol. 751, Amer. Math. Soc., Providence, RI, 2020, pp. 99–118, DOI 10.1090/conm/751/15112. MR4132086

[7] Lars Winther Christensen, Sergio Estrada, and Peder Thompson, *Gorenstein weak global dimension is symmetric*, Math. Nachr. **294** (2021), no. 11, 2121–2128, DOI 10.1002/mana.202100141. MR4371287

[8] Lars Winther Christensen, Fatih Köksal, and Li Liang, *Gorenstein dimensions of unbounded complexes and change of base (with an appendix by Driss Bennis)*, Sci. China Math. **60** (2017), no. 3, 401–420, DOI 10.1007/s11425-015-0798-2. MR3600932

[9] R. R. Colby, *Rings which have flat injective modules*, J. Algebra **35** (1975), 239–252, DOI 10.1016/0021-8693(75)90049-6. MR376763

[10] Nanqing Ding and Jianlong Chen, *Coherent rings with finite self-FP-injective dimension*, Comm. Algebra **24** (1996), no. 9, 2963–2980, DOI 10.1080/00927879608825724. MR1396867

[11] E. Enochs, S. Estrada, and A. Iacob, *Rings with finite Gorenstein global dimension*, Math. Scand. **102** (2008), no. 1, 45–58, DOI 10.7146/math.scand.a-15050. MR2420678

[12] Edgar E. Enochs and Overtoun M. G. Jenda, *Relative homological algebra*, De Gruyter Expositions in Mathematics, vol. 30, Walter de Gruyter & Co., Berlin, 2000, DOI 10.1515/9783110803662. MR1753146

[13] Sergio Estrada, Alina Iacob, and Katelyn Yeomans, *Gorenstein projective precovers*, Mediterr. J. Math. **14** (2017), no. 1, Paper No. 33, 10, DOI 10.1007/s00009-016-0822-5. MR3598789

[14] James Gillespie, *Model structures on modules over Ding-Chen rings*, Homology Homotopy Appl. **12** (2010), no. 1, 61–73. MR2607410

[15] James Gillespie, *On Ding injective, Ding projective and Ding flat modules and complexes*, Rocky Mountain J. Math. **47** (2017), no. 8, 2641–2673, DOI 10.1216/RMJ-2017-47-8-2641. MR3760311

[16] Pedro A. Guil Asensio and Ivo Herzog, *Sigma-cotorsion rings*, Adv. Math. **191** (2005), no. 1, 11–28, DOI 10.1016/j.aim.2004.01.006. MR2102841

[17] Henrik Holm, *Gorenstein homological dimensions*, J. Pure Appl. Algebra **189** (2004), no. 1-3, 167–193, DOI 10.1016/j.jpaa.2003.11.007. MR2038564

[18] Henrik Holm, *Rings with finite Gorenstein injective dimension*, Proc. Amer. Math. Soc. **132** (2004), no. 5, 1279–1283, DOI 10.1090/S0002-9939-03-07466-5. MR2053331

[19] Zhaoyong Huang, *Gorenstein injective and strongly cotorsion modules*, Israel J. Math. **198** (2013), no. 1, 215–228, DOI 10.1007/s11856-013-0033-8. MR3096637

[20] Alina Iacob, *Gorenstein injective envelopes and covers over two sided noetherian rings*, Comm. Algebra **45** (2017), no. 5, 2238–2244, DOI 10.1080/00927872.2016.1233193. MR3582857

[21] C. U. Jensen, *On the vanishing of* $\varprojlim^{(i)}$, J. Algebra **15** (1970), 151–166, DOI 10.1016/0021-8693(70)90071-2. MR260839

[22] Najib Mahdou, Mohammed Tamekkante, and Siamak Yassemi, *On (strongly) Gorenstein von Neumann regular rings*, Comm. Algebra **39** (2011), no. 9, 3242–3252, DOI 10.1080/00927872.2010.501773. MR2845571

[23] Tsutomu Nakamura and Peder Thompson, *Minimal semi-flat-cotorsion replacements and cosupport*, J. Algebra **562** (2020), 587–620, DOI 10.1016/j.jalgebra.2020.07.001. MR4127282

[24] R. S. Pierce, *The global dimension of Boolean rings*, J. Algebra **7** (1967), 91–99, DOI 10.1016/0021-8693(67)90069-5. MR229695

[25] Jan Trlifaj, *Infinite dimensional tilting modules and cotorsion pairs*, Handbook of tilting theory, London Math. Soc. Lecture Note Ser., vol. 332, Cambridge Univ. Press, Cambridge, 2007, pp. 279–321, DOI 10.1017/CBO9780511735134.011. MR2384614

[26] Junpeng Wang, *Ding projective dimension of Gorenstein flat modules*, Bull. Korean Math. Soc. **54** (2017), no. 6, 1935–1950, DOI 10.4134/BKMS.b160547. MR3733774

[27] Yueming Xiang, *Weak dimensions and Gorenstein weak dimensions of group rings*, Czechoslovak Math. J. **71(146)** (2021), no. 3, 803–816, DOI 10.21136/CMJ.2021.0102-20. MR4295246

[28] Jinzhong Xu, *Flat covers of modules*, Lecture Notes in Mathematics, vol. 1634, Springer-Verlag, Berlin, 1996, DOI 10.1007/BFb0094173. MR1438789

[29] Takeshi Yoshizawa, *On Gorenstein injectivity of top local cohomology modules*, Proc. Amer. Math. Soc. **140** (2012), no. 6, 1897–1907, DOI 10.1090/S0002-9939-2011-11059-1. MR2888177

Texas Tech University, Lubbock, TX 79409, U.S.A.
Email address: lars.w.christensen@ttu.edu
URL: http://www.math.ttu.edu/~lchriste

Universidad de Murcia, Murcia 30100, Spain
Email address: sestrada@um.es

Niagara University, Niagara University, New York 14109
Email address: peder.thompson@mdu.se
Current address: Mälardalen University, Västerås 72123, Sweden

Contemporary Mathematics
Volume **785**, 2023
https://doi.org/10.1090/conm/785/15777

Quasi-Self-Dual Codes over a Non-Unital Ring of Order 4

Steven T. Dougherty and Serap Şahinkaya

Dedicated to Professor Syed Tariq Rizvi for his contribution to Ring and Module Theory.

ABSTRACT. We relate quasi-self-dual codes over a non-unital ring to additive self-dual codes over the additive group of the ring. We show that the Hamming weight enumerator of a quasi-self-dual code over this non-unital ring must belong to the same ring of invariants that self-dual codes over the Frobenius rings with unities belong. Additionally, the Hamming weight enumerator of a Type IV code over E must reside in the same ring of invariants as a Type IV code over the field of order 4.

1. Introduction

Self-dual codes are a widely studied object in algebraic coding theory. Originally they were defined over finite fields, but within the past thirty years they have been defined over finite Frobenius rings with unity as well. These codes have found connections in numerous areas like unimodular lattices and invariant theory. More recently, codes which are self-dual or similar to self-dual codes have been defined over finite non-unital rings. In this situation, many of the techniques of classical coding theory no longer apply which forces us to look for other techniques to study these codes. Here we shall examine a way in which some of the classical techniques can be applied in this situation.

Codes that are self-dual that have weights divisible by a constant (divisible codes) have been one of the most intensely studied families of self-dual codes. Classically, there are four main families of divisible codes. They are Type I codes which are binary self-dual codes with weights divisible by 2, Type II codes which are binary self-dual codes with weights divisible by 4, Type III codes which are ternary codes with weights divisible by 3, and Type IV codes which are self-dual codes over the field of order 4 with weights divisible by 2. Later, in [**7**], the definition of Type IV codes was extended to include codes over the rings with unity of order 4, namely $R_1 = \mathbb{F}_2[u]/\langle u^2\rangle$, $A_1 = \mathbb{F}_2[v]/\langle v^2+v\rangle$, and $\mathbb{Z}_4$. (The rings R_1 and A_1 are part of the larger families of rings R_k and A_k, see [**4**] and [**5**] for a description of these rings and their uses in coding theory.) Later this definition was extended even

2020 *Mathematics Subject Classification.* Primary 11T71, Secondary 94B05.
Key words and phrases. Code, group, ring, orthogonality.
Serap Şahinkaya would like to thank T.U.B.I.T.A.K (Scientific and Technological Research Council of Turkey) for their support while writing this paper.

further to consider non-unital rings see [**1**], [**2**], [**3**] and [**12**]. In this setting, the results for these codes are quite different given that many of the classical foundations are lost.

We shall make a nomenclature convention of referring to rings with unity as rings and rings without unity as rngs (that is a ring without the identity "i"). The difficulty in extending codes to rngs or to non-Frobenius rings is that for the standard Euclidean duality, we do not necessarily have the double annihilator condition $(C^\perp)^\perp = C$, nor do we have the cardinality condition $|C||C^\perp| = |R^n|$, nor do we have MacWilliams relations which relate the weight enumerator of a code to the weight enumerator of its orthogonal. This has specific consequence to self-dual codes, since self-dual codes over Frobenius rings have weight enumerators that are held invariant by the action of the MacWilliams relations. This allows for the application of invariant theory which gives a ring of invariants where the weight enumerator of a self-dual code must reside. For codes over rngs or non-Frobenius rings we lose this important tool. For a complete description of the use of Frobenius rings in coding theory see [**6**].

In this paper, we shall examine self-dual codes over the rng E, as given in [**10**]. Codes over this rng were studied in [**1**], [**2**], [**3**], [**12**], [**15**] and [**16**]. We shall take a different approach than the approach given in these papers. But we shall show that we can get the results of invariant theory for codes over E even though, in the setting described, there are no MacWilliams relations. In essence, we show that the Hamming weight enumerator of a quasi-self-dual code over the ring E must reside in the same ring of invariants as the Hamming weight enumerator of a self-dual code over the field of order 4. Moreover, if the code satisfies the condition that all weights are even, then the Hamming weight enumerator of Type IV code over E must reside in the same ring of invariants as a Type IV code over the field of order 4. This is accomplished by using additive dualities and considering additive codes rather than using the Euclidean duality and linear codes (submodules of the ambient space).

2. Definitions

2.1. The Rng E. In [**10**], Fine gives a classification of all rngs of order p^2, where p is a prime. In that paper, the rng E is defined as

$$E = \langle a, b \mid 2a = 2b = 0, a^2 = a, b^2 = b, ab = a, ba = b\rangle.$$

We let c denote the element $a + b$. Then the addition and multiplication tables of the rng are as follows.

TABLE 1. Addition Table of E

+	0	a	b	c
0	0	a	b	c
a	a	0	c	b
b	b	c	0	a
c	c	b	a	0

TABLE 2. Multiplication Table of E

×	0	a	b	c
0	0	0	0	0
a	0	a	a	0
b	0	b	b	0
c	0	c	c	0

We note that the additive group of the rng E is the group $\mathbb{Z}_2 \times \mathbb{Z}_2$.

The rng E is non-commutative and has no multiplicative identity. It has a unique maximal ideal $\{0, c\}$ and $E/\{0, c\} = \mathbb{F}_2$. A left linear code over E is defined as a left E-submodule of E^n and a right linear code over E is defined as a right E-submodule of E^n. We attach the following inner-product to the ambient space: $[\mathbf{v}, \mathbf{w}] = \sum v_i w_i$. The left dual of a code C is defined as

$$\mathcal{L}(C) = \{\mathbf{v} \mid \mathbf{v} \in E^n,\ [\mathbf{v}, \mathbf{w}] = 0, \forall \mathbf{w} \in C\}.$$

The right dual of a code C is defined as

$$\mathcal{R}(C) = \{\mathbf{v} \mid \mathbf{v} \in E^n,\ [\mathbf{w}, \mathbf{v}] = 0, \forall \mathbf{w} \in C\}.$$

In general, we do not have $|C||\mathcal{R}(C)| = 4^n$ nor do we have $|C||\mathcal{L}(C)| = 4^n$. Moreover, we do not have $\mathcal{R}(\mathcal{L}(C)) = C$. For example, the code of length 1, $C = \{0, c\}$ is both left and right linear. Its left orthogonal is $\mathcal{L}(C) = E$. Here $4(2) \neq 4$.

A left linear code over E is said to be self-orthogonal if $C \subseteq \mathcal{L}(C)$. In [**1**], [**2**], [**3**] and [**12**], the authors define a quasi-self-dual code over E as a self-orthogonal code with cardinality $4^{\frac{n}{2}}$. See these references for a complete description of these codes.

The following appears in [**1**].

LEMMA 2.1. *Let B be a self-orthogonal binary code. The code C over E defined by $C = aB \oplus cB^\perp$ is a quasi-self-dual code. Moreover, every quasi-self-dual code over E is of this form.*

We shall exploit this lemma to give an important result about the weight enumerator of these codes.

2.2. Additive Codes. We shall now give the definitions to discuss additive codes. Let G be a finite abelian group. An additive code over a group G is a subgroup of G^n. We note that we are not assuming any ring structure at all, only an additive group. However, as every rng has an additive abelian group structure these codes can be defined over any rng.

We now give a description of the orthogonality for additive codes. A character of the group G is a homomorphism from G to the multiplicative group of the complex numbers $\mathbb{C}^*$. In some texts, characters are defined with images in $\mathbb{Q}/\mathbb{Z}$, but we shall use the version where the image is the set of non-zero Complex numbers. We denote the set of all characters of G by

$$\widehat{G} = \{\chi \mid \chi \text{ is a character of the group } G\}.$$

We have that $\widehat{G}$ is a group and is isomorphic to the group G, but not in a canonical way. In particular, the isomorphism we choose between $\widehat{G}$ and G gives rise to the orthogonality. This isomorphism gives a character table M as follows. Let $\{g_i\}$ denote the elements of the group G and let $\phi : G \to \widehat{G}$ be the isomorphism. Let the character χ_{g_i} be the image of g_i under the isomorphism ϕ, that is $\chi_{g_i} = \phi(g_i)$. The rows of the character table M are indexed by $\phi(g_i)$ and columns are indexed by elements of the group. The element placed in i-th row and j-th column of the table M corresponds to an element $\chi_{g_i}(g_j)$. We can now define the orthogonality given by this isomorphism.

Fix a duality M of the finite abelian group G, namely an isomorphism of G and $\widehat{G}$. Let C be a code over G, then the orthogonal of C with respect to this duality is defined as

$$C^M = \{(g_1, g_2, \ldots, g_n) \mid \prod_{i=1}^{n} \chi_{g_i}(c_i) = 1 \text{ for all } (c_1, c_2, \ldots, c_n) \in C\}.$$

The code C^M is associated with the set

$$\{\chi \in \widehat{G^n} : \chi(c) = 1 \text{ for all } c \in C\},$$

by $\widehat{G}^n = \widehat{G^n}$. Here we shall only consider symmetric dualities, namely we want $\chi_a(b) = \chi_b(a)$. For symmetric dualities we have that $M = M^T$. This means that

$$\begin{aligned} &\{(g_1, g_2, \ldots, g_n) \mid \prod_{i=1}^{n} \chi_{g_i}(c_i) = 1 \text{ for all } (c_1, c_2, \ldots, c_n) \in C\} \\ &= \{(g_1, g_2, \ldots, g_n) \mid \prod_{i=1}^{n} \chi_{c_i}(g_i) = 1 \text{ for all } (c_1, c_2, \ldots, c_n) \in C\}. \end{aligned}$$

In this way, we do not need to consider a left and a right dual, but rather a single dual to the code.

It follows that the code C^M is additive, whether or not C is additive. Moreover, if C is an additive code in G^n, then $|C||C^M| = |G|^n$. Proofs for these facts and a full account of the MacWilliams relations in this case can be found in [**6**].

This setup gives rise to an inner-product. Namely, if M is a duality then

$$\langle \mathbf{v}, \mathbf{w} \rangle_M = \prod_{i=1}^{n} \chi_{v_i}(w_i).$$

Since the duality is symmetric we have $\langle \mathbf{v}, \mathbf{w} \rangle_M = \langle \mathbf{w}, \mathbf{v} \rangle_M$.

Let $G = \{e = g_0, g_1, g_2, \ldots, g_{s-1}\}$. For a code C over G, the complete weight enumerator of the code is defined as

$$cwe(x_{g_0}, x_{g_1}, x_{g_2}, \ldots, x_{g_{s-1}}) = \sum_{\mathbf{c} \in C} \prod x_i^{n_i(\mathbf{c})}$$

where there are $n_i(\mathbf{c})$ occurrences of g_i in the vector $\mathbf{c}$. Let $M \cdot (x_{g_0}, x_{g_1}, x_{g_2}, \ldots, x_{g_{s-1}}) = (M(x_{g_0}, x_{g_1}, x_{g_2}, \ldots, x_{g_{s-1}})^T)^T$. The following MacWilliams relations are shown in [**6**]. Let C be an additive code over the group G, then

$$cwe_{C^M}(x_{g_0}, x_{g_1}, x_{g_2}, \ldots, x_{g_{s-1}}) = \frac{1}{|C|} cwe(M \cdot (x_{g_0}, x_{g_1}, x_{g_2}, \ldots, x_{g_{s-1}})).$$

It is worth noting that if the code over G is also a linear code over a finite Frobenius ring, then any duality M will provide MacWilliams relations for the complete weight enumerator for the classical case. Namely, the MacWilliams relations are not unique for a given ring. For example, over the field $\mathbb{F}_3$ one can take the matrix

$$\begin{pmatrix} 1 & 1 & 1 \\ 1 & \xi & \xi^2 \\ 1 & \xi^2 & \xi \end{pmatrix}$$

where $\xi = e^{\frac{2\pi i}{3}}$, for the MacWilliams relations or one can take the matrix

$$\begin{pmatrix} 1 & 1 & 1 \\ 1 & \xi^2 & \xi \\ 1 & \xi & \xi^2 \end{pmatrix}.$$

for the MacWilliams relations. The computations will give the same weight enumerator for the orthogonal. In either case, these matrices collapse to give the same MacWilliams relations for codes with respect to the Hamming weight enumerator as we now describe.

The Hamming weight enumerator is defined as $W_C(x,y) = cwe(x,y,y,\dots,y)$. The MacWilliams relations for the Hamming weight enumerator follow from the MacWilliams relations for the complete weight enumerator. Namely, if C is an additive code over G, then

$$W_{C^M}(x,y) = \frac{1}{|C|} W_C(x+(s-1)y, x-y).$$

It follows easily from the MacWilliams relations (by setting $x = y = 1$) that if C is an additive code then $|C||C^M| = |G^n|$.

If a code satisfies $C \subseteq C^M$ then the code is said to be self-orthogonal with respect to M. If the code satisfies $C = C^M$, then we say that the code is self-dual with respect to M. If $W_C(x,y) = W_{C^M}(x,y)$ then the code is said to be formally self-dual with respect to M.

We have the following lemma.

LEMMA 2.2. *Let C and D be self-dual codes over an abelian group G, with respect to M of length n and m respectively. Then $C \times D$ is a self-dual code with respect to M of length $n+m$.*

PROOF. It is immediate that $C^M \times D^M \subseteq (C \times D)^M$. Then since $|(C\times D)^M||C\times D| = |G|^n$ and $|C^M||C||D^M||D| = |G|^n$ we have that $C^M \times D^M = (C\times D)^M$. Then since $C = C^M$ and $D = D^M$ we have $C \times D = C^M \times D^M$ and we have the result. □

3. Dualities Over $\mathbb{Z}_2 \times \mathbb{Z}_2$

A simple counting argument gives that over the group $\mathbb{Z}_2 \times \mathbb{Z}_2$, there are four possible symmetric dualities that we name M_1, M_2, M_3 and M_4. There are also non-symmetric dualities over these rings but we ignore them since they do not satisfy the double annihilator condition. They are given in the following tables.

TABLE 3. Character Table of M_1

M_1	00	01	10	11
00	1	1	1	1
01	1	-1	1	-1
10	1	1	-1	-1
11	1	-1	-1	1

TABLE 4. Character Table of M_2

M_2	00	01	10	11
00	1	1	1	1
01	1	1	-1	-1
10	1	-1	1	-1
11	1	-1	-1	1

For each of these we make the association:

$$\begin{aligned} (0,0) &\to 0, \\ (0,1) &\to a, \\ (1,0) &\to b, \\ (1,1) &\to c. \end{aligned}$$

TABLE 5. Character Table of M_3

M_3	00	01	10	11
00	1	1	1	1
01	1	1	-1	-1
10	1	-1	-1	1
11	1	-1	1	-1

TABLE 6. Character Table of M_4

M_4	00	01	10	11
00	1	1	1	1
01	1	-1	-1	1
10	1	-1	1	-1
11	1	1	-1	-1

Recall that a code C is said to be right nice if $|C||\mathcal{R}(C)| = |E|^n$ and C is said to be left nice if $|C||\mathcal{L}(C)| = |E|^n$, see [**3**].

THEOREM 3.1. *If C is a linear code and $\mathcal{R}(C) = C^M$ for some M then C is right nice. If C is a linear code and $\mathcal{L}(C) = C^M$ for some M then C is left nice.*

PROOF. For any duality M we have $|C||C^M| = |E|^n$, this gives the result. □

We shall now investigate some elementary invariant theory in our case. For details, see Chapter 19 of [**14**]. Any code that satisfies $C = C^M$ for any duality M must have a weight enumerator that is held invariant by action of the MacWilliams relations, namely

$$G_1 = \frac{1}{2}\begin{pmatrix} 1 & 3 \\ 1 & -1 \end{pmatrix}.$$

The matrix G_1 generates a group $\mathcal{G}_4$, of order 2, with Molien series:

$$\Phi(\lambda) = \frac{1}{(1-\lambda)(1-\lambda^2)}.$$

The ring of invariants is

$$\mathcal{R}_4 = \mathbb{C}[x+y, y(x-y)].$$

This gives the following theorem.

THEOREM 3.2. *Let C be an additive formally self-dual code with respect to any symmetric duality over the ring E. Then $W_C(x,y) \in \mathcal{R}_4$.*

PROOF. Any additive formally self-dual code must be held invariant by the action of G_1. □

If, in addition, the code has only even Hamming weights then its weight enumerator is also held invariant by the action of the matrix

$$G_2 = \begin{pmatrix} 1 & 0 \\ 0 & -1 \end{pmatrix}.$$

Let $\mathcal{G}_{IV} = \langle G_1, G_2 \rangle$. The Molien series of this group is

$$\Phi(\lambda) = \frac{1}{(1-\lambda^2)(1-\lambda^6)}.$$

It is immediate that a Type IV code over any duality exists only when the length is even. The group $\mathcal{G}_{IV}$ is the dihedral group D_{12}. It has a ring of invariants

$$\mathcal{R}_{IV} = \mathbb{C}[x^2+3y^2, y^2(x^2-y^2)^2].$$

We shall say that any code that is self-dual over a duality M for the additive group of E and has only even weights is a Type IV code over E.

This gives the following theorem.

THEOREM 3.3. *Let C be a Type IV additive formally self-dual code with respect to any symmetric duality over the ring E. Then $W_C(x,y) \in \mathcal{R}_{IV}$.*

PROOF. Any additive formally self-dual code must be held invariant by the action of G_1. If it has only even weights, it must also be held invariant by the action of G_2. This gives that its Hamming weight enumerator must be an element of $\mathcal{R}_{IV}$. □

It is well known that if C is a code with weight enumerator in $\mathcal{R}_{IV}$ then the code satisfies the bound $d_H(C) \leq 2(1 + \lfloor \frac{n}{6} \rfloor)$.

We shall now show that additive self-dual codes with respect to any duality exist for all lengths and Type IV codes over E exist for all dualities for all even lengths. We begin by showing the existence of self-dual codes for length 1 over E for any duality.

(1) The additive code of length 1, $\{0, c\}$ is self-dual with respect to M_1 and M_2. It is also left linear over E.
(2) The additive code of length 1, $\{0, a\}$ is self-dual with respect to M_2 and M_3, but is not left linear since $ca = c \notin C$.
(3) The additive code of length 1, $\{0, b\}$ is self-dual with respect to M_2 and M_4, but is not left linear since $cb = c \notin C$.

All three of these codes are right linear since $xc = 0$ or all $x \in E$.

This gives the following.

THEOREM 3.4. *Additive self-dual codes over E exist for all symmetric dualities for all lengths.*

PROOF. We have shown that taking additive self-dual codes of length 1 exist for all symmetric dualities. Then by Lemma 2.2, we have that self-dual additive codes exist for all symmetric dualities for all lengths. □

The additive code of length 2, $\{00, aa, bb, cc\}$ is a self-dual code with respect to any symmetric duality M. Moreover, this code has only even weights.

THEOREM 3.5. *Type IV additive self-dual codes over E exist for all symmetric dualities for all even lengths.*

PROOF. We have shown that taking additive self-dual codes with only even weights of length 2 exist for all symmetric dualities. It is immediate that if C and D both have only even weights that $C \times D$ has only even weights. Then by Lemma 2.2, Type IV additive self-dual codes over E exist for all symmetric dualities for all even lengths. □

We note that the invariant theory gives that there are no Type IV additive self-dual codes for odd lengths. This is because in the Molien series the coefficient of λ^i is 0 for all odd i.

4. Quasi-Self-Dual Codes and Additive Self-Dual Codes

We shall now relate quasi-self-dual codes over the rng E to self-dual codes with respect to symmetric dualities.

THEOREM 4.1. *Let C be a quasi-self-dual code over the ring E. Then C is a self-dual code with respect to dualities M_1 and M_2.*

PROOF. For any two binary vectors $\mathbf{v}, \mathbf{w}$ let $|\mathbf{v} \wedge \mathbf{w}| = |\{i \mid v_i = w_i = 1\}|$. Let M be either M_1 or M_2.

The code C must be of the form $aB \oplus cB^{\perp}$ by Lemma 2.1, where B is a self-orthogonal binary code. Let $\mathbf{v}, \mathbf{v}' \in B$ and $\mathbf{w}, \mathbf{w}' \in B^{\perp}$. Then we have that

$$\begin{aligned}\langle a\mathbf{v} + c\mathbf{w}, a\mathbf{v}' + c\mathbf{w}'\rangle_M &= \langle a\mathbf{v}, a\mathbf{v}'\rangle_M \langle a\mathbf{v}, c\mathbf{w}'\rangle_M \langle c\mathbf{w}, a\mathbf{v}'\rangle_M \langle c\mathbf{w}, c\mathbf{w}'\rangle_M \\ &= \chi_a(a)^{|\mathbf{v}\wedge\mathbf{v}'|}\chi_a(c)^{|\mathbf{v}\wedge\mathbf{w}'|}\chi_c(a)^{|\mathbf{v}'\wedge\mathbf{w}|}\chi_c(c)^{|\mathbf{w}\wedge\mathbf{w}'|}.\end{aligned}$$

Since B is self-orthogonal we have that $|\mathbf{v} \wedge \mathbf{v}'| \equiv 0 \pmod 2$. Since B and $B^{\perp}$ orthogonal codes we have $|\mathbf{v} \wedge \mathbf{w}'| \equiv |\mathbf{v}' \wedge \mathbf{w}| \equiv 0 \pmod 2$. This gives that $\chi_a(a)^{|\mathbf{v}\wedge\mathbf{v}'|}\chi_a(c)^{|\mathbf{v}\wedge\mathbf{w}'|}\chi_c(a)^{|\mathbf{v}'\wedge\mathbf{w}|} = 1(1)(1) = 1$. Finally, for dualities M_1 and M_2 we have that $\chi_c(c) = 1$. This gives that $C \subseteq C^{M_1}$ and $C \subseteq C^{M_2}$.

Since C is quasi-self-dual over E we have $|C| = 4^{\frac{n}{2}}$ where n is the length of the code. We also know that $|C||C^M| = 4^n$ for any duality M. This gives $C = C^{M_1}$ and $C = C^{M_2}$. □

This leads to the most important result of the paper.

COROLLARY 4.2. *Let C be a quasi-self-dual code over the ring E. Then $W_C(x, y) \in \mathcal{R}_4$. Let C be a quasi-self-dual code over the ring E with only even weights. Then $W_C(x, y) \in \mathcal{R}_{IV}$.*

PROOF. We have that if C is a quasi-self-dual code then it is an additive self-dual code with respect to dualities M_1 and M_2 by Theorem 4.1. Theorem 3.2 gives that any additive formally self-dual code with respect to any symmetric duality over the ring E has its weight enumerator in $\mathcal{R}_4$. Theorem 3.3 gives that a Type IV additive formally self-dual code with respect to any symmetric duality over the ring E has weight enumerator in $\mathcal{R}_{IV}$. □

EXAMPLE 4.3. Consider the additive code $\{0, c\}$. This code is a quasi-self-dual code of length 1. It is of the form $a\{0\} + c\{1\}$. It is also self-dual with respect to M_1 and M_2. Its weight enumerator is $x + y \in \mathcal{R}_4$.

EXAMPLE 4.4. Consider the additive code $\{00, aa, bb, cc\}$. This code is a quasi-self-dual code of length 2. It is of the form $a\{00, 11\} + c\{00, 11\}$. It is self-dual with respect to M_1, M_2, M_3, M_4. Moreover, all its weights are even. It weight enumerator is $x^2 + 3y^2$ which is a weight enumerator in $\mathcal{R}_{IV}$.

We can use a similar technique to construct additive self-dual codes over various dualities. These self-dual codes may not be quasi-self-dual.

THEOREM 4.5. *Let C be a code over the ring E of the form $xB + yB^{\perp}$ where B is a binary self-orthogonal code and $x, y \in E$. If $\chi_y(y) = 1$ in the duality M then the code C is a self-orthogonal code with respect to M. If $x \neq y$ then C is a self-dual code with respect to M.*

PROOF. Let C be a code of the form $xB + yB^{\perp}$ where B is a self-orthogonal code. Let $\mathbf{v}, \mathbf{v}' \in B$ and $\mathbf{w}, \mathbf{w}' \in B^{\perp}$. Then we have that

$$\begin{aligned}\langle x\mathbf{v} + y\mathbf{w}, x\mathbf{v}' + y\mathbf{w}'\rangle_M &= \langle x\mathbf{v}, x\mathbf{v}'\rangle_M \langle x\mathbf{v}, y\mathbf{w}'\rangle_M \langle y\mathbf{w}, x\mathbf{v}'\rangle_M \langle y\mathbf{w}, y\mathbf{w}'\rangle_M \\ &= \chi_x(x)^{|\mathbf{v}\wedge\mathbf{v}'|}\chi_x(y)^{|\mathbf{v}\wedge\mathbf{w}'|}\chi_y(x)^{|\mathbf{v}'\wedge\mathbf{w}|}\chi_y(y)^{|\mathbf{w}\wedge\mathbf{w}'|}.\end{aligned}$$

Since B is self-orthogonal we have that $|\mathbf{v} \wedge \mathbf{v}'| \equiv 0 \pmod 2$. Since B and $B^{\perp}$ orthogonal codes we have $|\mathbf{v} \wedge \mathbf{w}'| \equiv |\mathbf{v}' \wedge \mathbf{w}| \equiv 0 \pmod 2$. This gives that $\chi_x(x)^{|\mathbf{v}\wedge\mathbf{v}'|}\chi_x(y)^{|\mathbf{v}\wedge\mathbf{w}'|}\chi_y(x)^{|\mathbf{v}'\wedge\mathbf{w}|} = 1(1)(1) = 1$. When $\chi_y(y) = 1$ the code is self-orthogonal.

If x and y are distinct then $xB + yB^{\perp}$ has cardinality $2^n = 4^{\frac{n}{2}}$ and we have the result. □

We note that if x and y are not distinct, we still get a self-orthogonal code, but $xB + xB^{\perp} = x(B + B^{\perp})$ which has cardinality $2^{\frac{n}{2}}$.

COROLLARY 4.6. *Let B be a self-orthogonal binary code. Then $xB + yB^{\perp}$ is a self-dual code with respect to duality M_i in the given cases:*

x	y	M_i
a	c	M_1
b	c	M_1

x	y	M_i
a	b	M_2
a	c	M_2
b	a	M_2
b	c	M_2
c	a	M_2
c	b	M_2

x	y	M_i
b	a	M_3
c	a	M_3

x	y	M_i
a	b	M_4
c	b	M_4

PROOF. Follows immediately from Theorem 4.5 noticing that in these cases the x and y are distinct and y is a self-orthogonal element with respect to the duality. □

Of course, these codes, while self-dual additive codes, are not, in general, quasi-self-dual codes. For example, if $B = \{000, 110\}$ then $B^{\perp}$ contains 001 and so for M_4 the code $aB + bB^{\perp}$ contains $00b$. This vector however, is not self-orthogonal in terms of the Euclidean inner-product on the ring R.

5. Conclusion

The results of this paper have shown that even though the codes identified in previous works as generalizations of Type IV code over commutative rings do not satisfy many of the foundational coding results, they still must have a weight enumerator that is of the same form as the classical Type IV codes. In particular, this gives importance to these codes as they truly are an important generalization of Type IV codes. In particular, there may be codes in this setting with weight enumerators for which there are no Type IV codes over the commutative Frobenius rings of order 4.

Additionally, we have shown that there is an even broader class of Type IV codes which are simply additive codes over the Klein-4 group with a variety of dualities. In these cases, it is also possible that there are codes with weight enumerators in the ring of invariants for which there are no codes over the commutative Frobenius rings of order 4.

References

[1] Adel Alahmadi, Alaa Altassan, Widyan Basaffar, Hatoon Shoaib, Alexis Bonnecaze, and Patrick Solé, *Type IV codes over a non-unital ring*, J. Algebra Appl. **21** (2022), no. 7, Paper No. 2250142, 16, DOI 10.1142/S0219498822501420. MR4448425

[2] Adel Alahmadi, Alaa Altassan, Widyan Basaffar, Alexis Bonnecaze, Hatoon Shoaib, and Patrick Solé, *Quasi Type IV codes over a non-unital ring*, Appl. Algebra Engrg. Comm. Comput. **32** (2021), no. 3, 217–228, DOI 10.1007/s00200-021-00488-6. MR4263587

[3] Adel Alahmadi, Amani Alkathiry, Alaa Altassan, Widyan Basaffar, Alexis Bonnecaze, Hatoon Shoaib, and Patrick Solé, *Type IV codes over a non-local non-unital ring*, Proyecciones **39** (2020), no. 4, 963–978. MR4156460

[4] Yasemin Cengellenmis, Abdullah Dertli, and S. T. Dougherty, *Codes over an infinite family of rings with a Gray map*, Des. Codes Cryptogr. **72** (2014), no. 3, 559–580, DOI 10.1007/s10623-012-9787-y. MR3217509

[5] Steven T. Dougherty, Bahattin Yildiz, and Suat Karadeniz, *Codes over R_k, Gray maps and their binary images*, Finite Fields Appl. **17** (2011), no. 3, 205–219, DOI 10.1016/j.ffa.2010.11.002. MR2793245

[6] Steven T. Dougherty, *Algebraic coding theory over finite commutative rings*, SpringerBriefs in Mathematics, Springer, Cham, 2017, DOI 10.1007/978-3-319-59806-2. MR3675844

[7] Steven T. Dougherty, Philippe Gaborit, Masaaki Harada, Akihiro Munemasa, and Patrick Solé, *Type IV self-dual codes over rings*, IEEE Trans. Inform. Theory **45** (1999), no. 7, 2345–2360, DOI 10.1109/18.796375. MR1725122

[8] Steven T. Dougherty, Jon-Lark Kim, and Nari Lee, *Additive self-dual codes over fields of even order*, Bull. Korean Math. Soc. **55** (2018), no. 2, 341–357, DOI 10.4134/BKMS.b160842. MR3789447

[9] S. T. Dougherty and C. Fernández-Córdoba, *Additive G-codes over $\mathbb{F}_q$ and their dualities*, Finite Fields Appl. **72** (2021), Paper No. 101821, 21, DOI 10.1016/j.ffa.2021.101821. MR4222304

[10] Benjamin Fine, *Classification of finite rings of order p^2*, Math. Mag. **66** (1993), no. 4, 248–252, DOI 10.2307/2690742. MR1240670

[11] W. Cary Huffman and Vera Pless, *Fundamentals of error-correcting codes*, Cambridge University Press, Cambridge, 2003, DOI 10.1017/CBO9780511807077. MR1996953

[12] Jon-Lark Kim and Dong Eun Ohk, *DNA codes over two noncommutative rings of order four*, J. Appl. Math. Comput. **68** (2022), no. 3, 2015–2038, DOI 10.1007/s12190-021-01598-7. MR4428971

[13] Vera Pless, *Coding constructions*, Handbook of coding theory, Vol. I, II, North-Holland, Amsterdam, 1998, pp. 141–176. MR1667938

[14] F. J. MacWilliams and N. J. A. Sloane, *The theory of error-correcting codes. II*, North-Holland Mathematical Library, Vol. 16, North-Holland Publishing Co., Amsterdam-New York-Oxford, 1977. MR0465510

[15] Minjia Shi, Shitao Li, Jon-Lark Kim, and Patrick Solé, *LCD and ACD codes over a noncommutative non-unital ring with four elements*, Cryptogr. Commun. **14** (2022), no. 3, 627–640, DOI 10.1007/s12095-021-00545-4. MR4416224

[16] M. Shi, S. Wang, J.L. Kim, P. Sole, *S*elf-orthogonal codes over a non-unital ring and combinatorial matrices, Des. Codes and Cryptog., https://link.springer.com/article/10.1007/s10623-021-00948-7.

University of Scranton, Scranton, PA 18518, USA
Email address: prof.steven.dougherty@gmail.com

Tarsus University, Faculty of Engineering, Department of Natural and Mathematical Sciences, Mersin, Turkey
Email address: serap@tarsus.edu.tr

Contemporary Mathematics
Volume **785**, 2023
https://doi.org/10.1090/conm/785/15778

Codes from the Skew Ring $M_2(\mathbb{F}_2) \rtimes_\varphi G$

S. T. Dougherty, Serap Şahinkaya, and Deniz Ustun

Dedicated to Professor Syed Tariq Rizvi for his contribution to Ring and Module Theory.

ABSTRACT. In this work, we study codes generated by elements in the skew group matrix ring $M_k(R) \rtimes_\varphi G$, where R is a finite commutative Frobenius ring, G is an arbitrary finite group, and φ is a group homomorphism from G to $Aut(M_k(R))$. We then determine all possible group homomorphisms $\varphi : G \to Aut(M_2(\mathbb{F}_2))$, for the cases where G is a cyclic group and a dihedral group. Finally, by using skew generator matrices we provide examples of binary self-dual codes and also binary linear optimal codes.

1. Introduction

Linear codes, specifically binary linear codes, are the fundamental structure of algebraic coding theory. Certain families of linear codes have been widely studied because of their algebraic structure and their many applications both in and outside of coding theory. The most important such family is the class of cyclic codes, which are codes held invariant by the action of the cyclic shift. These codes were generalized to such families as negacyclic codes, constacyclic codes, and polycyclic codes. One of the most important generalizations, because of it success in constructing good linear codes, is the family of skew cyclic codes. One of the first papers in this direction was the the work of Boucher et al.[**4**]. Following this paper, in [**5**], they obtained some skew codes with Hamming distance larger than the known linear codes with the same parameters. Following these important results, numerous papers have been written generalizing these codes over various alphabets, see for example, [**1**, **2**, **6**, **7**, **13**] and the references therein.

Another generalization of cyclic codes is the family of G-codes, where G is a finite group. These codes were defined in [**10**] as ideals in the group ring RG, where R is a finite ring and G is a finite group of order n. These codes are held invariant by the action of the group G, that is, their automorphism group contains the arbitrary group G as a subgroup. Recently, G-codes have been generalized in two different ways. First, it has been defined over the matrix ring $M_k(R)$, see [**11**]. The advantage of this construction is that it does not only depend on the choice of the group G, but it also depends on the form of the elements from the matrix ring $M_k(R)$, that is, the form of the $k \times k$ matrices over R. Then, in [**13**], G-codes

2020 *Mathematics Subject Classification.* Primary 11T71, 94B05; Secondary 16Z05.
Key words and phrases. Skew codes, optimal codes, G-codes, self-dual codes.
Serap Sahinkaya would like to thank T.U.B.I.T.A.K (Scientific and Technological Research Council of Turkey) for their support while writing this paper.

have been generalized to ideals in the skew group ring $R \rtimes_\varphi G$, where R is a finite commutative Frobenius ring and $\varphi : G \to Aut(R)$ is a group homomorphism. These codes generalize many of the well-known classes of codes such as cyclic, quasicyclic, constacyclic, skew cyclic, skew quasicyclic, and skew constacyclic codes.

Motivated by these, we study codes generated by elements from the skew matrix ring $M_k(R) \rtimes_\varphi G$, where R is a finite commutative Frobenius ring, G is an arbitrary finite group, and φ is a group homomorphism from G to $Aut(M_k(R))$. We mainly focus on the skew ring $M_2(\mathbb{F}_2) \rtimes_\varphi G$. We then determine all possible group homomorphisms from cyclic and dihedral groups to the automorphism group of $M_2(\mathbb{F}_2)$. As an application, we obtain binary self-dual codes and also optimal binary linear codes from skew generator matrices.

2. Preliminaries

We begin by recalling the standard definitions and notions that we use throughout this paper.

Let R be a finite Frobenius ring. A code of length n over R is a subset of R^n. If the subset is also a submodule then the code is said to be linear. Attached to the ambient space is the usual inner-product, namely $[\mathbf{v}, \mathbf{w}] = \sum v_i w_i$. The orthogonal is defined as $C^\perp = \{\mathbf{v} \mid [\mathbf{v}, \mathbf{w}] = 0, \forall \mathbf{w} \in C\}$. This code is linear, even if C is not. We require the ring R to be Frobenius so that the MacWilliams relations hold, see [**9**] for details. It follows from these relations that $|C||C^\perp| = |R|^n$. A code is said to be self-orthogonal if $C \subseteq C^\perp$ and self-dual if $C = C^\perp$.

2.1. Skew Rings.

The following defines skew group rings and the class of skew G-codes.

DEFINITION 2.1. Let R be a finite commutative Frobenius ring and let $G = \{g_1, g_2, \ldots, g_n\}$ be a fixed listing of elements of a finite group G. Let $\varphi : G \to Aut(R)$ be a group homomorphism. The skew group ring of G over R induced by φ is the ring of formal sums

$$R \rtimes_\varphi G = \left\{ \sum_{i=1}^{n} \alpha_i g_i : \alpha_i \in R \right\},$$

where the addition operation is component-wise and multiplication is given by

$$\alpha_i g_i *_\varphi \beta_j g_j = \alpha_i \varphi(g_i)(\beta_j) g_i g_j$$

and then extending linearly. A skew G-code is an ideal (left or right) in the skew group ring $R \rtimes_\varphi G$.

We first note that the group G need not be commutative. Moreover, it is clear that when φ is the trivial group homomorphism then $R \rtimes_\varphi G$ is the usual group ring RG. In the remainder of the paper, we will denote $\varphi(g_i)(\beta_j)$ by $\varphi_i(\beta_j)$ and $\varphi(g_i^{-1} g_k)(\beta_j)$ by $\varphi_{i^{-1}k}(\beta_j)$.

2.2. Generator Matrices from the Group Matrix Ring.

Let $v = A_{g_1} g_1 + A_{g_2} g_2 + \cdots + A_{g_n} g_n \in M_k(R) \rtimes_\varphi G$, that is, each A_{g_i} is a $k \times k$ matrix with entries from the ring R. We define the block matrix $\varphi(\tau_k(v)) \in (M_k(R))_n$ as:

$$\varphi(\tau_k(v)) = \begin{pmatrix} \varphi_1(A_{g_1^{-1}g_1}) & \varphi_1(A_{g_1^{-1}g_2}) & \varphi_1(A_{g_1^{-1}g_3}) & \dots & \varphi_1(A_{g_1^{-1}g_n}) \\ \varphi_2(A_{g_2^{-1}g_1}) & \varphi_2(A_{g_2^{-1}g_2}) & \varphi_2(A_{g_2^{-1}g_3}) & \dots & \varphi_2(A_{g_2^{-1}g_n}) \\ \vdots & \vdots & \vdots & \vdots & \vdots \\ \varphi_n(A_{g_n^{-1}g_1}) & \varphi_n(A_{g_n^{-1}g_2}) & \varphi_n(A_{g_n^{-1}g_3}) & \dots & \varphi_n(A_{g_n^{-1}g_n}) \end{pmatrix}, \tag{2.1}$$

for $\varphi : G \to Aut(M_k(R))$. We note that, $\varphi_i(A_{g_i^{-1}g_j})$ is a $k \times k$ matrix which is obtained from the matrix $A_{g_i^{-1}g_j}$ by applying φ_i to the each entries.

THEOREM 2.2. *Let R be a finite commutative Frobenius ring, k a positive integer and G a finite group of order n. Let $v \in M_k(R) \rtimes_\varphi G$ and C be the corresponding skew G-code in $M_k(R)$. Let $I_k(v)$ be the set of elements of $M_k(R) \rtimes_\varphi G$ such that $\sum A_i g_i \in I_k(v)$ if and only if there exists $i \in \{1, 2, \dots, n\}$, such that $(\varphi_i(A_1), \varphi_i(A_2), \dots, \varphi_i(A_n)) \in C_k(v)$, where $C_k(v)$ is the left ideal of the skew group ring generated by the element v. Then $I_k(v)$ is a left ideal in $M_k(R) \rtimes_\varphi G$.*

PROOF. Each row of $\varphi(\tau_k(v))$ corresponds to an element of the form $g_j \sum A_i g_i = \sum \varphi_j(A_i) g_j g_i$ in $M_k(R) \rtimes_\varphi G$, where g_j is any element of G. That is, the multiplication by g_j is done from left. The sum of any two elements in $I_k(v)$ corresponds exactly to the sum of the corresponding elements in $C_k(v)$ and so $I_k(v)$ is closed under addition.

Now we shall show when the product of an element in $M_k(R) \rtimes_\varphi G$ and an element in $I_k(v)$ is in $I_k(v)$. Let $w_1 = \sum B_i g_i \in M_k(R) \rtimes_\varphi G$, where B_i are the $k \times k$ matrices. Then if w_2 is a row in $C_k(v)$, it is of the form $\sum C_j h_j v$. Then $w_1 w_2 = \sum B_i g_i \sum C_j h_j v = \sum B_i \varphi_i(C_j) g_i h_j v$ which corresponds to an element in $C_k(v)$, which gives that the element is in $I_k(v)$. □

Next, we shall take elements in the skew ring $M_k(R) \rtimes_\varphi G$ and use them to generate codes over the ring R, not only over the ring $M_k(R)$. This is an extremely important distinction since R is a commutative ring and $M_k(R)$ is not commutative in general. The reasoning for this is that we want to be able to generate codes that are not reached by traditional techniques. In other words, we want to find codes that would be missed using generator matrices formed by the usual constructions. Recall the matrix construction $\tau_k(v)$ from [**11**]. Let $v = A_{g_1}g_1 + A_{g_2}g_2 + \dots + A_{g_n}g_n \in M_k(R)G$, where $M_k(R)$ is a non-commutative Frobenius matrix ring, G is a finite group of order n, and each A_{g_i} is a $k \times k$ matrix with entries from the ring R. We note that the rows of the matrices $A_{g_i^{-1}g_j}$ form rows of this matrix. That is, the elements of the matrix are elements of the ring, not a matrix ring. We have used the matrices A_{g_i} to form a larger matrix over R. We define the block matrix $\tau_k(v) \in (M_k(R))_n$ to be

$$\tau_k(v) = \begin{pmatrix} A_{g_1^{-1}g_1} & A_{g_1^{-1}g_2} & A_{g_1^{-1}g_3} & \dots & A_{g_1^{-1}g_n} \\ A_{g_2^{-1}g_1} & A_{g_2^{-1}g_2} & A_{g_2^{-1}g_3} & \dots & A_{g_2^{-1}g_n} \\ \vdots & \vdots & \vdots & \vdots & \vdots \\ A_{g_n^{-1}g_1} & A_{g_n^{-1}g_2} & A_{g_n^{-1}g_3} & \dots & A_{g_n^{-1}g_n} \end{pmatrix}. \tag{2.2}$$

Let

$$B_k(v) = \langle \tau_k(v) \rangle. \tag{2.3}$$

Here the code $B_k(v)$ is formed by taking all linear combinations of the rows of the matrix with coefficients in R. In this case the ring over which the code is defined is commutative so it is both a left linear and right linear code.

3. Generator Matrices of Self-Dual Codes

In this section, we investigate constructions of self-dual codes from the matrix $\varphi(\tau_k(v))$. These codes will be over the commutative ring R, not over the matrix ring.

LEMMA 3.1. *Let G be a group of order n and $v \in M_k(R) \rtimes_\varphi G$. The matrix $[I_{kn}|\varphi(\tau_k(v))]$ generates a self-dual code over R if and only if*

$$\varphi(\tau_k(v)) *_\varphi \varphi(\tau_k(v))^T = -I_{kn}.$$

PROOF. The proof is similar to the standard proof that $(I_m \mid A)$ generates a self-dual code of length $2m$ if and only if $AA^T = -I_m$. Namely, the fact that $\varphi(\tau_k(v)) *_\varphi \varphi(\tau_k(v))^T = -I_{kn}$ guarantees that the inner-product of any two rows is 0 and that the self-orthogonal code has maximum free rank. □

Recall from [**13**] that the skew canonical involution (or φ-canonical involution) for the skew group ring $R \rtimes_\varphi G$ is given by $w^* = \sum_{i=1}^n \alpha_i g_i^{-1}$ for $w = \sum_{i=1}^n \alpha_i g_i \in R \rtimes_\varphi G$. This notion can be extended to the skew group matrix ring $M_k(R) \rtimes_\varphi G$. Namely, let $* : M_k(R) \rtimes_\varphi G \to M_k(R) \rtimes_\varphi G$ be the canonical involution on the group matrix ring $M_k(R) \rtimes_\varphi G$ given by $v^* = \sum_g A_g g^{-1}$, for $v = \sum_g A_g g \in M_k(R) \rtimes_\varphi G$ where A_g are the $k \times k$ blocks.

THEOREM 3.2. *Let R be a finite Frobenius ring and G be a group of order n and $w \in M_k(R) \rtimes_\varphi G$. If $\varphi(\tau_k(w)) = (\varphi(\tau_k(w)))^T$, $\varphi(\tau_k(w))\varphi(\tau_k(w)) = \mathbf{0}$ then the code C generated by w is a self-orthogonal code. If, additionally, $|C| = |R|^{\frac{kn}{2}}$ then C is a self-dual code of length kn over the commutative ring R.*

PROOF. Let $(\varphi(\tau_k(w)))^T = \varphi(\tau_k(w))$. We know that $(\varphi(\tau_k(w))\varphi(\tau_k(w)))_{ij}$ is the inner product of i-th and j-th rows of $\varphi(\tau_k(w))$. Since $\varphi(\tau_k(w))\varphi(\tau_k(w)) = \mathbf{0}$ any two rows of $\varphi(\tau_k(w))$ are orthogonal, hence the code C is a self-orthogonal code. For the second statement, since $|C| = |R|^{\frac{kn}{2}}$, we have that C is self-dual. □

THEOREM 3.3. *Let R be a finite Frobenius ring and G be a group of order n, $w \in M_k(R) \rtimes_\varphi G$. If $\varphi(\tau_k(w)) = (\varphi(\tau_k(w)))^T$, $\varphi(\tau_k(w))\varphi(\tau_k(w)) = -I_{kn}$ and C is the code generated by $[I_{kn}|\varphi(\tau_k(w))]$, then C is a self-dual code of length $2kn$.*

PROOF. If C denotes the code generated by $[I_{kn}|\varphi(\tau_k(w))]$, then the free rank of the code is kn. In this case, for the self-orthogonality of C, we need the inner products of the i-th row and j-th row to be $-\delta_{ij}$. This condition is satisfied by the assumptions $\varphi(\tau_k(w)) = (\varphi(\tau_k(w)))^T$ and $\varphi(\tau_k(w))\varphi(\tau_k(w)) = -I_{kn}$, which completes the proof. □

4. Group Homomorphisms

In this section, we determine all possible group homomorphisms

$$\varphi : G \to Aut(M_2(\mathbb{F}_2)),$$

where G is a finite group and $M_2(\mathbb{F}_2)$ is the ring of 2×2 matrices over the field $\mathbb{F}_2$. We first need to find the automorphism group of $M_2(\mathbb{F}_2)$. It is well-known that automorphisms of matrix rings over arbitrary fields are of the following form:

$$M \to PMP^{-1}, \tag{4.1}$$

where P is invertible matrix, and the number of $n \times n$ invertible matrices in $M_n(\mathbb{F}_q)$ is

$$(q^n - 1)(q^n - q) \dots (q^n - q^{n-1}). \tag{4.2}$$

There exists 6 invertible matrices in $M_2(\mathbb{F}_2)$ according to Equation 4.2:

$$P_1 = P_1^{-1} = \begin{pmatrix} 1 & 0 \\ 0 & 1 \end{pmatrix}, P_2 = P_2^{-1} = \begin{pmatrix} 0 & 1 \\ 1 & 0 \end{pmatrix}, P_3 = P_4^{-1} = \begin{pmatrix} 1 & 1 \\ 1 & 0 \end{pmatrix},$$
$$P_4 = P_3^{-1} = \begin{pmatrix} 0 & 1 \\ 1 & 1 \end{pmatrix}, P_5 = P_5^{-1} = \begin{pmatrix} 1 & 1 \\ 0 & 1 \end{pmatrix}, P_6 = P_6^{-1} = \begin{pmatrix} 1 & 0 \\ 1 & 1 \end{pmatrix}.$$

Therefore, $Aut(M_2(\mathbb{F}_2))$ has 6 elements and they can be written according to Equation 4.1 as follows:

$$\eta_i : M_2(\mathbb{F}_2) \to M_2(\mathbb{F}_2), \quad \text{where} \quad \eta_i(M) = P_i M P_i^{-1} \quad \text{for each} \quad M \in M_2(\mathbb{F}_2).$$

More precisely;

$$\iota : M_2(\mathbb{F}_2) \to M_2(\mathbb{F}_2), \quad \iota(M) = P_1 M P_1^{-1}, \iota(\begin{pmatrix} a & b \\ c & d \end{pmatrix}) = \begin{pmatrix} a & b \\ c & d \end{pmatrix}$$
$$\eta_2 : M_2(\mathbb{F}_2) \to M_2(\mathbb{F}_2), \quad \eta_2(M) = P_2 M P_2^{-1}, \eta_2(\begin{pmatrix} a & b \\ c & d \end{pmatrix}) = \begin{pmatrix} d & c \\ b & a \end{pmatrix}$$
$$\eta_3 : M_2(\mathbb{F}_2) \to M_2(\mathbb{F}_2), \quad \eta_3(M) = P_3 M P_3^{-1}, \eta_3(\begin{pmatrix} a & b \\ c & d \end{pmatrix}) = \begin{pmatrix} b+d & a+b+c+d \\ b & a+b \end{pmatrix}$$
$$\eta_4 : M_2(\mathbb{F}_2) \to M_2(\mathbb{F}_2), \quad \eta_4(M) = P_4 M P_4^{-1}, \eta_4(\begin{pmatrix} a & b \\ c & d \end{pmatrix}) = \begin{pmatrix} c+d & c \\ a+b+c+d & a+c \end{pmatrix}$$
$$\eta_5 : M_2(\mathbb{F}_2) \to M_2(\mathbb{F}_2), \quad \eta_5(M) = P_5 M P_5^{-1}, \eta_5(\begin{pmatrix} a & b \\ c & d \end{pmatrix}) = \begin{pmatrix} a+c & a+b+c+d \\ c & c+d \end{pmatrix}$$
$$\eta_6 : M_2(\mathbb{F}_2) \to M_2(\mathbb{F}_2), \quad \eta_6(M) = P_6 M P_6^{-1}, \eta_6(\begin{pmatrix} a & b \\ c & d \end{pmatrix}) = \begin{pmatrix} a+b & b \\ a+b+c+d & b+d \end{pmatrix}.$$

It can be deduced from the relations that

$$\eta_3^2 = \eta_4, \ \eta_3\eta_2 = \eta_5, \ \eta_3^2\eta_2 = \eta_6.$$

Moreover, we have that $Aut(M_2(\mathbb{F}_2))$ is a dihedral group of order 6 given by the following presentation:

$$\mathcal{D}_6 = \langle \eta_3, \eta_2 | \eta_3^3 = \eta_2^2 = \iota, \ \eta_2\eta_3\eta_2 = \eta_3^{-1} \rangle.$$

The non-trivial subgroups of $\mathcal{D}_6$ are $\{\iota, \eta_3, \eta_4\}$, $\{\iota, \eta_2\}$, $\{\iota, \eta_5\}$, and $\{\iota, \eta_6\}$, where $\{\iota, \eta_3, \eta_4\}$ is a normal subgroup.

4.1. Skew Group Matrix Ring $M_2(\mathbb{F}_2) \rtimes_\varphi G$. We now determine all possible group homomorphisms $\varphi : G \to Aut(M_2(\mathbb{F}_2))$ for two different families of groups, namely cyclic groups and dihedral groups.

4.1.1. *The Cyclic Group $\mathcal{C}_n$.* Let $\mathcal{C}_n = \langle g | g^{n-1} = e \rangle$ be the cyclic group order n and we consider ordering of the elements of the group as

$$\{1, g, g^2, \ldots, g^{n-1}\}.$$

We only consider cyclic groups $\mathcal{C}_n$ of order $n = 2k$ and $n = 3k$ for $M_2(\mathbb{F}_2) \rtimes_\varphi \mathcal{C}_n$ due to the following lemma.

LEMMA 4.1. *Let $\mathcal{C}_n$ be the cyclic group of order n and let $\mathcal{D}_6$ be the dihedral group of order 6. A non-trivial $\varphi : \mathcal{C}_n \to Aut(M_2(\mathbb{F}_2)) = \mathcal{D}_6$ exists if $n = 2k$ or $n = 3k$ for a positive integer k.*

PROOF. It is well-known, by a basic group homomorphism property, that the order of image of an element divides the order of the element. It is also easy to see that order of elements of $\mathcal{D}_6$ are either 2 or 3. Therefore, a non-trivial group homomorphism φ exits if 2 or 3 divides n, which implies $n = 2k$ or $n = 3k$ for a positive integer k. □

THEOREM 4.2. *Let $\mathcal{C}_n$ be the cyclic group of order n and let φ be a homomorphism from $\mathcal{C}_n = \langle g \rangle$ to $Aut(M_2(\mathbb{F}_2))$.*

- *If $n = 2k$, then φ must be of the form*

$$\varphi_{i+1} = \varphi(g^i) = \begin{cases} \iota, & \text{if } i \text{ is even} \\ \alpha_j, & \text{if } i \text{ is odd} \end{cases} \tag{4.3}$$

where $\alpha_j \in \{\eta_2, \eta_5, \eta_6\}$.

- *If $n = 3k$, then φ must be of the form*

$$\varphi_{i+1} = \varphi(g^i) = \begin{cases} \iota, & i \equiv 0 \pmod 3 \\ \alpha_j, & i \equiv 1 \pmod 3 \\ \beta_j, & i \equiv 2 \pmod 3 \end{cases} \tag{4.4}$$

where $(\alpha_j, \beta_j) \in \{(\eta_3, \eta_4), (\eta_4, \eta_3)\}$.

PROOF. It is clear from $\varphi(aa) = \varphi(a)\varphi(a)$ and also from the fact that for a finite cyclic group, every subgroup's order is a divisor of the cyclic group's order. □

PROPOSITION 4.3. *Let $w \in M_2(\mathbb{F}_2) \rtimes_\varphi \mathcal{C}_n$. Then the generator matrix $\varphi(\tau_k(w))$ has one of the following forms:*

- *If $n = 2k$, then there exist three possible generator matrices*

$$\varphi(\tau_k(w)) = \begin{pmatrix} A_1 & A_2 & A_3 & \cdots & A_{2n} \\ \eta_i(A_{2n}) & \eta_i(A_1) & \eta_i(A_2) & \cdots & \eta_i(A_{2n-1}) \\ A_{2n-1} & A_{2n} & A_1 & \cdots & A_{2n-2} \\ \vdots & \vdots & \vdots & \vdots & \vdots \\ A_3 & A_4 & A_5 & \cdots & A_2 \\ \eta_i(A_2) & \eta_i(A_3) & \eta_i(A_4) & \cdots & \eta_i(A_1) \end{pmatrix},$$

where $A_i \in M_2(\mathbb{F}_2)$ and $\eta_i \in \{\eta_2, \eta_5, \eta_6\}$.

- *If $n = 3k$, then there exist two possible generator matrices*

$$\varphi(\tau_2(w)) = \begin{pmatrix} A_1 & A_2 & A_3 & \cdots & A_{3n} \\ \eta_i(A_{3n}) & \eta_i(A_1) & \eta_i(A_2) & \cdots & \eta_i(A_{3n-1}) \\ \eta_j(A_{3n-1}) & \eta_j(A_{3n}) & \eta_j(A_1) & \cdots & \eta_j(A_2) \\ \vdots & \vdots & \vdots & \vdots & \vdots \\ A_4 & A_5 & A_6 & \cdots & A_3 \\ \eta_i(A_3) & \eta_i(A_4) & \eta_i(A_5) & \cdots & \eta_i(A_2) \\ \eta_j(A_2) & \eta_j(A_3) & \eta_j(A_4) & \cdots & \eta_j(A_1) \end{pmatrix},$$

where $A_i \in M_2(\mathbb{F}_2)$, and $(\eta_i, \eta_j) \in \{(\eta_3, \eta_4), (\eta_4, \eta_3)\}$.

PROOF. The result follows from Theorem 4.2. □

4.1.2. *The Dihedral Group $\mathcal{D}_{2n}$ of order $2n$.* We first recall some basic properties of the dihedral group $\mathcal{D}_{2n}$. The elements of order 2 in the group $\mathcal{D}_{2n}$ are reflections which are of the form $b^{i-1}a$, where $1 \leq i \leq n$ if n is odd. Normal subgroups of the dihedral group are given as follows.

- If n is odd, then normal groups are of the form $\langle b^d \rangle$, where d divides n.
- If n is even, then normal groups are of the form $\langle b^d \rangle$, $\langle b^2, a \rangle$, $\langle b^2, ba \rangle$, where d divides n.

Now we are ready to determine all possible group homomorphisms from $\mathcal{D}_{2n}$ to $Aut(M_2(\mathbb{F}_2))$.

LEMMA 4.4. *Let $\mathcal{D}_{2n}$ be the dihedral group of order $2n$. Non-trivial homomorphisms $\varphi : \mathcal{D}_{2n} \to Aut(M_2(\mathbb{F}_2))$ are as follows.*

(1) *If $n = 1$ then there exists 3 non-trivial homomorphisms.*
(2) *If $n = 2$ then there exists 9 non-trivial homomorphisms.*
(3) *For $n > 2$ and n odd, then there exists 9 non-trivial homomorphism φ for $n = 3k$, and 3 non-trivial homomorphism φ for $n \neq 3k$.*
(4) *For $n > 2$ and n even, then there exists 15 non-trivial homomorphisms φ for $n = 3k$, and 9 non-trivial homomorphism φ for $n \neq 3k$.*

PROOF. (1) If $n = 1$ then $Ker(\varphi) = \iota$ since $\mathcal{D}_2 \cong C_2$. This gives the result. For each φ, the images of the generator are given in Table 4.5.

(4.5)

Case 1		Case 2		Case 3	
$\varphi(e) =$	ι	$\varphi(e) =$	ι	$\varphi(e) =$	ι
$\varphi(a) =$	η_2	$\varphi(a) =$	η_5	$\varphi(a) =$	η_6

(2) If $n = 2$ then $\mathcal{D}_4 \cong C_2 \times C_2$. The proof follows from the fact that $Ker(\varphi) = \langle b \rangle$, $Ker(\varphi) = \langle a \rangle$, or $Ker(\varphi) = \langle ba \rangle$. For each φ, the images of the

generators are given in Table 4.6.

(4.6)

Case 1		Case 2		Case 3	
$\varphi(e) =$	ι	$\varphi(e) =$	ι	$\varphi(e) =$	ι
$\varphi(b) =$	ι	$\varphi(b) =$	ι	$\varphi(b) =$	ι
$\varphi(a) =$	η_2	$\varphi(a) =$	η_5	$\varphi(a) =$	η_6
$\varphi(ba) =$	η_2	$\varphi(ba) =$	η_5	$\varphi(ba) =$	η_6
Case 4		Case 5		Case 6	
$\varphi(e) =$	ι	$\varphi(e) =$	ι	$\varphi(e) =$	ι
$\varphi(b) =$	η_2	$\varphi(a) =$	η_5	$\varphi(a) =$	η_6
$\varphi(a) =$	ι	$\varphi(b) =$	ι	$\varphi(b) =$	ι
$\varphi(ba) =$	η_2	$\varphi(ba) =$	η_5	$\varphi(ba) =$	η_6
Case 7		Case 8		Case 9	
$\varphi(e) =$	ι	$\varphi(e) =$	ι	$\varphi(e) =$	ι
$\varphi(b) =$	η_2	$\varphi(a) =$	η_5	$\varphi(a) =$	η_6
$\varphi(a) =$	η_2	$\varphi(a) =$	η_5	$\varphi(a) =$	η_6
$\varphi(ba) =$	ι	$\varphi(b) =$	ι	$\varphi(b) =$	ι

(3) We only consider the image of generators b, a of the dihedral group $\mathcal{D}_n$ and the rest of the images can be obtained by using them. As n is odd and the order of $\varphi(b)$ divides the order of b, we have that the order of $\varphi(b)$ cannot be even. Therefore, there are either 3 choices or 1 choice for $\varphi(b)$. If $n = 3t$, for an odd t then $\varphi(b)$ can be $\{\eta_3, \eta_3^2 = \eta_4, \eta_3^3 = \iota\}$ otherwise $\varphi(b) = \iota$. We now consider the image of a. If $\varphi(a) = \iota$, then $\varphi(b^i a) = \varphi(b^i) = \eta_3^j$ for $j = 1, 2, 3$ or $j = 3$. But $|\eta_3| = |\eta_3^2| = 3$ and $|b^i a| = 2$ which is a contradiction. Therefore, if $\varphi(a) = \iota$, then φ must be trivial. The other choices for $\varphi(a)$ are $\{\eta_2, \eta_5, \eta_6\}$. We tabulate our results in Table 4.7.

(4.7)

Case 1		Case 2		Case 3	
$\varphi(e) =$	ι	$\varphi(e) =$	ι	$\varphi(e) =$	ι
$\varphi(b) =$	ι	$\varphi(b) =$	ι	$\varphi(b) =$	ι
$\varphi(a) =$	η_2	$\varphi(a) =$	η_5	$\varphi(a) =$	η_6
Case 4		Case 5		Case 6	
$\varphi(e) =$	ι	$\varphi(e) =$	ι	$\varphi(e) =$	ι
$\varphi(b) =$	η_3	$\varphi(b) =$	η_3	$\varphi(b) =$	η_3
$\varphi(a) =$	η_2	$\varphi(a) =$	η_5	$\varphi(a) =$	η_6
Case 7		Case 8		Case 9	
$\varphi(e) =$	ι	$\varphi(e) =$	ι	$\varphi(e) =$	ι
$\varphi(b) =$	η_4	$\varphi(b) =$	η_4	$\varphi(b) =$	η_4
$\varphi(a) =$	η_2	$\varphi(a) =$	η_5	$\varphi(a) =$	η_6

(4) As in the previous item, if $n = 3k$, then there exist 9 non-trivial homomorphisms and if $n \neq 3k$, then there exist 3 non-trivial homomorphisms. In addition, $\varphi(b)$ can be one of $\{\eta_2, \eta_5, \eta_6\}$. In this case, there are two choices for the image of a, $\varphi(a) = \varphi(b)$ or $\varphi(a) = \iota$. This gives 6 more choices for the cases $n = 3k$ and $n \neq 3k$. Therefore we have 15 non-trivial homomorphisms for $n = 3k$ and 9 non-trivial homomorphisms for $n \neq 3k$. We only tabulate these 6 choices in Table 4.9 since Table 4.7 gives the first 9 non-trivial homomorphisms for $n = 3k$ and Table 4.8 give the first 3 non-trivial homomorphisms for $n \neq 3k$.

(4.8)

Case 1		Case 2		Case 3	
$\varphi(e) =$	ι	$\varphi(e) =$	ι	$\varphi(e) =$	ι
$\varphi(b) =$	ι	$\varphi(b) =$	ι	$\varphi(b) =$	ι
$\varphi(a) =$	η_2	$\varphi(a) =$	η_5	$\varphi(a) =$	η_6

(4.9)

Case 1		Case 2		Case 3	
$\varphi(e) =$	ι	$\varphi(e) =$	ι	$\varphi(e) =$	ι
$\varphi(b) =$	η_2	$\varphi(b) =$	η_2	$\varphi(b) =$	η_5
$\varphi(a) =$	η_2	$\varphi(a) =$	ι	$\varphi(a) =$	η_5
Case 4		Case 5		Case 6	
$\varphi(e) =$	ι	$\varphi(e) =$	ι	$\varphi(e) =$	ι
$\varphi(b) =$	η_5	$\varphi(b) =$	η_6	$\varphi(b) =$	η_6
$\varphi(a) =$	ι	$\varphi(a) =$	η_6	$\varphi(a) =$	ι

□

Let $\mathcal{D}_n$ be the dihedral group of order $2n$, where the elements are ordered as follows:

$$\{1, b, b^2, \ldots, b^{n-1}, a, ab, \ldots, ab^{n-1}\}. \tag{4.10}$$

Then the generator matrix has the following form

$$\sigma(v) = \begin{pmatrix} A & B \\ B & A \end{pmatrix}, \tag{4.11}$$

where A is a circulant $n \times n$ matrix and B is a reverse circulant $n \times n$ matrix.

PROPOSITION 4.5. *Let $\mathcal{D}_{2n}$ be the dihedral group of order $2n$ for $n > 2$, and let $v_i \in M_2(\mathbb{F}_2) \rtimes_\varphi \mathcal{D}_n$ with the ordering given in Equation 4.10.*

(1) *Let n be an odd.*
- *If $n = 3k$ then the generator matrix is either*

$$\left(\begin{array}{cccc|cccc} A_1 & A_b & \cdots & A_{b^{n-1}} & A_a & A_{ab^{n-1}} & \cdots & A_{ab} \\ A_{b^{n-1}} & A_1 & \cdots & A_{b^{n-2}} & A_{ab^{n-1}} & A_{ab^{n-2}} & \cdots & A_a \\ \vdots & \vdots & \vdots & \vdots & \vdots & \vdots & \vdots & \vdots \\ A_b & A_{b^2} & \cdots & A_1 & A_{ab}) & A_a & \cdots & A_{ab^2} \\ \hline \eta_i(A_a) & \eta_i(A_{ab^{n-1}}) & \cdots & \eta_i(A_{ab}) & \eta_i(A_1) & \eta_i(A_b) & \cdots & \eta_i(A_{b^{n-1}}) \\ \eta_i(A_{ab^{n-1}}) & \eta_i(A_{ab^{n-2}}) & \cdots & \eta_i(A_a) & \eta_i(A_{b^{n-1}}) & \eta_i(A_1) & \cdots & \eta_i(A_{b^{n-2}}) \\ \vdots & \vdots & \vdots & \vdots & \vdots & \vdots & \vdots & \vdots \\ \eta_i(A_{ab}) & \eta_i(A_a) & \cdots & \eta_i(A_{ab^2}) & \eta_i(A_b) & \eta_i(A_{b^2}) & \cdots & \eta_i(A_1) \end{array}\right),$$

for $i \in \{2,5,6\}$, *or*

$$\left(\begin{array}{cccc|cccc}
\iota(A_1) & \iota(A_b) & \cdots & \iota(A_{b^{n-1}}) & \iota(A_a) & \iota(A_{ab^{n-1}}) & \cdots & \iota(A_{ab}) \\
\eta_i(A_{b^{n-1}}) & \eta_i(A_1) & \cdots & \eta_i(A_{b^{n-2}}) & \eta_i(A_{ab^{n-1}}) & \eta_i(A_{ab^{n-2}}) & \cdots & \eta_i(A_a) \\
\vdots & \vdots & \vdots & \vdots & \vdots & \vdots & \vdots & \vdots \\
\eta_j(A_b) & \eta_j(A_{b^2}) & \cdots & \eta_j(A_1) & \eta_j(A_{ab}) & \eta_j(A_a) & \cdots & \eta_j(A_{ab^2}) \\
\hline
\eta_k(A_a) & \eta_k(A_{ab^{n-1}}) & \cdots & \eta_k(A_{ab}) & \eta_k(A_1) & \eta_k(A_b) & \cdots & \eta_k(A_{b^{n-1}}) \\
\eta_m(A_{ab^{n-1}}) & \eta_m(A_{ab^{n-2}}) & \cdots & \eta_m(A_a) & \eta_m(A_{b^{n-1}}) & \eta_m(A_1) & \cdots & \eta_m(A_{b^{n-2}}) \\
\vdots & \vdots & \vdots & \vdots & \vdots & \vdots & \vdots & \vdots \\
\eta_n(A_{ab}) & \eta_n(A_a) & \cdots & \eta_n(A_{ab^2}) & \eta_n(A_b) & \eta_n(A_{b^2}) & \cdots & \eta_n(A_1)
\end{array}\right),$$

where

$$(i,j,k,m,n) \in \{(\eta_3,\eta_4,\eta_2,\eta_6,\eta_5),(\eta_3,\eta_4,\eta_5,\eta_2,\eta_6),(\eta_3,\eta_4,\eta_6,\eta_5,\eta_2),$$
$$(\eta_4,\eta_3,\eta_2,\eta_5,\eta_6),(\eta_4,\eta_3,\eta_5,\eta_6,\eta_2),(\eta_4,\eta_3,\eta_6,\eta_2,\eta_5)\}.$$

- *If* $n \neq 3k$, *then the generator matrix is in the following form:*

$$\left(\begin{array}{cccc|cccc}
A_1 & A_b & \cdots & A_{b^{n-1}} & A_a & A_{ab^{n-1}} & \cdots & A_{ab} \\
A_{b^{n-1}} & A_1 & \cdots & A_{b^{n-2}} & A_{ab^{n-1}} & A_{ab^{n-2}} & \cdots & A_a \\
\vdots & \vdots & \vdots & \vdots & \vdots & \vdots & \vdots & \vdots \\
A_b & A_{b^2} & \cdots & A_1 & A_{ab}) & A_a & \cdots & A_{ab^2} \\
\hline
\eta_i(A_a) & \eta_i(A_{ab^{n-1}}) & \cdots & \eta_i(A_{ab}) & \eta_i(A_1) & \eta_i(A_b) & \cdots & \eta_i(A_{b^{n-1}}) \\
\eta_i(A_{ab^{n-1}}) & \eta_i(A_{ab^{n-2}}) & \cdots & \eta_i(A_a) & \eta_i(A_{b^{n-1}}) & \eta_i(A_1) & \cdots & \eta_i(A_{b^{n-2}}) \\
\vdots & \vdots & \vdots & \vdots & \vdots & \vdots & \vdots & \vdots \\
\eta_i(A_{ab}) & \eta_i(A_a) & \cdots & \eta_i(A_{ab^2}) & \eta_i(A_b) & \eta_i(A_{b^2}) & \cdots & \eta_i(A_1)
\end{array}\right),$$

where $i \in \{2,5,6\}$,

(2) *If* n *is even, then the generator matrix either has one of the form of the nine cases for* $n = 3k$, *or one of the form of the three cases for* $n \neq 3k$ *that is given in previous item or it is of the following form:*

$$\left(\begin{array}{cccc|cccc}
\iota(A_1) & \iota(A_b) & \cdots & \iota(A_{b^{n-1}}) & \iota(A_a) & \iota(A_{ab^{n-1}}) & \cdots & \iota(A_{ab}) \\
\eta_i(A_{b^{n-1}}) & \eta_i(A_1) & \cdots & \eta_i(A_{b^{n-2}}) & \eta_i(A_{ab^{n-1}}) & \eta_i(A_{ab^{n-2}}) & \cdots & \eta_i(A_a) \\
\vdots & \vdots & \vdots & \vdots & \vdots & \vdots & \vdots & \vdots \\
\eta_i(A_b) & \eta_i(A_{b^2}) & \cdots & \eta_i(A_1) & \eta_i(A_{ab}) & \eta_i(A_a) & \cdots & \eta_i(A_{ab^2}) \\
\hline
\eta_i(A_a) & \eta_i(A_{ab^{n-1}}) & \cdots & \eta_i(A_{ab}) & \eta_i(A_1) & \eta_i(A_b) & \cdots & \eta_i(A_{b^{n-1}}) \\
\iota(A_{ab^{n-1}}) & \iota(A_{ab^{n-2}}) & \cdots & \iota(A_a) & \iota(A_{b^{n-1}}) & \iota(A_1) & \cdots & \iota(A_{b^{n-2}}) \\
\vdots & \vdots & \vdots & \vdots & \vdots & \vdots & \vdots & \vdots \\
\iota(A_{ab}) & \iota(A_a) & \cdots & \iota(A_{ab^2}) & \iota(A_b) & \iota(A_{b^2}) & \cdots & \iota(A_1)
\end{array}\right),$$

where $i \in \{2,5,6\}$, *or*

$$\left(\begin{array}{cccc|cccc}
\iota(A_1) & \iota(A_b) & \cdots & \iota(A_{b^{n-1}}) & \iota(A_a) & \iota(A_{ab^{n-1}}) & \cdots & \iota(A_{ab}) \\
\eta_i(A_{b^{n-1}}) & \eta_i(A_1) & \cdots & \eta_i(A_{b^{n-2}}) & \eta_i(A_{ab^{n-1}}) & \eta_i(A_{ab^{n-2}}) & \cdots & \eta_i(A_a) \\
\vdots & \vdots & \vdots & \vdots & \vdots & \vdots & \vdots & \vdots \\
\eta_i(A_b) & \eta_i(A_{b^2}) & \cdots & \eta_i(A_1) & \eta_i(A_{ab}) & \eta_i(A_a) & \cdots & \eta_i(A_{ab^2}) \\
\hline
\iota(A_a) & \iota(A_{ab^{n-1}}) & \cdots & \iota(A_{ab}) & \iota(A_1) & \iota(A_b) & \cdots & \iota(A_{b^{n-1}}) \\
\eta_i(A_{ab^{n-1}}) & \eta_i(A_{ab^{n-2}}) & \cdots & \eta_i(A_a) & \eta_i(A_{b^{n-1}}) & \eta_i(A_1) & \cdots & \eta_i(A_{b^{n-2}}) \\
\vdots & \vdots & \vdots & \vdots & \vdots & \vdots & \vdots & \vdots \\
\eta_i(A_{ab}) & \eta_i(A_a) & \cdots & \eta_i(A_{ab^2}) & \eta_i(A_b) & \eta_i(A_{b^2}) & \cdots & \eta_i(A_1)
\end{array}\right)$$

where $i \in \{2,5,6\}$.

PROOF. Let $v \in M_2(\mathbb{F}_2) \rtimes_\varphi \mathcal{D}_n$. Then

$$\varphi(\tau_2(v)) = \left(\begin{array}{cccc|cccc}
\varphi_1(A_1) & \varphi_1(A_b) & \cdots & \varphi_1(A_{b^{n-1}}) & \varphi_1(A_a) & \varphi_1(A_{ab^{n-1}}) & \cdots & \varphi_1(A_{ab}) \\
\varphi_b(A_{b^{n-1}}) & \varphi_b(A_1) & \cdots & \varphi_b(A_{b^{n-2}}) & \varphi_b(A_{ab^{n-1}}) & \varphi_b(A_{ab^{n-2}}) & \cdots & \varphi_b(A_a) \\
\vdots & \vdots & \vdots & \vdots & \vdots & \vdots & \vdots & \vdots \\
\varphi_{b^{n-1}}(A_b) & \varphi_{b^{n-1}}(A_{b^2}) & \cdots & \varphi_{b^{n-1}}(A_1) & \varphi_{b^{n-1}}(A_{ab}) & \varphi_{b^{n-1}}(A_a) & \cdots & \varphi_{b^{n-1}}(A_{ab^2}) \\
\hline
\varphi_a(A_a) & \varphi_a(A_{ab^{n-1}}) & \cdots & \varphi_a(A_{ab}) & \varphi_a(A_1) & \varphi_a(A_b) & \cdots & \varphi_a(A_{b^{n-1}}) \\
\varphi_{ab}(A_{ab^{n-1}}) & \varphi_{ab}(A_{ab^{n-2}}) & \cdots & \varphi_{ab}(A_a) & \varphi_{ab}(A_{b^{n-1}}) & \varphi_{ab}(A_1) & \cdots & \varphi_{ab}(A_{b^{n-2}}) \\
\vdots & \vdots & \vdots & \vdots & \vdots & \vdots & \vdots & \vdots \\
\varphi_{ab^{n-1}}(A_{ab}) & \varphi_{ab^{n-1}}(A_a) & \cdots & \varphi_{ab^{n-1}}(A_{ab^2}) & \varphi_{ab^{n-1}}(A_b) & \varphi_{ab^{n-1}}(A_{b^2}) & \cdots & \varphi_{ab^{n-1}}(A_1)
\end{array}\right).$$

We now consider the 9 cases in Table 4.7 and denote the generator matrix of the i-th case of Table 4.7 by $\varphi(\tau_k(v_i))$.

(1) The first three cases can be obtained easily by the basic group property that $\varphi_{ab} = \varphi_a\varphi_b$ as follows.

(a) The first case is:

$$\varphi(\tau_2(v_1)) = \left(\begin{array}{cccc|cccc}
A_1 & A_b & \cdots & A_{b^{n-1}} & A_a & A_{ab^{n-1}} & \cdots & A_{ab} \\
A_{b^{n-1}} & A_1 & \cdots & A_{b^{n-2}} & A_{ab^{n-1}} & A_{ab^{n-2}} & \cdots & A_a \\
\vdots & \vdots & \vdots & \vdots & \vdots & \vdots & \vdots & \vdots \\
A_b & A_{b^2} & \cdots & A_1 & A_{ab}) & A_a & \cdots & A_{ab^2} \\
\hline
\eta_2(A_a) & \eta_2(A_{ab^{n-1}}) & \cdots & \eta_2(A_{ab}) & \eta_2(A_1) & \eta_2(A_b) & \cdots & \eta_2(A_{b^{n-1}}) \\
\eta_2(A_{ab^{n-1}}) & \eta_2(A_{ab^{n-2}}) & \cdots & \eta_2(A_a) & \eta_2(A_{b^{n-1}}) & \eta_2(A_1) & \cdots & \eta_2(A_{b^{n-2}}) \\
\vdots & \vdots & \vdots & \vdots & \vdots & \vdots & \vdots & \vdots \\
\eta_2(A_{ab}) & \eta_2(A_a) & \cdots & \eta_2(A_{ab^2}) & \eta_2(A_b) & \eta_2(A_{b^2}) & \cdots & \eta_2(A_1)
\end{array}\right),$$

(b) The second case is:

$$\varphi(\tau_2(v_2)) = \left(\begin{array}{cccc|cccc}
A_1 & A_b & \cdots & A_{b^{n-1}} & A_a & A_{ab^{n-1}} & \cdots & A_{ab} \\
A_{b^{n-1}} & A_1 & \cdots & A_{b^{n-2}} & A_{ab^{n-1}} & A_{ab^{n-2}} & \cdots & A_a \\
\vdots & \vdots & \vdots & \vdots & \vdots & \vdots & \vdots & \vdots \\
A_b & A_{b^2} & \cdots & A_1 & A_{ab}) & A_a & \cdots & A_{ab^2} \\
\hline
\eta_5(A_a) & \eta_5(A_{ab^{n-1}}) & \cdots & \eta_5(A_{ab}) & \eta_5(A_1) & \eta_5(A_b) & \cdots & \eta_5(A_{b^{n-1}}) \\
\eta_5(A_{ab^{n-1}}) & \eta_5(A_{ab^{n-2}}) & \cdots & \eta_5(A_a) & \eta_5(A_{b^{n-1}}) & \eta_5(A_1) & \cdots & \eta_5(A_{b^{n-2}}) \\
\vdots & \vdots & \vdots & \vdots & \vdots & \vdots & \vdots & \vdots \\
\eta_5(A_{ab}) & \eta_5(A_a) & \cdots & \eta_5(A_{ab^2}) & \eta_5(A_b) & \eta_5(A_{b^2}) & \cdots & \eta_5(A_1)
\end{array}\right),$$

(c) The third case is:

$$\varphi(\tau_2(v_3)) = \left(\begin{array}{cccc|cccc}
A_1 & A_b & \cdots & A_{b^{n-1}} & A_a & A_{ab^{n-1}} & \cdots & A_{ab} \\
A_{b^{n-1}} & A_1 & \cdots & A_{b^{n-2}} & A_{ab^{n-1}} & A_{ab^{n-2}} & \cdots & A_a \\
\vdots & \vdots & \vdots & \vdots & \vdots & \vdots & \vdots & \vdots \\
A_b & A_{b^2} & \cdots & A_1 & A_{ab}) & A_a & \cdots & A_{ab^2} \\
\hline
\eta_6(A_a) & \eta_6(A_{ab^{n-1}}) & \cdots & \eta_6(A_{ab}) & \eta_6(A_1) & \eta_6(A_b) & \cdots & \eta_6(A_{b^{n-1}}) \\
\eta_6(A_{ab^{n-1}}) & \eta_6(A_{ab^{n-2}}) & \cdots & \eta_6(A_a) & \eta_6(A_{b^{n-1}}) & \eta_6(A_1) & \cdots & \eta_6(A_{b^{n-2}}) \\
\vdots & \vdots & \vdots & \vdots & \vdots & \vdots & \vdots & \vdots \\
\eta_6(A_{ab}) & \eta_6(A_a) & \cdots & \eta_6(A_{ab^2}) & \eta_6(A_b) & \eta_6(A_{b^2}) & \cdots & \eta_6(A_1)
\end{array}\right).$$

For the remaining six cases in Table 4.7, we consider

$$\sigma(v) = \begin{pmatrix} A & B \\ B & A \end{pmatrix}, \tag{4.12}$$

where

$$A = \begin{pmatrix}
A_1 & A_b & A_{b^2} & \cdots & A_{b^{n-1}} \\
A_{b^{n-1}} & A_1 & A_b & \cdots & A_{b^{n-2}} \\
\vdots & \vdots & \vdots & \vdots & \vdots \\
A_b & A_{b^2} & A_{b^3} & \cdots & A_1
\end{pmatrix} \text{ and } B = \begin{pmatrix}
A_a & A_{ab^{n-1}} & A_{ab^{n-2}} & \cdots & A_{ab} \\
A_{ab^{n-1}} & A_{ab^{n-2}} & A_{ab^{n-3}} & \cdots & A_a \\
\vdots & \vdots & \vdots & \vdots & \vdots \\
A_{ab} & A_a & A_{ab^{n-1}} & \cdots & A_{ab^2}
\end{pmatrix}.$$

Then generator matrices $\varphi(\tau_2(v_i))$ for $i = 4, \ldots, 9$ are computed as follows:

- $\varphi(\tau_2(v_4)) = \begin{pmatrix} A_1 & B_1 \\ B_2 & A_2 \end{pmatrix}$, where A_1, B_1 are obtained by applying ι to the $(1+3k)$-th rows, applying η_3 to the $(2+3k)$-th rows and applying η_4 to the $(3+3k)$-th rows of A and B, where $0 \leq k < \frac{n}{3}$. The matrices A_2, B_2 are obtained by applying η_2 to the$(1+3k)$-th rows, applying η_6 to the $(2+3k)$-th rows, and applying η_5 to the $(3+3k)$-th rows to the rows of A and B for $\frac{n}{3} \leq k < \frac{2n}{3}$.
- $\varphi(\tau_2(v_5)) = \begin{pmatrix} A_1 & B_1 \\ B_2 & A_2 \end{pmatrix}$, where A_1, B_1 are obtained by applying ι to the $(1+3k)$-th rows, applying η_3 to the $(2+3k)$-th rows, applying η_4 to the $(3+3k)$-th rows of A and B for $0 \leq k < \frac{n}{3}$. The matrices A_2 and B_2 are obtained by applying η_5 to the $(1+3k)$-th rows, η_2 to the $(2+3k)$-th rows and η_6 to the $(3+3k)$-th rows of A and B for $\frac{n}{3} \leq k < \frac{2n}{3}$.
- $\varphi(\tau_2(v_6)) = \begin{pmatrix} A_1 & B_1 \\ B_2 & A_2 \end{pmatrix}$, where A_1, B_1 are obtained by applying ι to the $(1+3k)$-th rows, η_3 to the $(2+3k)$-th rows and η_4 to the $(3+3k)$-th rows of A and B for $0 \leq k < \frac{n}{3}$. The matrices A_2, B_2 are obtained by applying η_6 to the $(1+3k)$-th rows, η_5 to the $(2+3k)$-th rows, η_2 to the $(3+3k)$-th rows of A and B for $\frac{n}{3} \leq k < \frac{2n}{3}$.
- $\varphi(\tau_2(v_7)) = \begin{pmatrix} A_1 & B_1 \\ B_2 & A_2 \end{pmatrix}$, where A_1, B_1 are obtained by applying ι to the $(1+3k)$-th rows, η_4 to the $(2+3k)$-th rows and η_3 to the $(3+3k)$-th rows of A and B for $0 \leq k < \frac{n}{3}$. The matrices A_2, B_2 are obtained by applying η_2 to the $(1+3k)$-th rows, η_5 to the $(2+3k)$-th rows, η_6 to the $(3+3k)$-th rows of A and B for $\frac{n}{3} \leq k < \frac{2n}{3}$.
- $\varphi(\tau_2(v_8)) = \begin{pmatrix} A_1 & B_1 \\ B_2 & A_2 \end{pmatrix}$, where A_1, B_1 are obtained by applying ι to the $(1+3k)$-th rows, η_4 to the $(2+3k)$-th rows, η_3 to the $(3+3k)$-th rows of A and B for $0 \leq k < \frac{n}{3}$. The matrices A_2, B_2 are obtained by applying η_5 to the $(1+3k)$-th rows, η_6 to the $(2+3k)$-th rows, η_2 to the $(3+3k)$-th rows of A and B for $\frac{n}{3} \leq k < \frac{2n}{3}$.
- $\varphi(\tau_2(v_9)) = \begin{pmatrix} A_1 & B_1 \\ B_2 & A_2 \end{pmatrix}$, where A_1, B_1 are obtained by applying ι to the $(1+3k)$-th rows, η_4 to the $(2+3k)$-th rows, η_3 to the $(3+3k)$-th rows of A and B for $0 \leq k < \frac{n}{3}$. The matrices A_2, B_2 are obtained by applying η_6 to the$(1+3k)$-th rows, η_2 to the $(2+3k)$-th rows, η_5 to the $(3+3k)$-th rows of A and B for $\frac{n}{3} \leq k < \frac{2n}{3}$.

For $n \neq 3k$, we consider 3 cases in Table 4.8. But these 3 cases are the same as the first three cases of Table 4.7 so the generator matrices are the same.

(2) It is enough to write generator matrices from the Table 4.9 since we already wrote the generator matrices from Table 4.7 and 4.8 in the previous item.

- $\varphi(\tau_2(v_1)) = \begin{pmatrix} A_1 & B_1 \\ B_2 & A_2 \end{pmatrix}$, where A_1, B_1 are obtained by applying ι to the $(1+2k)$-th rows, η_2 to the $(2+2k)$-th rows of A and B for $0 \le k < \frac{n}{2}$. The matrices A_2, B_2 are obtained by applying η_2 to the $(1+2k)$-th rows, ι to the $(2+2k)$-th rows of A and B for $\frac{n}{2} \le k < n$.
- $\varphi(\tau_2(v_2)) = \begin{pmatrix} A_1 & B_1 \\ B_2 & A_2 \end{pmatrix}$, where A_1, B_1 are obtained by applying ι to the $(1+2k)$-th rows, η_2 to the $(2+2k)$-th rows of A and B for $0 \le k < \frac{n}{2}$. The matrices A_2, B_2 are obtained by applying ι to the $(1+2k)$-th rows, η_2 to the $(2+2k)$-th rows of A and B for $\frac{n}{2} \le k < n$.
- $\varphi(\tau_2(v_3)) = \begin{pmatrix} A_1 & B_1 \\ B_2 & A_2 \end{pmatrix}$, where A_1, B_1 are obtained by applying ι to the $(1+2k)$-th rows, η_5 to the $(2+2k)$-th rows of A and B for $0 \le k < \frac{n}{2}$. The matrices A_2, B_2 are obtained by applying η_5 to the $(1+2k)$-th rows, ι to the $(2+2k)$-th rows of A and B for $\frac{n}{2} \le k < n$.
- $\varphi(\tau_2(v_4)) = \begin{pmatrix} A_1 & B_1 \\ B_2 & A_2 \end{pmatrix}$, where A_1, B_1 are obtained by applying ι to the $(1+2k)$-th rows, η_5 to the $(2+2k)$-th rows of A and B for $0 \le k < \frac{n}{2}$. The matrices A_2, B_2 are obtained by applying ι to the $(1+2k)$-th rows, η_5 to the $(2+2k)$-th rows of A and B for $\frac{n}{2} \le k < n$.
- $\varphi(\tau_2(v_5)) = \begin{pmatrix} A_1 & B_1 \\ B_2 & A_2 \end{pmatrix}$, where A_1, B_1 are obtained by applying ι to the $(1+2k)$-th rows, η_6 to the $(2+2k)$-th rows of A and B for $0 \le k < \frac{n}{2}$. A_2, B_2 are obtained by applying η_6 to the $(1+2k)$-th rows, ι to the $(2+2k)$-th rows of A and B for $\frac{n}{2} \le k < n$.
- $\varphi(\tau_2(v_6)) = \begin{pmatrix} A_1 & B_1 \\ B_2 & A_2 \end{pmatrix}$, where A_1, B_1 are obtained by applying ι to the $(1+2k)$-th rows, η_6 to the $(2+2k)$-th rows of A and B for $0 \le k < \frac{n}{2}$. A_2, B_2 are obtained by applying ι to the $(1+2k)$-th rows, η_6 to the $(2+2k)$-th rows of A and B for $\frac{n}{2} \le k < n$.

□

5. Examples of good codes from $M_2(\mathbb{F}_2) \rtimes_\varphi G$

In this section, we perform our search using the software package MAGMA ([**3**]).

5.1. Binary self-dual codes. Let $v_1 \in M_2(\mathbb{F}_2) \rtimes_\varphi \mathcal{D}_{2n}$. By using the generator matrices $\varphi(\tau_2(v_1))$ from the proof of Proposition 4.5, we obtained binary self-dual codes of different lengths. We employ a generator matrix of the form $[I_n \mid \varphi(\tau_2(v))]$, where I_n is the $n \times n$ identity matrix, and search for binary self-dual codes with parameters $[32, 16, 8]$, $[40, 20, 8]$, $[56, 28, 10]$, $[64, 32, 12]$ and $[72, 36, 12]$. We tabulated the parameters of the code, order of automorphism group, and the first row of the matrices A_i, where the A_i are 2×2 circulant matrices.

We know by [**8**] that there exists two possible weight enumerators for binary self-dual codes of length 32 which are:

$$W_1 = 1 + 620y^8 + 13888y^{12} + 36518y^{16} + \cdots$$
$$W_2 = 1 + 364y^8 + 2048y^{10} + 6720y^{12} + 14336y^{14} + 18598y^{16} + \cdots.$$

TABLE 1. Binary Self-Dual Codes [32,16,8]

$(A_1, A_b, \cdots, A_{b^{n-1}})$	$(A_a, A_{ab^{n-1}}, \cdots, A_{ab})$	W_i	$\lvert Aut(C_i)\rvert$
$(0,0,0,1,0,0,1,0)$	$(1,1,0,1,1,0,0,1)$	W_1	23040
$(0,0,0,0,1,1,1,1)$	$(1,0,1,1,1,0,1,0)$	W_2	294912

We recall from [**8**, **16**] that any singly-even binary self-dual $[40, 20, 8]$ code has the following weight enumerator:

$$W(y) = 1 + (125 + 16\beta)y^8 + (1664 - 64\beta)y^{10} + \dots.$$

TABLE 2. Binary Self-Dual Codes [40,20,8]

$(A_1, A_b, \cdots, A_{b^{n-1}})$	$(A_a, A_{ab^{n-1}}, \cdots, A_{ab})$	β	$\lvert Aut(C_i)\rvert$
$(0,1,1,1,0,1,0,0,1,0)$	$(0,0,1,1,0,0,1,0,0,1)$	0	40
$(1,1,0,1,1,1,1,0,1,0)$	$(0,0,0,0,0,0,1,0,1,0)$	10	245760

The possible weight enumerators of a singly-even binary self-dual $[56, 28, 10]$ code are given in [**15**] as

$$W_{56,1} = 1 + (308 + 4\alpha)x^{10} + (4246 - 8\alpha)x^{12} + \dots,$$
$$W_{56,2} = 1 + (308 + 4\alpha)x^{10} + (3990 - 8\alpha)x^{12} + \dots.$$

TABLE 3. Singly-even Binary Self-Dual Codes $[56, 28, 10]$

$(A_1, A_b, \cdots, A_{b^{n-1}})$	$(A_a, A_{ab^{n-1}}, \cdots, A_{ab})$	α	*Type*	$\lvert Aut(C_i)\rvert$
$(1,0,1,0,1,0,0,1,1,0,1,0,1,1)$	$(0,0,0,1,1,1,1,0,1,0,0,0,0,0)$	-14	$W_{56,2}$	28
$(1,1,1,1,1,0,0,1,1,1,1,0,0,0)$	$(0,1,0,1,0,1,1,0,1,0,1,1,1,0)$	0	$W_{56,2}$	28
$(0,0,0,0,0,0,1,1,1,1,1,0,1,1)$	$(1,0,0,0,1,1,1,0,0,0,0,1,1,0)$	-28	$W_{56,2}$	28
$(1,1,0,0,1,1,0,0,0,0,0,0,1,0)$	$(0,0,0,1,1,0,1,0,1,1,1,1,0,1)$	-42	$W_{56,2}$	28

The possible weight enumerators of a singly-even binary self-dual $[64, 32, 12]$ code are given in [**8**, **17**] as

$$W_{64,1} = 1 + (1312 + 16\alpha)x^{12} + (22016 - 64\alpha)x^{14} + \dots$$
$$W_{64,2} = 1 + (1312 + 16\alpha)x^{12} + (23040 - 64\alpha)x^{14} + \dots$$

TABLE 4. Singly-even Binary Self-Dual Codes $[64, 32, 12]$

$(A_1, A_b, \cdots, A_{b^{n-1}})$	$(A_a, A_{ab^{n-1}}, \cdots, A_{ab})$	α	*Type*	$\lvert Aut(C_i)\rvert$
$(0,0,1,0,0,1,0,1,1,0,0,1,0,0,1,0)$	$(0,0,1,0,1,1,1,0,1,1,1,1,1,0,1,1)$	32	$W_{64,2}$	32
$(0,1,0,0,1,0,0,0,1,1,1,1,1,1,0,0)$	$(1,0,0,1,1,1,1,0,0,1,0,0,1,0,1,1)$	16	$W_{64,2}$	64
$(0,0,1,0,0,1,1,0,1,0,0,0,0,0,0,0)$	$(1,1,0,1,0,1,0,1,0,0,0,0,1,0,1,0)$	104	$W_{64,2}$	32
$(1,0,1,0,1,1,1,1,0,1,0,1,0,1,0,0)$	$(0,0,1,0,0,1,1,0,1,0,0,0,0,0,0,0)$	0	$W_{64,2}$	32
$(0,0,0,1,0,1,1,1,1,1,1,0,1,1,0,0)$	$(0,0,0,1,1,0,1,0,1,0,0,0,1,1,1,1)$	48	$W_{64,2}$	32
$(1,1,1,0,0,1,0,0,1,1,0,0,1,0,0,1)$	$(1,0,1,0,0,0,1,1,1,1,0,0,1,1,0,1)$	80	$W_{64,2}$	128

The possible weight enumerators for a Type I $[72, 36, 12]$ codes are as follows ([**12**]):

$$W_{72,1} = 1 + 2\beta y^{12} + (8640 - 64\gamma)y^{14} + (124281 - 24\beta + 384\gamma)y^{16} + \dots$$

$$W_{72,2} = 1 + 2\beta y^{12} + (7616 - 64\gamma)y^{14} + (134521 - 24\beta + 384\gamma)y^{16} + \dots$$

where β and γ are parameters. The possible weight enumerators for Type II $[72, 36, 12]$ codes are ([**12**]):

$$1 + (4398 + \alpha)y^{12} + (197073 - 12\alpha)y^{16} + (18396972 + 66\alpha)y^{20} + \dots$$

where α is a parameter.

TABLE 5. Some Examples of Binary Type I Self-Dual Codes of Length 72

$(A_1, A_b, \cdots, A_{b^8})$	$(A_a, A_{ab^8}, \cdots, A_{ab})$	γ	β	$\lvert Aut(C_i)\rvert$
$(0,1,0,0,0,0,1,1,0,0,1,1,1,0,0,0,0,1)$	$(0,1,0,0,1,0,0,0,1,0,0,0,1,0,0,1,1,0)$	0	354	36
$(0,1,1,1,1,1,1,0,0,0,1,0,0,0,1,1,1,0)$	$(1,1,0,0,1,0,1,1,1,0,1,0,1,0,0,1,1,1)$	0	297	36
$(0,1,1,0,0,0,1,0,1,1,1,1,1,0,0,1,1,0)$	$(0,0,1,0,0,0,1,0,0,0,1,1,1,1,0,0,0,1)$	0	201	36
$(0,0,0,1,1,0,1,1,1,0,0,0,1,0,0,1,0,0)$	$(1,1,0,1,0,1,1,0,0,0,0,1,0,0,0,0,0,0)$	18	300	36
$(1,1,1,0,0,0,1,1,0,0,1,0,1,1,0,1,0,0)$	$(1,0,0,0,1,1,1,0,1,0,1,0,0,0,1,0,0,1)$	0	330	36
$(1,0,1,1,0,0,1,1,1,0,0,1,0,1,0,1,1,1)$	$(1,1,0,1,1,1,0,1,1,1,0,0,1,0,0,1,0,0)$	0	327	36
$(0,0,0,1,0,0,0,0,0,0,0,1,1,1,0,0,1,0)$	$(1,0,0,1,1,1,0,0,0,1,0,1,1,0,0,0,1,0)$	0	222	36
$(0,1,0,1,0,1,1,0,1,0,0,1,0,0,1,1,1,1)$	$(1,1,1,1,1,1,0,1,1,0,0,0,1,1,1,0,0,0)$	0	189	36
$(1,1,1,1,1,1,0,0,1,0,0,0,0,1,0,0,1,0)$	$(1,1,1,0,1,1,0,1,1,0,1,0,1,0,1,0,1,1)$	0	246	36
$(0,1,0,1,0,0,1,0,1,1,0,0,0,0,1,0,1,1)$	$(0,1,0,0,0,1,0,1,0,1,1,1,1,1,0,0,0,1)$	0	252	36

5.2. Binary optimal codes. Let $v_i \in M_2(\mathbb{F}_2) \rtimes_\varphi \mathcal{D}_{2n}$. By using the generator matrices from Proposition 4.5, we have provided several examples of binary optimal codes according to database [**14**]. Generator matrices $\varphi(\tau_2(v_i))$ in Table 5.2 are the same generator matrices as in the proof of Proposition 4.5. Here, the A_i are 2×2 circulant matrices. We tabulated the first row of the matrices A_i, the parameters of the code, and also the generator matrix.

TABLE 6. Examples of Binary Linear Optimal Codes of Different Lengths

$(A_1, A_b, \cdots, A_{b^{n-1}})$	$(A_a, A_{ab^{n-1}}, \cdots, A_{ab})$	$[n, k, d]$	Generator Matrix
(0, 0, 0, 1, 0, 1)	(1, 1, 1, 0, 1, 0)	[12, 6, 4]	$\varphi(\tau_2(v_1))$
(1, 1, 0, 1, 0, 0)	(1, 0, 0, 1, 1, 0)	[12, 10, 2]	$\varphi(\tau_2(v_1))$
(1, 1, 1, 1, 1, 1)	(1, 1, 0, 1, 0, 1)	[12, 9, 2]	$\varphi(\tau_2(v_1))$
(1, 0, 0, 0, 0, 1)	(0, 1, 1, 0, 0, 0)	[12, 5, 4]	$\varphi(\tau_2(v_1))$
(1, 1, 1, 0, 1, 0)	(1, 0, 0, 0, 0, 1)	[12, 7, 4]	$\varphi(\tau_2(v_1))$
(1, 1, 1, 1, 1, 1)	(1, 1, 1, 1, 1, 1)	[12, 1, 12]	$\varphi(\tau_2(v_1))$
(0, 0, 1, 1, 1, 1)	(1, 1, 0, 0, 1, 1)	[12, 2, 8]	$\varphi(\tau_2(v_1))$
(1, 0, 1, 0, 1, 0)	(0, 1, 1, 1, 1, 0)	[12, 12, 1]	$\varphi(\tau_2(v_1))$
(0, 1, 0, 0, 0, 0, 0, 0)	(0, 0, 0, 0, 1, 1, 1, 0)	[16, 10, 4]	$\varphi(\tau_2(v_1))$
(0, 1, 0, 0, 1, 1, 1, 0)	(1, 1, 1, 0, 1, 1, 1, 1)	[16, 16, 1]	$\varphi(\tau_2(v_1))$
(0, 1, 1, 0, 1, 0, 0, 1)	(0, 1, 1, 1, 0, 0, 0, 1)	[16, 12, 2]	$\varphi(\tau_2(v_1))$
(0, 0, 1, 1, 0, 1, 1, 0)	(0, 0, 1, 0, 1, 0, 1, 1)	[16, 9, 4]	$\varphi(\tau_2(v_1))$
(0, 0, 0, 0, 1, 1, 1, 1)	(1, 1, 0, 0, 0, 0, 1, 1)	[16, 3, 8]	$\varphi(\tau_2(v_1))$
(1, 1, 1, 0, 1, 1, 1, 0)	(1, 0, 1, 1, 0, 1, 0, 0)	[16, 6, 6]	$\varphi(\tau_2(v_1))$
(1, 0, 0, 0, 0, 1, 1, 1)	(0, 0, 0, 1, 1, 1, 1, 0)	[16, 4, 8]	$\varphi(\tau_2(v_1))$
(1, 0, 0, 1, 0, 1, 1, 0)	(1, 1, 1, 1, 0, 0, 0, 0)	[16, 5, 8]	$\varphi(\tau_2(v_1))$
(0, 1, 1, 0, 0, 1, 1, 1, 0, 1)	(1, 1, 1, 1, 0, 0, 1, 0, 1, 1)	[20, 16, 2]	$\varphi(\tau_2(v_1))$
(0, 1, 0, 0, 1, 1, 1, 0, 0, 1)	(1, 1, 0, 1, 0, 0, 0, 0, 1, 1)	[20, 18, 2]	$\varphi(\tau_2(v_1))$
(1, 0, 1, 1, 1, 0, 0, 0, 1, 0)	(1, 0, 0, 1, 1, 0, 0, 1, 0, 0)	[20, 20, 1]	$\varphi(\tau_2(v_1))$
(1, 1, 1, 1, 0, 0, 1, 0, 1, 0)	(0, 0, 0, 0, 0, 0, 0, 1, 0, 1)	[20, 12, 4]	$\varphi(\tau_2(v_1))$
(1, 1, 1, 0, 1, 0, 1, 0, 1, 0)	(0, 1, 0, 1, 0, 0, 0, 1, 1, 0)	[20, 13, 4]	$\varphi(\tau_2(v_1))$
(1, 0, 1, 1, 1, 0, 0, 0, 0, 0)	(1, 1, 0, 1, 0, 1, 1, 1, 1, 1)	[20, 17, 2]	$\varphi(\tau_2(v_1))$
(0, 1, 1, 0, 0, 1, 0, 0, 1, 0)	(1, 0, 1, 0, 0, 1, 0, 1, 0, 0)	[20, 8, 8]	$\varphi(\tau_2(v_1))$
(0, 1, 0, 1, 1, 1, 1, 1, 1, 1, 1, 0)	(0, 1, 1, 1, 1, 0, 1, 1, 0, 0, 1, 0)	[24, 20, 2]	$\varphi(\tau_2(v_1))$
(0, 0, 1, 1, 1, 1, 0, 0, 0, 0, 1, 1)	(0, 1, 0, 0, 0, 0, 1, 0, 1, 1, 0, 1)	[24, 24, 1]	$\varphi(\tau_2(v_1))$
(1, 1, 0, 0, 1, 1, 0, 0, 0, 1, 0, 1)	(0, 0, 1, 1, 0, 1, 1, 1, 0, 1, 1, 1)	[24, 16, 4]	$\varphi(\tau_2(v_1))$
(1, 0, 0, 1, 0, 0, 0, 0, 0, 1, 1, 0)	(1, 0, 1, 0, 1, 0, 1, 0, 1, 1, 1, 1)	[24, 9, 8]	$\varphi(\tau_2(v_1))$
(1, 1, 1, 0, 0, 1, 0, 0, 1, 0, 1, 0)	(1, 0, 1, 0, 1, 1, 0, 1, 0, 1, 1, 1)	[24, 8, 8]	$\varphi(\tau_2(v_1))$
(0, 1, 1, 1, 1, 1, 1, 1, 0, 1, 1, 1)	(0, 0, 1, 0, 1, 1, 0, 0, 1, 1, 1, 0)	[24, 10, 8]	$\varphi(\tau_2(v_1))$
(0, 0, 1, 0, 0, 0, 0, 1, 0, 1, 0, 1, 1, 0, 0, 1)	(0, 1, 0, 0, 0, 0, 1, 0, 1, 1, 1, 1, 0, 1, 1)	[32, 32, 1]	$\varphi(\tau_2(v_1))$
(1, 0, 1, 1, 1, 0, 1, 1, 0, 0, 0, 0, 1, 0, 1, 1)	(0, 1, 0, 0, 0, 1, 1, 0, 1, 1, 1, 0, 0, 0, 1, 1)	[32, 30, 2]	$\varphi(\tau_2(v_1))$
(0, 1, 0, 1, 0, 0, 1, 0, 0, 0, 0, 1, 0, 0, 0, 0)	(1, 1, 1, 1, 0, 0, 0, 0, 1, 0, 0, 1, 1, 1, 1, 1)	[32, 16, 8]	$\varphi(\tau_2(v_1))$
(1, 1, 0, 0, 1, 1, 1, 0, 0, 0, 1, 1, 1, 1, 1, 1, 1, 0)	(1, 1, 0, 1, 0, 0, 0, 0, 1, 0, 1, 0, 1, 0, 1, 0, 0, 1)	[36, 17, 8]	$\varphi(\tau_2(v_1))$
(0, 1, 0, 0, 0, 0, 0, 0, 1, 1, 1, 0, 1, 1, 0, 1, 1, 0)	(1, 0, 1, 1, 0, 0, 0, 0, 1, 1, 0, 1, 0, 1, 0, 1, 1, 1)	[36, 18, 8]	$\varphi(\tau_2(v_1))$
(0, 1, 1, 1, 0, 1, 1, 1, 1, 0, 0, 0, 1, 1, 1, 0, 0, 0)	(0, 0, 1, 1, 1, 1, 1, 1, 1, 0, 1, 1, 0, 1, 1, 0, 0, 1)	[36, 12, 2]	$\varphi(\tau_2(v_1))$
(1, 1, 0, 1, 0, 1, 0, 1, 1, 1, 0, 0, 0, 1, 0, 0, 0, 0)	(0, 0, 1, 1, 0, 0, 1, 0, 0, 0, 1, 1, 1, 0, 1, 1, 1, 1)	[36, 26, 4]	$\varphi(\tau_2(v_2))$
(0, 1, 1, 1, 0, 1, 0, 1, 0, 1, 1, 1, 0, 0, 0, 0, 1, 0)	(1, 1, 0, 1, 0, 0, 0, 0, 1, 0, 1, 0, 1, 0, 1, 0, 0, 1)	[36, 31, 2]	$\varphi(\tau_2(v_2))$
(0, 0, 1, 0, 0, 1, 0, 1, 0, 1, 0, 1, 0, 1, 0, 1, 0, 1)	(1, 1, 0, 0, 1, 0, 1, 1, 1, 1, 1, 0, 0, 1, 0, 0, 0, 1)	[36, 34, 2]	$\varphi(\tau_2(v_3))$
(0, 0, 0, 1, 1, 0, 0, 1, 1, 1, 1, 1, 0, 1, 1, 0, 1, 1, 0, 0)	(1, 1, 1, 1, 0, 0, 0, 1, 1, 1, 1, 0, 0, 0, 0, 0, 0, 1, 0, 1)	[40, 40, 1]	$\varphi(\tau_2(v_1))$
(1, 1, 1, 1, 1, 0, 0, 1, 0, 0, 0, 1, 0, 0, 0, 0, 0, 0, 1, 1)	(1, 1, 0, 1, 0, 0, 0, 0, 0, 0, 1, 1, 1, 1, 0, 0, 0, 0, 1, 1)	[40, 36, 2]	$\varphi(\tau_2(v_1))$
(0, 1, 1, 0, 1, 0, 1, 1, 1, 1, 0, 1, 0, 1, 1, 1, 1, 0, 0, 0)	(0, 0, 0, 0, 1, 0, 1, 0, 1, 0, 0, 0, 0, 1, 1, 1, 1, 0, 1, 0)	[40, 21, 8]	$\varphi(\tau_2(v_1))$

References

[1] Taher Abualrub, Ali Ghrayeb, Nuh Aydin, and Irfan Siap, *On the construction of skew quasi-cyclic codes*, IEEE Trans. Inform. Theory **56** (2010), no. 5, 2081–2090, DOI 10.1109/TIT.2010.2044062. MR2723473

[2] Maheshanand Bhaintwal, *Skew quasi-cyclic codes over Galois rings*, Des. Codes Cryptogr. **62** (2012), no. 1, 85–101, DOI 10.1007/s10623-011-9494-0. MR2873109

[3] Wieb Bosma, John Cannon, and Catherine Playoust, *The Magma algebra system. I. The user language*, J. Symbolic Comput. **24** (1997), no. 3-4, 235–265, DOI 10.1006/jsco.1996.0125. MR1484478

[4] D. Boucher, W. Geiselmann, and F. Ulmer, *Skew-cyclic codes*, Appl. Algebra Engrg. Comm. Comput. **18** (2007), no. 4, 379–389, DOI 10.1007/s00200-007-0043-z. MR2322946

[5] Delphine Boucher and Felix Ulmer, *Coding with skew polynomial rings*, J. Symbolic Comput. **44** (2009), no. 12, 1644–1656, DOI 10.1016/j.jsc.2007.11.008. MR2553570

[6] D. Boucher and F. Ulmer, *Linear codes using skew polynomials with automorphisms and derivations*, Des. Codes Cryptogr. **70** (2014), no. 3, 405–431, DOI 10.1007/s10623-012-9704-4. MR3160738

[7] Delphine Boucher, Patrick Solé, and Felix Ulmer, *Skew constacyclic codes over Galois rings*, Adv. Math. Commun. **2** (2008), no. 3, 273–292, DOI 10.3934/amc.2008.2.273. MR2429458

[8] J. H. Conway and N. J. A. Sloane, *A new upper bound on the minimal distance of self-dual codes*, IEEE Trans. Inform. Theory **36** (1990), no. 6, 1319–1333, DOI 10.1109/18.59931. MR1080819

[9] Steven T. Dougherty, *Algebraic coding theory over finite commutative rings*, SpringerBriefs in Mathematics, Springer, Cham, 2017, DOI 10.1007/978-3-319-59806-2. MR3675844

[10] Steven T. Dougherty, Joseph Gildea, Rhian Taylor, and Alexander Tylyshchak, *Group rings, G-codes and constructions of self-dual and formally self-dual codes*, Des. Codes Cryptogr. **86** (2018), no. 9, 2115–2138, DOI 10.1007/s10623-017-0440-7. MR3816217

[11] S. T. Dougherty, A. Korban, S. Sahinkaya and D. Ustun, "Group Matrix Ring Codes and Constructions of Self-Dual Codes", Applicable Algebra in Engineering, Communication and Computing, (2021), https://doi.org/10.1007/s00200-021-00504-9.

[12] Steven T. Dougherty, Joseph Gildea, Adrian Korban, Abidin Kaya, Alexander Tylyshchak, and Bahattin Yildiz, *Bordered constructions of self-dual codes from group rings and new extremal binary self-dual codes*, Finite Fields Appl. **57** (2019), 108–127, DOI 10.1016/j.ffa.2019.02.004. MR3915399

[13] S. T. Dougherty, S. Sahinkaya, B. Yildiz B, "Skew G-Codes", to appear in J. Algebra Appl.

[14] M. Grassl Code Tables, bounds on the parameters of various types of codes. `http://www.codetables.de/`. Accessed 30 November 2021.

[15] Masaaki Harada and Ken Saito, *Singly even self-dual codes constructed from Hadamard matrices of order 28*, Australas. J. Combin. **70** (2018), 288–296. MR3737113

[16] Masaaki Harada and Akihiro Munemasa, *Some restrictions on weight enumerators of singly even self-dual codes*, IEEE Trans. Inform. Theory **52** (2006), no. 3, 1266–1269, DOI 10.1109/TIT.2005.864416. MR2238092

[17] W. Cary Huffman, *On the classification and enumeration of self-dual codes*, Finite Fields Appl. **11** (2005), no. 3, 451–490, DOI 10.1016/j.ffa.2005.05.012. MR2158773

University of Scranton, Scranton, Pennsylvania 18518
Email address: `prof.steven.dougherty@gmail.com`

Faculty of Engineering, Department of Natural and Mathematical Sciences, Mersin, Turkey
Email address: `serap@tarsus.edu.tr`

Tarsus University, Faculty of Engineering, Department of Computer Engineering, Mersin, Turkey
Email address: `denizust@gmail.com`

Contemporary Mathematics
Volume **785**, 2023
https://doi.org/10.1090/conm/785/15779

The underlying vector spaces of certain endomorphism rings

Gabriella D'Este

Dedicated to Professor Syed Tariq Rizvi for his contribution to Ring and Module Theory

1. Introduction

The Baer-Kaplansky Theorem [**3**, Theorem 108.1] says that two torsion abelian groups with isomorphic endomorphism rings are actually isomorphic. In the sequel, following [**4**], we say that a class of modules, say $\mathcal{C}$, is a *Baer-Kaplansky class* if two modules in $\mathcal{C}$ whose endomorphism rings are isomorphic (as rings) are necessarily isomorphic.

Assuming that the Generalized Continuum Hypothesis holds, one can construct classes of vector spaces (of finite dimension or of reasonably large infinite dimension) with the property that two vector spaces in these classes with endomorphism rings of the same dimension are necessarily isomorphic [**2**, Theorem 2.2]. Under the same hypotheses, a class of divisible abelian groups with the property that two groups are isomorphic if their endomorphism rings have isomorphic underlying additive groups contains both very small and very large divisible groups [**2**, Corollary 4.6 and Proposition 4.7].

In this note we characterize Baer-Kaplansky classes $\mathcal{C}$ with two indecomposable modules, defined over a K–algebra A, such that the endomorphism rings of two non isomorphic modules of $\mathcal{C}$ are non isomorphic vector spaces over K. Let U and V denote the two indecomposable and finite dimensional modules of $\mathcal{C}$. Assume that the following conditions hold:

- $\dim_K \operatorname{End}_A(U)$ is different from $\dim_K \operatorname{End}_A(V)$;
- $\dim_K \operatorname{End}_A(U) - \dim_K \operatorname{Hom}_A(U,V)$ is different from $\dim_K \operatorname{End}_A(V) - \dim_K \operatorname{Hom}_A(V,U)$.

Then the class $\operatorname{add}(U \oplus V)$, formed by all modules of the form $U^m \oplus V^n$ with $m, n \in \mathbb{N}$, is a Baer-Kaplansky class [**2**, Lemma 3.1].

We will show (Theorem 1) that if the vector space $\operatorname{Hom}_A(U,V) \oplus \operatorname{Hom}_A(V,U)$ is neither too small nor too large, then the endomorphism rings of two non isomorphic modules in $\operatorname{add}(U \oplus V)$ with the same number of indecomposable summands are non isomorphic vector spaces over K. It is easy to see that the class of finite dimensional projective (resp. injective) modules may satisfy (Example 2) the arithmetic condition (i) of Theorem 1. However this does not always happen (Example 3). We will also see that very special classes with three indecomposable modules contain

2020 *Mathematics Subject Classification.* 16G20, 16D99.

many non isomorphic modules with endomorphisms rings of the same dimension over K (Example 4).

In the following K is a field and modules are left modules over finite dimensional K-algebras A usually given by quivers ([**1**] or [**6**]). If $1, \dots, n$ are the vertices of a quiver, then the simple modules corresponding to these vertices are denoted by $1, \dots, n$ respectively. Finally, pictures of the form $\begin{smallmatrix}1\\2\\1\end{smallmatrix}$, $\begin{smallmatrix}2\\1\,2\\2\end{smallmatrix}$, $\begin{smallmatrix}1\\1\,2\,2\,3\end{smallmatrix}$ describe in an obvious way indecomposable modules of dimension 3, 4 and 5 respectively.

2. Theorem and Examples

The next theorem shows that only very rarely a Baer-Kaplansky class $\mathcal{C}$ containing two indecomposable modules has the property that the endomorphism rings of two non isomorphic modules belonging to $\mathcal{C}$ and with the same number of indecomposable summands are non isomorphic vector spaces.

THEOREM 1. *Let A be a K-algebra, let U and V be indecomposable finite dimensional A-modules. Let $a = \dim_K \operatorname{End}_A(U)$, $b = \dim_K \operatorname{End}_A(V)$, $c = \dim_K \operatorname{Hom}_A(U, V)$ and $d = \dim_K \operatorname{Hom}_A(V, U)$. Assume $a < b$ and $a - c \neq b - d$. Then* $\operatorname{add}(U \oplus V)$ *is a Baer-Kaplansky class and the following conditions are equivalent:*

(i) $2a \leq c + d \leq 2b$.

(ii) *If $m, n, s \in \mathbb{N}$, $s \leq m$, $X = U^m \oplus V^n$, $Y = U^{m-s} \oplus V^{n+s}$ and $\dim_K \operatorname{End}_A(X) = \dim_K \operatorname{End}_A(Y)$, then X is isomorphic to Y.*

PROOF. Our hypotheses on a, b, c, d and [**2**, Lemma 3.1] guarantee that $\operatorname{add}(U \oplus V)$ is a Baer-Kaplansky class. We first note that

(1) $\dim_K \operatorname{End}_A(U^m \oplus V^n) = (m\ n) \begin{pmatrix} a & d \\ c & b \end{pmatrix} \begin{pmatrix} m \\ n \end{pmatrix}$ for all $m, n \in \mathbb{N}$.

Let $m, n, s \in \mathbb{N}$, $s \leq m$ and let X and Y denote the following modules: $X = U^m \oplus V^n$ and $Y = U^{m-s} \oplus V^{n+s}$. Then we deduce from (1) that

(2) $\dim_K \operatorname{End}_A(X) = am^2 + bn^2 + (c+d)mn$ and
$\dim_K \operatorname{End}_A(Y) = a(m-s)^2 + b(n+s)^2 + (c+d)(m-s)(n+s)$.

Consequently, we have

(3) $\dim_K \operatorname{End}_A(Y) - \dim_K \operatorname{End}_A(X) = st$ where $t = (c+d-2a)m + (2b - c - d)n + (a+b-c-d)s$.

Suppose first $c+d < 2a$. In this case let $m = a+b-c-d$, $n = 0$ and $s = 2a-c-d$. Then we have $m, n, s \in \mathbb{N}$, $0 < s < m$ and we deduce from (3) that

(4) $\dim_K \operatorname{End}_A(Y) - \dim_K \operatorname{End}_A(X) = st$ where $t = (c+d-2a)(a+b-c-d) + (a+b-c-d)(2a-c-d) = 0$.

Assume now $c+d > 2b$. In this case, let $m = s = c+d-2b$ and $n = b-a$. Then we have $m, n, s \in \mathbb{N}$, $0 < s = m$. Hence we deduce from (3) and from our hypotheses of m, n, s that

(5) $\dim_K \operatorname{End}_A(Y) - \dim_K \operatorname{End}_A(X) = st$ where
$t = (c+d-2a)(c+d-2b) + (2b-c-d)(b-a) + (a+b-c-d)(c+d-2b) = (c+d-2b)(c+d-2a-b+a+a+b-c-d) = 0$.

Putting together (4) and (5), we conclude that

(6) If (i) does not hold, then (ii) does not hold.

Finally, assume $2a \leq c+d \leq 2b$. Next, let m, n, s be natural numbers such that $0 < s \leq m$. As before, let $X = U^m \oplus V^n$, $Y = U^{m-s} \oplus V^{n+s}$. These hypotheses and (3) imply that

(7) $\dim_K \mathrm{End}_A Y - \dim_K \mathrm{End}_A X = st$ where $t = (c+d-2a)m + (2b-c-d)n + (a+b-c-d)s \geq (c+d-2a)s - (c+d-a-b)s \geq (b-a)s > 0$.

Hence we deduce from (7) that

(8) (i) implies (ii).

This observation and (6) guarantee that (i) holds if and only if (ii) holds. □

We give an example of an algebra with many modules satisfying the hypotheses and condition (i) of Theorem 1. Following the definition given in [**5**, page 167], we recall that a *tilting* (resp. *cotilting*) *module* T over an algebra A with n simple modules is a module of projective (resp. injective) dimension ≤ 1, such that $\mathrm{Ext}^1_A(T,T) = 0$ and the number of isomorphism classes of the indecomposable summands of T is equal to n.

EXAMPLE 2. There is a K-algebra A admitting indecomposable modules U_1, U_2, U_3, V such that, for any $j = 1, 2, 3$, the modules U_j and V satisfy condition (i) of Theorem 1 and the following facts hold:

(a) $U_1 \oplus V$ is a projective generator.
(b) $U_2 \oplus V$ is an injective cogenerator.
(c) $U_3 \oplus V$ is a tilting and cotilting module of projective and injective dimension equal to one.

CONSTRUCTION. Let A be the algebra given by the quiver

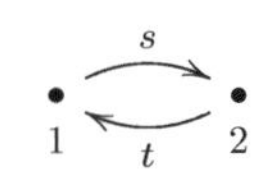

with relation $st = 0$. Then the Auslander-Reiten quiver is of the form

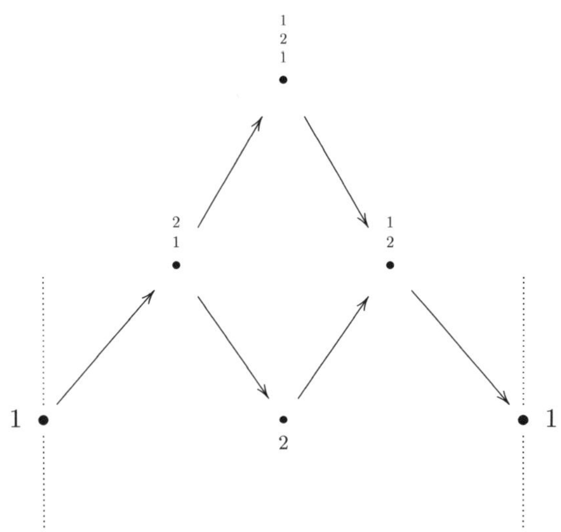

Consequently the modules $U_1 = \begin{smallmatrix}2\\1\end{smallmatrix}$, $U_2 = \begin{smallmatrix}1\\2\end{smallmatrix}$, $U_3 = 1$ and $V = \begin{smallmatrix}1\\2\\1\end{smallmatrix}$ have all the desired properties. □

EXAMPLE 3. There are a K-algebra A and indecomposable projective (resp. injective) modules U and V with the following properties:

(a) $\operatorname{End}_A(U) \simeq K$, $\operatorname{End}_A(V) \simeq K[x]/(x^3)$ and $\operatorname{add}(U \oplus V)$ is a Baer-Kaplansky class.
(b) If X and Y are two non isomorphic projective (resp. injective) modules which are the sum of $n \leq 2$ indecomposable modules, then $\dim_K \operatorname{End}_A(X) \neq \dim_K \operatorname{End}_A(Y)$.
(c) $\operatorname{End}_A(U^3)$ and $\operatorname{End}_A(U^2 \oplus V)$ are isomorphic vector spaces over K.

CONSTRUCTION. Let A be the algebra given by the quiver 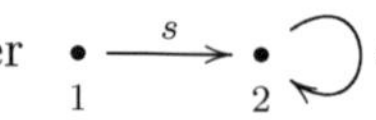

with relations $ts = 0$ and $t^3 = 0$. Let $U = \begin{smallmatrix} 1 \\ 2 \end{smallmatrix}$ and $V = \begin{smallmatrix} 2 \\ 2 \\ 2 \end{smallmatrix}$ $\Big($resp. $U = 1$ and $V = \begin{smallmatrix} & 2 \\ 1 & 2 \\ & 2 \end{smallmatrix}\Big)$. Then we have $\operatorname{End}_A(U) \simeq K$, $\operatorname{End}_A(V) \simeq K[x]/(x^3)$, $\operatorname{Hom}_A(U,V) = 0$ and $\operatorname{Hom}_A(V,U) \simeq K$. Hence (a) follows from Theorem 1 and it is easy to check that also (b) holds. We finally note that $\dim_K \operatorname{End}_A(U^3) = \begin{pmatrix} 3 & 0 \end{pmatrix} \begin{pmatrix} 1 & 1 \\ 0 & 3 \end{pmatrix} \begin{pmatrix} 3 \\ 0 \end{pmatrix} = 9$ and $\dim_K \operatorname{End}_A(U^2 \oplus V) = \begin{pmatrix} 2 & 1 \end{pmatrix} \begin{pmatrix} 1 & 1 \\ 0 & 3 \end{pmatrix} \begin{pmatrix} 2 \\ 1 \end{pmatrix} = 9$. Hence (c) holds. □

In the next example we investigate a class containing three indecomposable modules such that any two of them satisfy the hypotheses and condition (i) of Theorem 1.

EXAMPLE 4. There are a K-algebra A and indecomposable modules P_1, P_2, P_3 with the following properties:

(a) If $1 \leq i < j \leq 3$, then the modules $U = P_i$ and $V = P_j$ satisfy the hypotheses and condition (i) of Theorem 1.
(b) There exist non isomorphic projective modules P and Q, which are direct sums of $n \geq 5$ indecomposable modules, such that $\operatorname{End}_A(P)$ and $\operatorname{End}_A(Q)$ are isomorphic vector spaces and one of the following conditions hold:
   (1) $P = P_2^n$ and $Q = P_1 \oplus P_2 \oplus P_3 \oplus \ldots$.
   (2) $P = P_1^r \oplus P_3^{n-r}$ and $Q = P_2^s \oplus P_3^{n-s}$ for some $0 < r, s < n$.
   (3) $P = P_1^r \oplus P_2^{n-r}$ for some $0 < r < n$ and $Q = P_1 \oplus P_2 \oplus P_3 \oplus \ldots$.
   (4) $P = P_1^r \oplus P_3^{n-r}$ for some $0 < r < n$ and $Q = P_1 \oplus P_2 \oplus P_3 \oplus \ldots$.
   (5) $P = P_1 \oplus P_2 \oplus P_3 \oplus \ldots$ and $Q = P_1 \oplus P_2 \oplus P_3 \oplus \ldots$.
   (6) $P = P_1 \oplus P_2 \oplus P_3 \oplus \ldots$ and $Q = P_2^r \oplus P_3^{n-r}$ for some $0 < r < n$.

CONSTRUCTION. Let A be the algebra given by the quiver

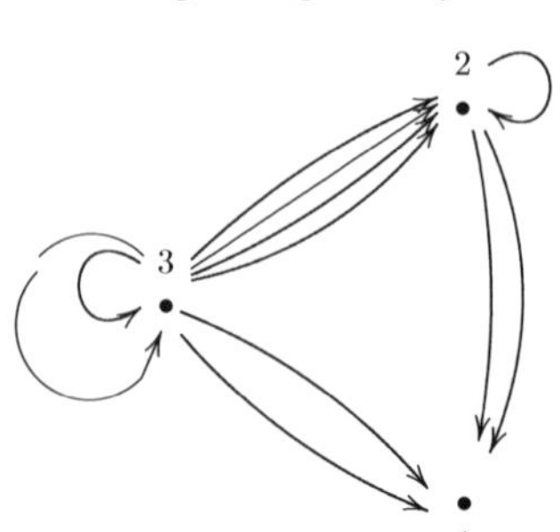

with relations such that the indecomposable projective modules are

$$P_1 = 1, P_2 = \begin{smallmatrix} & 2 & \\ 1 & 1 & 2 \end{smallmatrix}, P_3 = \begin{smallmatrix} & & & 3 & & & & \\ 1 & 1 & 2 & 2 & 2 & 2 & 3 & 3 \end{smallmatrix}.$$

Hence we have $\mathrm{End}_A(P_1) \simeq K$, $\mathrm{End}_A(P_2) \simeq K[x]/(x^2)$, and $\mathrm{End}_A(P_3)$ is isomorphic to the algebra of all matrices of the form $\left(\begin{smallmatrix} x & 0 & 0 \\ y & x & 0 \\ z & 0 & x \end{smallmatrix}\right)$ with $x, y, z \in K$. Moreover, we have $\mathrm{Hom}_A(P_1, P_2) \simeq K^2$, $\mathrm{Hom}_A(P_1, P_3) \simeq K^2$, $\mathrm{Hom}_A(P_2, P_3) \simeq K^4$ and $\mathrm{Hom}_A(P_i, P_j) = 0$ if $i > j$. Consequently it is easy to see that (a) holds. To prove (b), let $D = (d_{ij})$ be the matrix $\left(\begin{smallmatrix} 1 & 0 & 0 \\ 2 & 2 & 0 \\ 2 & 4 & 3 \end{smallmatrix}\right)$ which satisfies $d_{ij} = \dim_k \mathrm{End}_A(P_j, P_i)$ for all i, j. Then we have $(\,x\;y\;z\,)\, D \left(\begin{smallmatrix} x \\ y \\ z \end{smallmatrix}\right) = \dim_K \mathrm{End}_A(P_1^x \oplus P_2^y \oplus P_3^z)$ for all $x, y, z \in \mathbb{N}$. Since $(\,0\;5\;0\,)\, D \left(\begin{smallmatrix} 0 \\ 5 \\ 0 \end{smallmatrix}\right) = 50 = (\,1\;1\;3\,)\, D \left(\begin{smallmatrix} 1 \\ 1 \\ 3 \end{smallmatrix}\right)$, it follows that the modules $P = P_2^5$ and $Q = P_1 \oplus P_2 \oplus P_3^3$ satisfy (1) with $n = 5$. We next observe that $(\,2\;0\;5\,)\, D \left(\begin{smallmatrix} 2 \\ 0 \\ 5 \end{smallmatrix}\right) = 99 = (\,0\;6\;1\,)\, D \left(\begin{smallmatrix} 0 \\ 6 \\ 1 \end{smallmatrix}\right)$. Consequently the modules $P = P_1^2 \oplus P_3^5$ and $Q = P_2^6 \oplus P_3$ satisfy (2) with $n = 7$. Since $(\,3\;5\;0\,)\, D \left(\begin{smallmatrix} 3 \\ 5 \\ 0 \end{smallmatrix}\right) = 89 = (\,4\;1\;3\,)\, D \left(\begin{smallmatrix} 4 \\ 1 \\ 3 \end{smallmatrix}\right)$ (resp. $(\,5\;5\;0\,)\, D \left(\begin{smallmatrix} 5 \\ 5 \\ 0 \end{smallmatrix}\right) = 125 = (\,6\;1\;3\,)\, D \left(\begin{smallmatrix} 6 \\ 1 \\ 3 \end{smallmatrix}\right)$), it follows that the modules $P = P_1^3 \oplus P_2^5$ and $Q = P_1^4 \oplus P_2 \oplus P_3^3$ (resp. $P = P_1^5 \oplus P_2^5$ and $Q = P_1^6 \oplus P_2 \oplus P_3^3$) satisfy (3) with $n = 8$ (resp. $n = 10$). We also note that $(\,4\;0\;5\,)\, D \left(\begin{smallmatrix} 4 \\ 0 \\ 5 \end{smallmatrix}\right) = 131 = (\,2\;6\;1\,)\, D \left(\begin{smallmatrix} 2 \\ 6 \\ 1 \end{smallmatrix}\right)$ (resp. $(\,5\;0\;5\,)\, D \left(\begin{smallmatrix} 5 \\ 0 \\ 5 \end{smallmatrix}\right) = 150 = (\,3\;6\;1\,)\, D \left(\begin{smallmatrix} 3 \\ 6 \\ 1 \end{smallmatrix}\right)$). Consequently the modules $P = P_1^4 \oplus P_3^5$ and $Q = P_1^2 \oplus P_2^6 \oplus P_3$ (resp. $P = P_1^5 \oplus P_3^5$ and $Q = P_1^3 \oplus P_2^6 \oplus P_3$) satisfy (4) with $n = 9$ (resp. $n = 10$). We finally observe that $(\,3\;3\;4\,)\, D \left(\begin{smallmatrix} 3 \\ 3 \\ 4 \end{smallmatrix}\right) = 165 = (\,2\;7\;1\,)\, D \left(\begin{smallmatrix} 2 \\ 7 \\ 1 \end{smallmatrix}\right)$ (resp. $(\,3\;1\;6\,)\, D \left(\begin{smallmatrix} 3 \\ 1 \\ 6 \end{smallmatrix}\right) = 185 = (\,1\;7\;2\,)\, D \left(\begin{smallmatrix} 1 \\ 7 \\ 2 \end{smallmatrix}\right)$). It follows that $P = P_1^3 \oplus P_2^3 \oplus P_3^4$ and $Q = P_1^2 \oplus P_2^7 \oplus P_3$ (resp. $P = P_1^3 \oplus P_2 \oplus P_3^6$ and $Q = P_1 \oplus P_2^7 \oplus P_3^2$) satisfy (5) with $n = 10$. Since $(\,1\;3\;4\,)\, D \left(\begin{smallmatrix} 1 \\ 3 \\ 4 \end{smallmatrix}\right) = 129 = (\,0\;7\;1\,)\, D \left(\begin{smallmatrix} 0 \\ 7 \\ 1 \end{smallmatrix}\right)$, it follows that $P = P_1 \oplus P_2^3 \oplus P_3^4$ and $Q = P_2^7 \oplus P_3$ satisfy (6) with $n = 8$.

□

References

[1] Maurice Auslander, Idun Reiten, and Sverre O. Smalø, *Representation theory of Artin algebras*, Cambridge Studies in Advanced Mathematics, vol. 36, Cambridge University Press, Cambridge, 1997. Corrected reprint of the 1995 original. MR1476671

[2] G. D'Este and D. Keskin Tütüncü and R. Tribak, *Baer–Kaplansky classes of vector spaces and modules determined by numerical invariants,* Communications in Algebra, Vol. 51(3), (2023), 1089–1104.

[3] László Fuchs, *Infinite abelian groups. Vol. II*, Pure and Applied Mathematics. Vol. 36-II, Academic Press, New York-London, 1973. MR0349869

[4] George Ivanov and Peter Vámos, *A characterization of FGC rings*, Proceedings of the Second Honolulu Conference on Abelian Groups and Modules (Honolulu, HI, 2001), Rocky Mountain J. Math. **32** (2002), no. 4, 1485–1492, DOI 10.1216/rmjm/1181070035. MR1987620

[5] Claus Michael Ringel, *Tame algebras and integral quadratic forms*, Lecture Notes in Mathematics, vol. 1099, Springer-Verlag, Berlin, 1984, DOI 10.1007/BFb0072870. MR774589

[6] Ralf Schiffler, *Quiver representations*, CMS Books in Mathematics/Ouvrages de Mathématiques de la SMC, Springer, Cham, 2014, DOI 10.1007/978-3-319-09204-1. MR3308668

Department of Mathematics, University of Milano, Milano, Italy
Email address: gabriella.deste@unimi.it

Contemporary Mathematics
Volume **785**, 2023
https://doi.org/10.1090/conm/785/15780

Algebraic structures from the point of view of complete multiplicative lattices

Alberto Facchini

Dedicated to Professor Syed Tariq Rizvi for his contribution to Ring and Module Theory

ABSTRACT. General results on multiplicative lattices found recently by Facchini, Finocchiaro and Janelidze have been studied in the particular case of groups by Facchini, de Giovanni and Trombetti. In this paper we prove that these results hold not only for the multiplicative lattices of all normal subgroups of a group, but also for much more general multiplicative lattices. Therefore they can be applied to other algebraic structures, for instance to braces.

1. Introduction

In the history of Mathematics there have been at least two very successful attempts in trying to put in order algebraic structures. One has been with the creation of Universal Algebra, developed in particular by Garrett Birkhoff, and the other has been with the discovery of Category Theory first [**12**], and then the development of Categorical Algebra. Smith [**30**], Hagemann-Hermann [**18**] and Freese and McKenzie [**22**] introduced a generalization to any algebra in a congruence-permutable variety (or congruence-modular variety, respectively) of the commutator for groups. This immediately opened the possibility of applying concepts like solvability, nilpotence and the center, to very general settings (see [**9**, p. 283]). In this paper we follow this direction, stressing how the point of view of multiplicative lattices also allows to reorganize several concepts relative to algebraic structures. Of course, our project is not so ambitious like creating a new point of view comparable to the extremely powerful introduction of Universal Algebra and Category Theory!

DEFINITION 1.1. A multiplicative lattice is a complete lattice $(L, \vee, \wedge)$ equipped with a further binary operation $\cdot\colon L \times L \to L$ (multiplication) satisfying $x \cdot y \leqslant x \wedge y$ for all $x, y \in L$.

2020 *Mathematics Subject Classification.* Primary 06B75; Secondary 16T25, 20A99, 22F05.
Partially supported by Ministero dell'Istruzione, dell'Università e della Ricerca (Progetto di ricerca di rilevante interesse nazionale "Categories, Algebras: Ring-Theoretical and Homological Approaches (CARTHA)"), Fondazione Cariverona (Research project "Reducing complexity in algebra, logic, combinatorics - REDCOM" within the framework of the programme Ricerca Scientifica di Eccellenza 2018), and the Department of Mathematics "Tullio Levi-Civita" of the University of Padua (Research programme DOR1828909 "Anelli e categorie di moduli").

As usual, we will often write xy instead of $x \cdot y$. The smallest and the largest elements of the complete lattice L will be denoted by 0 and 1, respectively. Notice that 1 is not an identity for the multiplication $\cdot$, in general.

Multiplicative lattices are an algebraic structure to which little attention has been devoted, but which already appeared in Krull [**25**], and has been studied by M. Ward [**33, 34**], M. Ward and R. P. Dilworth [**35**], D. D. Anderson [**2, 3**], E. W. Johnson, and J. A. Johnson [**19, 20**]. In all these papers, further axioms are required: associativity or commutativity of multiplication, distributivity of multiplication with respect to the join $\vee$, existence of an identity, compatibility of multiplication and partial order, the multiplication is the meet $\wedge$, and so on. We don't require any of these further conditions.

It is better to view a multiplicative lattice L not as a complete lattices like in Definition 1.1, but as a complete join-semilattice $(L, \vee)$ necessarily with 0 and 1. Let's be more precise. A *join-semilattice* is a partially ordered set that has a join (a least upper bound) for any nonempty finite subset. Equivalently, a join-semilattice is an algebraic structure $(L, \vee)$ for which $\vee$ is a binary operation that is associative, commutative and in which every element of L is idempotent. A *complete* join-semilattice is a partially ordered set (define $x \le y$ if and only if $x \vee y = y$) that has a join (a least upper bound) for any subset. In a complete join-semilattice a meet (a greatest lower bound) also exists for any subset, so that complete join-semilattices are exactly complete lattices. Every complete join-semilattice has a least element 0 (the least upper bound of the empty subset) and a greatest element 1 (the least upper bound of the improper subset). If we view a complete join-semilattice as an algebraic structure $(L, \vee, 0, 1)$, then $(L, \vee, 0)$ is a commutative monoid in which every element is idempotent and in which the identity for the operation $\vee$ is 0, and $(L, \vee, 0, 1)$ is a commutative monoid with zero in which every element is idempotent, the identity for the operation $\vee$ is 0, and the zero is 1 (because $x \vee 1 = 1$ for every $x \in L$). This argument could seem annoying, but it is necessary in order to explain the notion of morphism of multiplicative lattices: there is no difference between complete lattices and complete join-semilattices, but the difference is in morphisms: a morphism $f: L \to M$ of complete join-semilattices is a mapping such that $f(\bigvee X) = \bigvee f(X)$ for every subset X of L.

It is better not to look at multiplicative lattices as complete lattices $(L, \vee, \wedge)$ with a further third operation $\cdot$: it is better to view them as complete join-semilattices $(L, \vee)$ with a further second operation $\cdot$ for which $x \cdot y \le x$ and $x \cdot y \le y$ for every $x, y \in L$. In some sense, the operation $\cdot$ in multiplicative lattices is a replacement of the operation $\wedge$. That is, in some sense, multiplicative lattices are deformations of lattices: given a lattice $(L, \vee, \wedge)$, the operation $\vee$ remains, but the operation $\wedge$ is replaced by the multiplication $\cdot$, and $xy \le x$, $xy \le y$ for every $x, y \in L$.

LEMMA 1.2. *If $f: L \to M$ is a morphism of complete join-semilattices, then there is a unique morphism $u: M \to L$ of complete meet-semilattices such that, for every $x \in L$ and every $y \in M$, $f(x) \le y \Leftrightarrow x \le u(y)$.*

Considering the complete lattices L and M as categories, u is a right adjoint of f. The category CML of complete multiplicative lattices is the category whose morphisms are the morphisms $f: L \to M$ of complete join-semilattices such that $f(x)f(x') \leqslant f(xx')$ for all $x, x' \in L$. Thus a morphism of complete multiplicative lattice is a mapping $f: L \to M$ such that $f(\bigvee X) = \bigvee f(X)$ for every subset X of L, $f(x)f(x') \leqslant f(xx')$ for all $x, x' \in L$, and $f(1) = 1$.

This paper strongly depends on the papers [**13**] and [**14**].

2. Prime elements and Zariski spectrum of a multiplicative lattice

The content of this section is taken from [**13**]. Let L be a multiplicative lattice. An element p of L is a *prime element* of L if $p \neq 1$ and

$$xy \leqslant p \Rightarrow (x \leqslant p \text{ or } y \leqslant p)$$

for every $x, y \in L$. The set $\operatorname{Spec}(L)$ of all prime elements of L is the *Zariski spectrum* (or the *prime spectrum*) of L.

Let $\mathcal{P}(\operatorname{Spec}(L))$ be the power set of $\operatorname{Spec}(L)$, that is, the set of all subsets of $\operatorname{Spec}(L)$. We can define two mappings

$$V: L \to \mathcal{P}(\operatorname{Spec}(L))$$

by

$$V(x) = \{ p \in \operatorname{Spec}(L) \mid x \leqslant p \}$$

for every $x \in L$, and

$$I: \mathcal{P}(\operatorname{Spec}(L)) \to L,$$

defined by $I(X) = \bigwedge X$ for every subset X of $\operatorname{Spec}(L)$.

If we view V as a mapping $V: L \to \mathcal{P}(\operatorname{Spec}(L))^{\mathrm{op}}$, then V turns out to be a morphism of complete join-semilattices (i.e., $V(\bigvee X) = \bigcap_{x \in X} V(x)$ for every subset X of L), and $I: \mathcal{P}(\operatorname{Spec}(L))^{\mathrm{op}} \to L$ is the right adjoint of V in the sense of Lemma 1.2. The mapping $V: L \to \mathcal{P}(\operatorname{Spec}(L))$ has two further important properties:

(1) V transforms the multiplication in L into the union in $\mathcal{P}(\operatorname{Spec}(L))$, that is, V is a magma morphism of the magma $(L, \cdot)$ into the magma (the commutative monoid) $(\mathcal{P}(\operatorname{Spec}(L)), \cup)$:

$$V(xy) = V(x) \cup V(y) \qquad \text{for every } x, y \in L,$$

and $V(1) = \varnothing$. Hence V is a morphism of multiplicative lattices of $(L, \vee, \cdot)$ into the multiplicative lattice $(\mathcal{P}(\operatorname{Spec}(L))^{\mathrm{op}}, \cap, \cup)$.

(2) The image $V(L)$ of the mapping V satisfies the axioms for the closed sets of a topology on $\operatorname{Spec}(L)$, called the *Zariski topology*, and $\operatorname{Spec}(L)$ with this topology turns out to be a sober topological space [**13**, Lemma 2.6].

From the fact that I is a right adjoint of V, we get that V and I induce mutually inverse antiisomorphisms of partially ordered sets between the partially ordered subset $\operatorname{Rad}(L) := I(\mathcal{P}(\operatorname{Spec}(L)))$ of L and the partially ordered subset $V(L)$ of $\mathcal{P}(\operatorname{Spec}(L))$. Hence, we have a lattice isomorphism between $\operatorname{Rad}(L)$, the lattice of all elements of L that are intersections of primes, and $\Omega(\operatorname{Spec}(L))$, the lattice of all open subsets of the Zariski space $\operatorname{Spec}(L)$. We will come back to the elements of L that are intersections of primes in Proposition 4.10.

Since $\Omega(\operatorname{Spec}(L))$ is a sublattice of the distributive lattice $\mathcal{P}(\operatorname{Spec}(L))$, it follows that the lattices $\operatorname{Rad}(L) \cong \Omega(\operatorname{Spec}(L))$ are also distributive lattices, which are complete (though joins of infinite subsets in $\operatorname{Rad}(L)$ are not the same as in L, and meets of infinite subsets of $\Omega(\operatorname{Spec}(L))$ are not the same as in $\mathcal{P}(\operatorname{Spec}(L))$, but are the interiors of the intersections).

THEOREM 2.1 ([**13**, Theorem 3.2]). *Let $f: L \to M$ be a morphism in* CML *and let $u: M \to L$ be its right adjoint. Then:*

(a) *If p is a prime element of M, then $u(p)$ is a prime element of L.*
(b) *The map* $\mathrm{Spec}(f): \mathrm{Spec}(M) \to \mathrm{Spec}(L)$ *defined by* $\mathrm{Spec}(f)(p) = u(p)$ *of every* $p \in \mathrm{Spec}(M)$ *is continuous. Moreover,* $(\mathrm{Spec}(f))^{-1}(\mathrm{V}(x)) = \mathrm{V}(f(x))$ *for every* $x \in L$.
(c) *The assignment* $L \mapsto \mathrm{Spec}(L)$ *defines a functor* $\mathrm{Spec}: \mathsf{CML}^{\mathrm{op}} \to \mathsf{STop}$, *where* STop *is the category of sober topological spaces.*
(d) [**13**, 12.3(e)] *The functor* $\mathrm{Spec}: \mathsf{CML}^{\mathrm{op}} \to \mathsf{STop}$ *is the right adjoint of the functor* $\Omega: \mathsf{STop} \to \mathsf{CML}^{\mathrm{op}}$, *that assigns to each sober space X the complete lattice $\Omega(X)$ of its open subsets with multiplication the intersection of two open subsets.*

Our main application of multiplicative lattices is, for an algebraic structure A, its lattice of congruences $\mathcal{L}(A)$. But on $\mathcal{L}(A)$ we must fix a multiplication, and the spectrum then depends on the multiplication chosen on $\mathcal{L}(A)$. For example, consider a discrete valuation ring R, for instance $\mathbb{Z}_p$. Its lattice of congruences $\mathcal{L}(R)$ is isomorphic to the complete lattice $\mathbb{Z}_{\le 0} \cup \{-\infty\}$. If on $\mathcal{L}(R)$ we fix the usual multiplication of ideals, we get that the spectrum of $\mathcal{L}(R)$ has two points (the zero ideal and the maximal ideal). But if we fix as multiplication on $\mathcal{L}(R)$ intersection of ideals, then all proper ideals are prime, so that the spectrum of $\mathcal{L}(R)$ is infinite. If we fix as multiplication on $\mathcal{L}(R)$ the constant mapping 0 (i.e., the product of any two ideals of R is always the zero ideal), then the the spectrum of $\mathcal{L}(R)$ is empty. A further possibility is to fix as product of two ideals I and J in $\mathcal{L}(R)$ for an arbitrary (noncommutative) ring R the two-sided ideal generated by all commutators $[i,j] = ij - ji$, where $i \in I$ and $j \in J$. Hence the spectrum doesn't depend only on the ring R and the lattice $\mathcal{L}(R)$, but also on the multiplication chosen on the lattice of its congruences.

2.1. Changing the operation. Another thing to which little attention has been devoted in Algebra until now is composition of operations. It has been studied in [**24**] and [**26**]. Given two binary operations $*$ and $\circ$ on a set X, it is possible to define a third binary operation $\Box$ setting $x \Box y := (x * y) \circ (y * x)$. For example, given an associative algebra A over a field k, it is possible to give A a Lie algebra structure taking as Lie bracket the commutator $[x,y] = xy - yx$. More precisely, for a set X, it is possible to consider the set $\mathcal{M}(X)$ of all binary operations on X, and define a binary operation $\lhd$ on $\mathcal{M}(X)$ setting, for every $x, y \in X$, $x(* \lhd \circ)y := (x*y)\circ(y*x)$. In this way, one gets that $\mathcal{M}(X)$ is a monoid with respect to the operation $\lhd$, which is non-commutative if X has at least two elements [**24, 26**]. The identity of the monoid $\mathcal{M}(X)$ is the operation $x * y := x$ for every $x, y \in X$.

We can apply this to our multiplicative lattices $(L, \vee, \cdot)$. Given a multiplicative lattice $(L, \vee, \cdot)$, we can define a new binary operation $\Box$ on L setting $x \Box y := (x \cdot y) \vee (y \cdot x)$. Then $(L, \vee, \Box)$ turns out to be a multiplicative lattice [**13**, Example 3.7]. A second example for this operation on $\mathcal{M}(X)$ is given by Jordan product. If A is an associative algebra over a field of characteristic $\neq 2$, one can construct a Jordan algebra A^+ with the same underlying vector space and the multiplication $x \circ y$ (the *Jordan product*) defined by

$$x \circ y := \frac{xy + yx}{2}.$$

The coefficient $\frac{1}{2}$ is convenient here because if any two elements x, y of A commute, then $x \circ y = x \cdot y$, and if e is any two-sided identity in A, then e is also a two-sided identity in A^+. This is not necessary for a multiplicative lattice $(L, \vee, \cdot)$, where the operation $\vee$ is always idempotent, so that if any two elements x, y of $(L, \cdot)$ commute, then $x \Box y = x \cdot y$, and if e is any two-sided identity in $(L, \cdot)$ then e is also a two-sided identity in $(L, \Box)$. Notice that, as for Jordan algebras, the operation $\Box$ on L is always commutative.

Assume that L satisfies the *monotonicity condition*, i.e., that $x \leq y$ and $x' \leq y'$ imply $xx' \leq yy'$ for every $x, y, x', y' \in L$. It can then be proved that an element $p \in L$ is prime in the multiplicative lattice $(L, \vee, \cdot)$ if and only if it is a prime element in the multiplicative lattice $(L, \vee, \Box)$, and that the spectra of these two multiplicative lattices are the same topological space [**13**, Example 3.7]. This shows that in the study of multiplicative lattices L with the monotonicity condition and their spectra $\mathrm{Spec}(L)$, it will be often possible to assume the operation $\cdot$ commutative.

3. Some further terminology in multiplicative lattices

Keeping in mind the example $\mathcal{N}(G)$ of the multiplicative lattice of all normal subgroups of a group G, with the commutator of normal subgroups as multiplication, the following terminology turns out to be very natural.

Let x be an element of a multiplicative lattice L. The *lower central series* (or *descending central series*) of x is the descending series

$$x = x_1 \geq x_2 \geq x_3 \geq \dots,$$

where $x_{n+1} := x_n \cdot x$ for every $n \geq 1$. If $x_n = 0$ for some $n \geq 1$, then x is *left nilpotent*. The element x is *idempotent* if $x_2 = x \cdot x = x$, and *abelian* if $x_2 = x \cdot x = 0$. Similarly, the element x is *right nilpotent* if ${}_nx = 0$ for some n, where now in the descending series the elements ${}_nx$ are defined recursively by ${}_{n+1}x := x \cdot {}_nx$. Clearly, if the multiplication in the multiplicative lattice L is either associative or commutative, then left nilpotency coincides with right nilpotency.

The *derived series* of x [**13**, Definition 6.1] is the descending series

$$x := x^{(0)} \geq x^{(1)} \geq x^{(2)} \geq \dots,$$

where $x^{(n+1)} := x^{(n)} \cdot x^{(n)}$ for every $n \geq 0$. The term $x' := x_2 = x \cdot x = x^{(1)}$ is the *derived element* of x. The element x of L is *solvable* if $x^{(n)} = 0$ for some integer $n \geq 0$.

If the multiplication on the lattice is associative, then left nilpotency, right nilpotency and solvability of an element $x \in L$ coincide.

For the multiplicative lattice $\mathcal{N}(G)$ of a group G with operation the commutator of two normal subgroups (which is commutative, but not associative), the element $1 = G$ of $\mathcal{N}(G)$ is left (=right) nilpotent as an element of the multiplicative lattice in the sense just described if and only if the group G is nilpotent, is a solvable element if and only if G is a solvable group, is an abelian element if and only if G is an abelian group, and is an idempotent element if and only if the group G is perfect. For the multiplicative lattice $\mathcal{L}(R)$ of a ring R with operation the product of two ideals (which is associative, but not commutative in general), the element $1 = R$ of $\mathcal{L}(R)$ is left nilpotent (=right nilpotent=solvable) as an element of the multiplicative lattice if and only if the ring R is nilpotent, and is an abelian element if and only if the ring R is an additive abelian group with the zero multiplication.

Recall that an element x of a lattice L is *meet-irreducible* if $x = y \wedge z$ implies $x = y$ or $x = z$ for every $y, z \in L$.

DEFINITION 3.1. An element s of a multiplicative lattice L is *semiprime* if $x^2 \leq s$ implies $x \leq s$ for every $x \in L$.

LEMMA 3.2. *In a multiplicative lattice L, every prime element is meet-irreducible and semiprime.*

PROOF. Clearly, prime elements are semiprime. Let $p \in P$ be a prime element and suppose $p = y \wedge z$. Then $yz \leq y \wedge z = p$ implies that $y \leq p$ or $z \leq p$. But $p = y \wedge z$ implies $p \leq y$ and $p \leq z$. Therefore either $p = y$ and $p = z$. □

THEOREM 3.3. *Let L be a multiplicative lattice that satisfies the monotonicity condition. Then all* $\mathrm{Spec}(L) = L \setminus \{1\}$ *if and only if L is linearly ordered and every element of L is idempotent.*

PROOF. Suppose $\mathrm{Spec}(L) = L \setminus \{1\}$. If L is not linearly ordered, there exist $x, y \in L$ such that $x \nleq y$ and $y \nleq x$. Then x and y are two elements of L with meet $x \wedge y$, and $x > x \wedge y$, $y > x \wedge y$. This shows that $x \wedge y$ is not meet-irreducible, hence is not prime (Lemma 3.2). But $x > x \wedge y$ shows that $x \wedge y \in L \setminus \{1\}$. A contradiction.

Now let x be an element of L that is not idempotent. Then $x^2 < x$, so that x^2 is not semiprime. In particular x^2 is not prime (Lemma 3.2), another contradiction.

Conversely, assume L linearly ordered and that every element of L is idempotent. Let p be any element of L, $p \neq 1$. In order to show that p is prime, suppose $x, y \in L$, $x \nleq p$ and $y \nleq p$. Then $x > p$ and $y > p$. As L is linearly ordered, it follows that $x \wedge y > p$. By the monotonicity condition, it follows that $xy \geq (x \wedge y)(x \wedge y) = x \wedge y > p$. This proves that p is prime. □

4. Hyperabelian multiplicative lattices

In the first two lemmas of this Section we will suppose that $(L, \vee, \cdot)$ is a multiplicative lattice, which is algebraic (every element of L is the join of compact elements) and satisfies the monotonicity condition. Moreover $C(L)$ will denote the set of all compact elements of L. One has that:

LEMMA 4.1 ([**13**, Lemma 6.16]). *An element $p \neq 1$ in L is prime if and only if*

$$xy \leqslant p \Rightarrow (x \leqslant p \text{ or } y \leqslant p)$$

for every $x, y \in C(L)$.

An *m-system* in L is a nonempty subset S of $C(L)$ such that for every $x, y \in S$ there exists $z \in S$ such that $z \leq xy$.

LEMMA 4.2. *An element $p \in L$ is prime if and only if $S_p := \{c \in C(L) \mid c \nleq p\}$ is an m-system in L.*

PROOF. Clearly, $p = 1$ if and only if $S_p = \varnothing$.

Assume p prime, and fix $x, y \in S_p$. Then $x, y \nleq p$, so that $xy \nleq p$ because p is prime. Since L is algebraic, there exists a compact element $z \in L$ such that $z \leq xy$ and $z \nleq p$. Therefore $z \in S_p$.

Conversely, suppose that S_p is an m-system. We already know that $p \neq 1$. By Lemma 4.1, in order to show that p is prime we must show that if x and y are compact elements of L with $xy \leq p$, then either $x \leq p$ or $y \leq p$. Assume by contradiction that $x, y \in C(L)$ and $xy \leq p$, but $x, y \in S_p$. Since S_p is an m-system, there exists $z \in S_p$ such that $z \leq xy$. Then $z \leq p$, so $z \notin S_p$, a contradiction. □

We will say that *m-distributivity* holds in a multiplicative lattice $(L, \vee, \cdot)$ if

$$(x \vee y)z = xz \vee yz \quad \text{and} \quad x(y \vee z) = xy \vee xz$$

for every $x, y, z \in L$. Notice that when the operation $\cdot$ is the meet $\wedge$, m-distributivity holds if and only if the lattice L is a distributive lattice. Hence multiplicative lattices in which m-distributivity holds can be viewed as a deformation of bounded distributive lattices.

LEMMA 4.3. *If m-distributivity holds in a multiplicative lattice L, then the monotonicity condition holds in L.*

PROOF. Assume that m-distributivity holds in L, and suppose $x \leq y$, $x' \leq y'$, $x, y, x', y' \in L$. Then $x \vee y = y$ and $x' \vee y' = y'$, so $(x \vee y)(x' \vee y') = yy$. By m-distributivity, $xx' \vee xy' \vee yx' \vee yy' = yy'$. Thus $xx' \vee xy' \vee yx' \leq yy'$. It follows that $xx' \leq yy'$, hence the monotonicity condition holds in L. □

REMARK 4.4. When $\cdot$ is associative and m-distributivity holds, the commutative operation $\square$ considered in Section 2.1 satisfies the Jordan identity $(x \square y) \square (x \square x) = x \square (y \square (x \square x))$. Jordan algebras are the commutative algebras that satisfy the Jordan identity.

PROPOSIITON 4.5 ([**13**, Section 5]). *Let L be an m-distributive lattice in which 1 is a compact element. The following conditions are equivalent:*

(a) $1 \cdot 1 = 1$ *in L.*
(b) *Every maximal element of $L \setminus \{1\}$ is prime*
(c) *L has enough primes, in the sense that for every $x \neq 1 \in L$, there exists a prime $p \in L$ with $x \leqslant p$.*

LEMMA 4.6. *Let $(L, \vee, \cdot)$ be a multiplicative lattice, which is algebraic and satisfies the monotonicity condition. An element $s \in L$ is semiprime if and only if $c^2 \leq s$ implies $c \leq s$ for every compact element c of L.*

PROOF. Suppose that $c^2 \leq s$ implies $c \leq s$ for every compact element $c \in L$, and let $x \in L$ be an element such that $x \nleq s$. Then there exists a compact element $c_0 \in L$ such that $c_0 \leq x$ and $c_0 \nleq s$. Then $c_0^2 \nleq s$. By the monotonicity condition, $c_0^2 \leq x^2$. Hence $x^2 \nleq s$. This proves that s is semiprime. □

For the rest of this Section we will suppose that $(L, \vee, \cdot)$ is a multiplicative lattice, which is algebraic (every element of L is the join of compact elements) and in which m-distributivity holds.

LEMMA 4.7. *Let L be an m-distributive multiplicative lattice. Then an element of L is prime if and only if it is a meet-irreducible semiprime element of L.*

PROOF. One implication has been already seen in Lemma 3.2. For the other, assume that m-distributivity holds in L, that p is a meet-irreducible semiprime element of L, that $a, b \in L$, $a \nleq p$ and $b \nleq p$. To prove that p is prime, we must show that $ab \nleq p$. Now $a \vee p > p$ and $b \vee p > p$. Since p is meet-irreducible, $(a \vee p) \wedge (b \vee p) > p$. As p is semiprime, it follows that $((a \vee p) \wedge (b \vee p))^2 \nleq p$. But $((a \vee p) \wedge (b \vee p))^2 = ((a \vee p) \wedge (b \vee p))((a \vee p) \wedge (b \vee p)) \leq (a \vee p)(b \vee p) = ab \vee ap \vee pb \vee p^2 \leq ab \vee p$. It follows that $ab \nleq p$. □

The next proposition is the analogue, for multiplicative lattices, of the fact that an ideal of a commutative ring R maximal with respect to the property of being disjoint from a multiplicatively closed subset of R, is a prime ideal of R.

PROPOSIITON 4.8. *Let S be an m-system in L and $p \in L$ be a maximal element in the set $\{ y \in L \mid S \cap \{ c \in C(L) \mid c \leq y \} = \varnothing \}$. Then p is a prime element of L.*

PROOF. Assume by contradiction that S is an m-system, that $p \in L$ is a maximal element in the set $T := \{ y \in L \mid S \cap \{ c \in C(L) \mid c \leq y \} = \varnothing \}$, but that p is not a prime element. By Lemma 4.3 we can apply Lemma 4.1, so that there exist $x, y \in C(L)$ such that $x \nleq p$, $y \nleq p$, but $xy \leq p$. Thus $x \vee p > p$ and $y \vee p > p$. By the maximality of p, one has that $S \cap \{ c \in C(L) \mid c \leq x \vee p \} \neq \varnothing$ and $S \cap \{ c \in C(L) \mid c \leq y \vee p \} \neq \varnothing$. Hence there exist $x', y' \in S$ such that $x' \leq x \vee p$ and $y' \leq y \vee p$. By the monotonicity condition, $x'y' \leq (x \vee p)(y \vee p)$, hence $x'y' \leq xy \vee py \vee xp \vee pp$, so $x'y' \leq p$. Since S is an m-system, there exists $z \in S$ such that $z \leq x'y'$. Thus $z \leq p$. Hence $z \in S \cap \{ c \in C(L) \mid c \leq p \}$, which is therefore not empty, so that p is not an element in the set T. This is a contradiction, because p was a maximal element in T. □

LEMMA 4.9. *Let I and V be the mappings defined at the beginning of Section 2, and let x be an element of L. If $x < IV(x)$, then x is not semiprime.*

PROOF. Let x be an element of L. Assume by contradiction that $s := IV(x) > x$ and x is semiprime, so that $y^2 \leq x$ implies $y \leq x$ for every element $y \in L$. From $x < s$ and the fact that L is an algebraic lattice, it follows that there exists an element $c_0 \in C(L)$ such that $c_0 \leq s$ and $c_0 \nleq x$. Then $c_0^2 \nleq x$, so that here exists an element $c_1 \in C(L)$ such that $c_1 \leq c_0^2$ and $c_1 \nleq x$. In this way it is possible to construct a descending chain $c_0 \geq c_1 \geq c_2 \geq \dots$ of elements of $C(L)$ such that $c_n \nleq x$ and $c_{n+1} \leq c_n^2$ for every $n \geq 0$. Set $S := \{ c_n \mid n \geq 0 \}$. The nonempty set S is an m-system in L, because if n, m are natural numbers and ℓ is the greatest between them, then $c_n c_m \geq c_\ell c_\ell \geq c_{\ell+1}$. Notice that x belongs to the set $T := \{ y \in L \mid S \cap \{ c \in C(L) \mid c \leq y \} = \varnothing \}$. It is possible to apply Zorn's Lemma to the nonempty set $U := \{ y \in T \mid x \leq y \}$, because if $\{ u_\lambda \mid \lambda \in \Lambda \}$ is a nonempty chain in U, then $u := \bigvee_{\lambda \in \Lambda} u_\lambda \in L$ is in U, because if $S \cap \{ c \in C(L) \mid c \leq u \} \neq \varnothing$, there exists $c \in S$ such that $c \leq u = \bigvee_{\lambda \in \Lambda} u_\lambda \in L$. But c is compact, hence there exists $\lambda_0 \in \Lambda$ such that $c \leq u_{\lambda_0}$. Thus $S \cap \{ c \in C(L) \mid c \leq u_{\lambda_0} \} \neq \varnothing$, which is a contradiction. This shows that it is possible to apply Zorn's Lemma to U, hence there is a maximal element p in U, which is a maximal element in T, hence p is prime by Proposition 4.8. Now $p \in U$ implies $x \leq p$, so $p \in V(x)$. Therefore $s \leq p$. Since $p \in U$, we know that $S \cap \{ c \in C(L) \mid c \leq p \} = \varnothing$. But $c_0 \leq s \leq p$ and $c_0 \in S$. This contradiction proves the Lemma. □

PROPOSIITON 4.10. *The following conditions are equivalent for an element t of L:*

(a) *$t = \bigwedge P$ for some set P of prime elements of L.*
(b) *t is semiprime.*

PROOF. Suppose that (a) holds, $x \in L$ and $x^2 \leq t$. Then $x^2 \leq p$ for every $p \in P$, so that $x \leq p$. It follows that $x \leq t$, as desired. This proves that t is semiprime.

Conversely, suppose that (b) holds. Let $P := V(x)$ be the set of all prime elements p of L such that $t \leq p$. By Lemma 4.9, we get from (b) that $t := \bigwedge P$. □

The *semiprime radical* of L is the meet of all prime elements of L. The multiplicative lattice L is *semisimple* if its semiprime radical is 0. The complete distributive lattice $\mathrm{Rad}(L)$ introduced in Section 2 consists of all semiprime elements of L.

THEOREM 4.11. *Let $(L, \vee, \cdot)$ be an algebraic multiplicative lattice in which m-distributivity holds. The following conditions are equivalent:*

(a) *1 is the unique semiprime element of L.*
(b) *For every $x \in L$, $x \neq 1$, there exists $y \in L$ such that $y > x$ and $y^2 \leq x$.*
(c) *There exists a strictly ascending chain*

$$0 := x_0 \leq x_1 \leq x_2 \leq \cdots \leq x_\omega \leq x_{\omega+1} \leq \cdots \leq x_\alpha := 1$$

in L indexed in the ordinal numbers less or equal to α for some ordinal α, such that $x_{\beta+1}^2 \leq x_\beta$ for every ordinal $\beta < \alpha$ and $x_\gamma = \bigvee_{\beta<\gamma} x_\beta$ for every limit ordinal $\gamma \leq \alpha$.
(d) *The lattice L has no prime elements, that is,* $\mathrm{Spec}(L) = \varnothing$.
(e) *The semiprime radical of L is 1.*
(f) *Every m-system of L contains 0.*

PROOF. (a) $\Rightarrow$ (b) Suppose that (a) holds. Let x be an element of L, $x \neq 1$. By (a), x is not semiprime. Hence there exists $z \in L$ such that $z^2 \leq x$ and $z \nleq x$. Then $y := z \vee x > x$ is such that $y^2 = z^2 \vee zx \vee xz \vee x^2 \leq z^2 \vee x = x$. This proves (b).

(b) $\Rightarrow$ (c) Suppose that (b) holds. The constructions of the strictly ascending chain of elements x_β of L is by transfinite induction. Set $x_0 := 0$. Assume that x_β has been defined. If $x_\beta = 1$, then β is the required ordinal α, and the construction stops. If $x_\beta < 1$, by (b) there exists $x_{\beta+1} \in L$ such that $x_{\beta+1} > x_\beta$ and $x_{\beta+1}^2 \leq x_\beta$. Finally, if γ is a limit ordinal, set $x_\gamma := \bigvee_{\beta<\gamma} x_\beta$.

(c) $\Rightarrow$ (d) Transfinite induction on the ordinal number α. If $\alpha = 0$, then $L = \{0\}$, and L has no prime elements.

Notice that, for every $x \in L$, the closed interval $[0, x] := \{y \in L \mid y \leq x\}$ is a multiplicative sublattice of L, because $[0, x]$ is a complete sublattice, which is closed under multiplication because if $y, y' \in [0, x]$, then $yy' \leq y \wedge y' \leq x \wedge x = x$. (But notice that the greatest element x of $[0, x]$ is in general different from the greatest element 1 of L.)

Suppose $p \in L$ is a prime element of L. There are two cases: α limit or α nonlimit.

If α is a limit ordinal, then $1 = \bigvee_{\beta<\alpha} x_\beta$. Since $p < 1$, there exists $\beta_0 < \alpha$ such that $x_{\beta_0} \nleq p$. Let $\beta_1 < \alpha$ be the smallest ordinal such that $x_{\beta_1} \nleq p$. Then $x_\gamma \leq p$ for every $\gamma < \beta_1$. Notice that β_1 must be necessarily a nonlimit ordinal, so $\beta_1 = \beta_2 + 1$ for some ordinal β_2. Thus $x_{\beta_2} \leq p$ and $x_{\beta_2+1} \nleq p$. But $x_{\beta_2+1}^2 \leq x_{\beta_2} \leq p$ implies $x_{\beta_2+1} \leq p$, a contradiction. This proves that α must be a nonlimit ordinal, $\alpha = \alpha_0 + 1$ say. Then $x_{\alpha_0} \nleq p$, otherwise $1 \cdot 1 = x_{\alpha_0+1} \cdot x_{\alpha_0+1} \leq x_{\alpha_0} \leq p$, hence $1 \leq p$, contradiction. Thus $x_{\alpha_0} \wedge p < x_{\alpha_0}$, and the element $x_{\alpha_0} \wedge p$ of the interval $[0, x_{\alpha_0}]$ is prime in $[0, x_{\alpha_0}]$, because if $a, b \in [0, x_{\alpha_0}]$ and $ab \leq p$, then either $a \leq p$ or $b \leq p$, so that either $a \leq x_{\alpha_0} \wedge p$ or $b \leq x_{\alpha_0} \wedge p$. This contradicts the inductive hypothesis.

(d) $\Rightarrow$ (a) Follows immediately from the characterization of semiprime elements as intersections of primes (Proposition 4.10).

(a) $\Rightarrow$ (e) Follows immediately from the definition of semiprime radical as the meet of all prime elements.

(e) $\Rightarrow$ (f) Suppose that there exists an m-system S in L such that $0 \notin S$. We will apply Proposition 4.8 showing that the set $T := \{ y \in L \mid S \cap \{ c \in C(L) \mid c \leq y \} = \varnothing \}$ has a maximal element, hence L has a prime element, which contradicts (e). Now $0 \in T$ because $0 \notin S$. Hence it suffices to apply Zorn's Lemma, that is, it suffices to show that T is inductive. The proof of this is exactly the same as the proof of Lemma 4.9.

(f) $\Rightarrow$ (d) If p is a prime element of L, then by Lemma 4.2 the set $S_p := \{ c \in C(L) \mid c \nleq p \}$ is an m-system in L, and $0 \notin S_p$. □

We say that a multiplicative lattice L is *hyperabelian* if it satisfies the equivalent conditions of Theorem 4.11.

4.1. Infinite m-distributivity. In this subsection we consider *a multiplicative lattice L in which $(\bigvee X)(\bigvee Y) = \bigvee_{(x,y)\in X\times Y} xy$ for any two subsets $X, Y \subseteq L$ (infinite m-distributivity both on the right and on the left).*

For such a multiplicative lattice L, it is possible to consider, for any element $x \in L$, the *right annihilator* $\mathrm{r.ann}_L(x)$ of x. It is the join of the set A of all elements $y \in L$ such that $xy = 0$. Clearly, $\mathrm{r.ann}_L(x)$ is the greatest element in this set A. Similarly for the *left annihilator* $\mathrm{l.ann}_L(x) := \bigvee\{ y \in L \mid yx = 0 \}$ of x. For instance, if $L := \mathcal{N}(G)$ is the multiplicative lattice of all normal subgroups of a group G, then, for every $N \in \mathcal{N}(G)$, $\mathrm{r.ann}_L(N) = \mathrm{l.ann}_L(N)$ is the centralizer of N in L. It is also possible to define the *right center* $rZ(x) := x \wedge \mathrm{r.ann}_L(x)$ and the *left center* $lZ(x) := x \wedge \mathrm{l.ann}_L(x)$ of any element $x \in L$.

The *left upper central series* (or *left ascending central series*) of the lattice L is the increasing sequence $0 := z_0 \leq z_1 \leq \dots \leq z_\omega \leq \dots$ of elements of L, indexed in the ordinal numbers, where, for every ordinal γ, $z_{\gamma+1}$ is the greatest of the elements $z \in L$ such that $z \cdot 1 \leq z_\gamma$. If γ is a limit ordinal, then z_γ is the join of the set of all the elements z_β with $\beta \leq \gamma$. The *left hypercenter* of L is the largest element of the left upper central series. The lattice L is *left hypercentral* if the left upper central series reaches 1.

REMARK 4.12. Notice that if L is a multiplicative lattice and $h \in L$, then the interval $[0,h]_L := \{ x \in L \mid x \leq h \}$ in L is a multiplicative sublattice of L. Clearly, 0 is a prime element of $[0,h]_L$ if and only if for every $a, b \in L$, $a \leq h$, $b \leq h, ab = 0 \Rightarrow a = 0$ or $b = 0$.

Let us prove that [**14**, Lemma 3.1] extends from groups to multiplicative lattices with infinite distributivity.

LEMMA 4.13. *Let L be a multiplicative lattice with infinite m-distributivity, $h \in L$, and $n \in [0,h]_L$. If 0 is a prime element of $[0,n]_L$, then $\mathrm{l.ann}_{[0,h]}(n)$ coincides with $\mathrm{r.ann}_{[0,h]}(n)$ and is the unique element $p \in \mathrm{Spec}([0,h]_L)$ such that $p \wedge n = 0$.*

PROOF. Consider the property "$p \wedge n = 0$" of an element p of L.

Step 1: $\mathrm{r.ann}_{[0,h]}(n)$ *has the property.*

In fact,

$$(\mathrm{r.ann}_{[0,h]}(n) \wedge n)^2 = (\mathrm{r.ann}_{[0,h]}(n) \wedge n)(\mathrm{r.ann}_{[0,h]}(n) \wedge n) \leq n \cdot \mathrm{r.ann}_{[0,h]}(n) = 0.$$

Since 0 is prime in $[0,n]_L$, it follows that $\mathrm{r.ann}_{[0,h]}(n) \wedge n = 0$.

Step 2: $\mathrm{r.ann}_{[0,h]}(n)$ *is prime in* $[0,h]_L$.

First of all notice that $\mathrm{r.ann}_{[0,h]}(n) < h$, otherwise $nh = 0$, so that $n^2 = 0$, so $n = 0$, hence 0 is not prime in $[0,n]_L$, a contradiction.

Suppose $a, b \in L$, $a, b \leq h$, and $ab \leq \mathrm{r.ann}_{[0,h]}(n)$. Then $(a \wedge n)(b \wedge n) \leq ab \wedge n \leq \mathrm{r.ann}_{[0,h]}(n) \wedge n = 0$ by Step 1. Since 0 is prime in $[0,n]_L$, it follows that either $a \wedge n = 0$ or $b \wedge n = 0$. If, for instance, $a \wedge n = 0$, then $na = 0$, so $a \leq \mathrm{r.ann}_{[0,h]}(n)$. Similarly if $b \wedge n = 0$.

Step 3: $\mathrm{r.ann}_{[0,h]}(n)$ *is the unique prime element of* $[0,h]_L$ *with the property.*

Let p be a prime element of $[0,h]_L$ with the property, i.e., such that $p \wedge n = 0$. Then $np = 0$, so $p \leq \mathrm{r.ann}_{[0,h]}(n)$. From $n \cdot \mathrm{r.ann}_{[0,h]}(n) = 0 \leq p$, it follows that either $n \leq p$ or $\mathrm{r.ann}_{[0,h]}(n) \leq p$. If $n \leq p$, then $n = p \wedge n = 0$, so 0 is not prime in $[0,n]_L$, a contradiction. Thus $\mathrm{r.ann}_{[0,h]}(n) \leq p$. Since we have already seen that $p \leq \mathrm{r.ann}_{[0,h]}(n)$, we get that $p = \mathrm{r.ann}_{[0,h]}(n)$, which concludes Step 3.

Step 4: $\mathrm{r.ann}_{[0,h]}(n) = \mathrm{l.ann}_{[0,h]}(n)$.

We have seen that $\mathrm{r.ann}_{[0,h]}(n)$ is the unique element $p \in \mathrm{Spec}([0,h]_L)$ such that $p \wedge n = 0$. This is a right/left symmetric property, so that $\mathrm{l.ann}_{[0,h]}(n)$ is also the unique element $p \in \mathrm{Spec}([0,h]_L)$ such that $p \wedge n = 0$. This concludes the proof of the Lemma. □

5. Some examples.

5.1. Groups. We have already mentioned above the example of groups. For any fixed group G, consider the complete modular lattice $\mathcal{N}(G)$ of all normal subgroups of G, i.e., in the language of Universal Algebra, the lattice of all congruences of G. It is possible to give $\mathcal{N}(G)$ a multiplicative lattice structure in a number of ways. The most interesting one is that of defining as a product of $A, B \in \mathcal{N}(G)$ the commutator $[A, B]$, which is a normal subgroup of G contained in A and B. This multiplication on $\mathcal{N}(G)$ is commutative, but not associative. But it is also possible to define as product of $A, B \in \mathcal{N}(G)$ the intersection $A \cap B$, or the trivial subgroup of G with one element. In this way we get three different structures of multiplicative lattice on $\mathcal{N}(G)$. Of course, it is not possible to take as a product in the multiplicative lattice $\mathcal{N}(G)$ the product AB of two normal subgroups $A, B \in \mathcal{N}(G)$, because it is not true that $AB \subseteq A \cap B$: in fact, AB is the join $A \vee B$ in the complete lattice $\mathcal{N}(G)$.

Notice that groups behave so similarly to commutative rings without identity because in the multiplicative lattice $\mathcal{N}(G)$ of any group G, the multiplication $[-,-]$ of $\mathcal{N}(G)$ is always a commutative operation, though it is not associative.

5.2. *G*-groups. Another immediate application of our theory is to G-groups, which we studied in [**1**,**4**,**5**,**15**,**16**]. The notion of G-group H is classical. Sometimes G is called an *operator group* on H [**31**, Definition 8.1]. For any two groups G and H we say that H is a *(left) G-group* if there is a fixed group homomorphism $\alpha: G \to \mathrm{Aut}(H)$. That is, a G-group is a pair (H, α), where H is a group and $\alpha: G \to \mathrm{Aut}(H)$ is a group morphism. Equivalently, a G-group can be defined as a group H endowed with a mapping $\cdot: G \times H \to H$, $(g, h) \mapsto gh$, called *left scalar multiplication*, such that

(a) $g(hh') = (gh)(gh')$,
(b) $(gg')h = g(g'h)$, and
(c) $1_G h = h$

for every $g, g' \in G$ and every $h, h' \in H$.

The reader will notice the analogy between the category G-Grp of G-groups and the category R-Mod of all left modules over a fixed ring R. The category R-Mod has as objects the pairs (M,λ), where M is an additive abelian group and $\lambda\colon R \to \mathrm{End}(M)$ is a ring morphism. In the category G-Grp the morphisms $f\colon(H,\varphi) \to (H',\varphi')$ are the group homomorphisms $f\colon H \to H'$ such that $f(gh) = gf(h)$ for every $g \in G$, $h \in H$, exactly as in the category R-Mod.

We say that H is an *abelian G-group* if H is a G-group and H is an abelian group. Most of the results in [**14**] hold not only for groups, but, more generally, for arbitrary G-groups.

The most important example of G-group is the regular left G-group, that is, the group G on which G acts via the inner automorphisms of G. More precisely, for any fixed group G, let $\alpha_g\colon G \to G$, $\alpha_g\colon h \mapsto ghg^{-1}$, be the *inner automorphism* of G determined by g, for every $g \in G$. There is a canonical group morphism $\alpha\colon G \to \mathrm{Aut}(G)$, $g \mapsto \alpha_g$, which makes G a G-group. It is called the *regular* left G-group, similarly to the left R-module ${}_RR$, which is called the regular left R-module. Other important examples of G-groups are the normal subgroups N of G (they are the G-subgroups of the regular left G-group G), and the factor groups G/N (=homomorphic images of the regular left G-group G). Notice that the lattice of all subgroups of G is the lattice of all $\{1\}$-subgroups of the $\{1\}$-group G, the lattice of all normal subgroups of G is the lattice of all G-subgroups of the regular left G-group G, the lattice of all characteristic subgroups of G is the lattice of all $\mathrm{Aut}(G)$-subgroups of the $\mathrm{Aut}(G)$-group G.

For any fixed G-group H, the associated multiplicative lattice is the lattice of all normal G-subgroups of H, i.e., the normal subgroups of H that are closed under the action of the elements of G. Multiplication in this lattice is given by the commutator subgroup.

5.3. Rings without identity. For rngs (associative rings possibly without identity) the associated multiplicative lattice is the same as for rings with identity. Since now we will be dealing with rngs (associative rings without identity), and their theory goes back to the times of Jacobson and Kaplansky, let us briefly recall the theory of Jacobson radical for rings not necessarily with identity. Let Rng denote the category of all rings, with or without identity, in which morphisms are the mappings that preserve addition and multiplication, and let Ring be the category of all rings with identity, in which morphisms are the mappings that preserve addition, multiplication and the identities. There is a further third category involved. It is the category Ring_a of all rings with augmentation, whose objects are the pairs (S,ε_S), where S is an object in Ring and $\varepsilon_S\colon S \to \mathbb{Z}$ is a morphism in Ring. Morphisms $f\colon(S,\varepsilon_S) \to (S',\varepsilon_{S'})$ in Ring_a are the morphisms $f\colon S \to S'$ in Ring such that $\varepsilon_{S'}f = \varepsilon_S$:

$$\begin{array}{ccccc} \mathbb{Z} & \longrightarrow & S & \xrightarrow{\varepsilon_S} & \mathbb{Z} \\ \| & & \downarrow f & & \| \\ \mathbb{Z} & \longrightarrow & S' & \xrightarrow{\varepsilon_{S'}} & \mathbb{Z}. \end{array}$$

There is an equivalence of categories $F\colon \mathsf{Rng} \to \mathsf{Ring}_a$ that associates with any object R of Rng the abelian group direct sum $R_+ := R \oplus \mathbb{Z}$ with multiplication defined by $(r,z)\cdot(r',z') = (rr'+zr'+z'r, zz')$ for every $(r,z),(r',z') \in R_+$. This ring R_+ is a ring with augmentation. Notice that there is a short exact sequence $0 \to R \to R_+ \to \mathbb{Z} \to 0$ in the category Rng, which is a pointed category with zero object the zero ring with one element. In particular, R is an ideal of the ring with identity R_+. The quasi-inverse of the category equivalence F associates with each object (S, ε_S) of Ring_a the kernel of ε_S.

The construction of the ring R_+ goes back to J. L. Dorroh ([**10**], but also see [**27**]). The right (left, two-sided) ideals of R are exactly the right (left, two-sided) ideals of R_+ contained in $R \oplus 0$. Conversely, the right (left, two-sided) ideals of R_+ are all of the form $K = \{ (-r, a) \mid r \in J,\ a \in A,\ \varphi(r) = a + Z \}$, where $Z \leq A$ are ideals of the ring of integers $\mathbb{Z}$, J is a right (left, two-sided, resp.) ideal of R, and $\varphi: J \to A/Z$ is a morphism in the category Rng.

Given any left module over a rng R, that is, a pair (M, λ), where M is an additive abelian group and $\lambda: R \to \mathrm{End}(M)$ is a morphism in Rng, then λ extends uniquely to a morphism $\lambda_+: R_+ \to \mathrm{End}(M)$ in Ring. This assignement defines a category isomorphism between the categories Mod-R and Mod-R_+.

A similar situation occurs in several different settings. For instance, let G be a group. Then we can associate with every G-group (H, φ) the semidirect product $H \rtimes G$, getting an equivalence between the category G-Grp of G-groups and the category G-Sdp of all G-semidirect products. The objects of G-**Sdp** are the triples (P, α, β), where P is a group and $\alpha: G \to P$, $\beta: P \to G$ are morphisms such that the composite mapping $\beta\alpha$ is the identity automorphism id_G of the group G. The morphisms from (P, α, β) to (P', α', β') in G-**Sdp** are all group morphisms $a: P \to P'$ which make the diagram

$$\begin{array}{ccccc} G & \xrightarrow{\alpha} & P & \xrightarrow{\beta} & G \\ \| & & \downarrow a & & \| \\ G & \xrightarrow{\alpha'} & P' & \xrightarrow{\beta'} & G \end{array}$$

commute. Here the vertical arrows on the right and on the left are the identity morphism id_G of G. Thus the diagram commutes if and only if $a\alpha = \alpha'$ and $\beta' a = \beta$. The category G-**Sdp** is the category of pointed objects over the G-group G (Borceux [**8**, page 28]; also see [**5**]). Let (H, φ) be a G-group. Let P be the semidirect product $H \rtimes G$, that is, the cartesian product $H \times G$ with the operation defined by $(h_1, g_1)(h_2, g_2) = (h_1 \varphi(g_1)(h_2), g_1 g_2)$ Let $\alpha: G \to P$ and let $\beta: P \to G$ be defined by $\alpha(g) = (1, g)$ and $\beta(h, g) = g$. Then α and β are group morphisms and $\beta\alpha = id_G$, so that (P, α, β) is an object of G-**Sdp**.

If (H, φ) and (H', φ') are G-groups, (P, α, β) and (P', α', β') are the corresponding objects in G-**Sdp** and $f: (H, \varphi) \to (H', \varphi')$ is a G-group morphism, define $\widetilde{f}: P = H \rtimes G \to P' = H' \rtimes G$ by $\widetilde{f}(h, g) = (f(h), g)$ for every $(h, g) \in P = H \rtimes G$. Then $\widetilde{f}$ is a morphism in the category G-**Sdp**. The assignments $(H, \varphi) \mapsto (P, \alpha, \beta)$ and $f \mapsto \widetilde{f}$ define a category equivalence $F: G$-$\mathsf{Grp} \to G$-**Sdp** (see [**8**, Proposition 5.7] or [**5**, 4.1]).

An even clearer analogy is with the category of k-algebras. Let k be a commutative ring with identity. Recall that a k-algebra R is a ring that is also a k-module R_k, and in which $(rs)\lambda = (r\lambda)s = r(s\lambda)$ for every $r, s \in R$ and $\lambda \in k$. When R is a ring with identity, this is equivalent to the existence of a morphism $\alpha: k \to R$ in Ring with $\alpha(k)$ contained in the center of R. An augmented k-algebra is a triple (R, α, β), where R is a ring with identity and $\alpha: k \to R$, $\beta: R \to k$ are morphisms in Ring with $\alpha(k) \subseteq Z(R)$ and $\beta\alpha = id_k$, the identity automorphism of the ring k.

For any k-algebra R (with or without identity) it is clear how to give $R_+ := R \oplus k$ a structure of augmented k-algebra with identity. There is a short exact sequence $0 \to R \to R_+ \to k \to 0$ in the pointed category of all k-algebras.

Let us go back to rngs. On any rng R it is possible to define the operation $x \circ y = x + y + xy$ [**23**, p. 85]. Then $(R, \circ)$ is a multiplicative monoid with the additive zero 0_R of the ring R as its identity. The Jacobson radical $J(R)$ of R is the set of all $x \in R$ such that yx is left invertible in $(R, \circ)$ for all $y \in R$ (equivalently, the set of all $x \in R$ such that xy is right invertible in $(R, \circ)$ for all $y \in R$). The ideal $J(R)$ is contained in the group of units of the monoid $(R, \circ)$. For any rng R, one has that $J(R_+) = J(R) \oplus 0$.

Clearly, the mapping $\psi: S \to S$, $\psi: x \mapsto 1 + x$, is a monoid isomorphism of the monoid $(S, \circ)$ onto the monoid $(S, \cdot)$ for every ring S. In particular, $\varphi(J(S)) \subseteq \varphi(U(S, \circ)) = U(S, \cdot)$ for any ring S, and $\varphi(J(R)) = \varphi(J(R_+)) \subseteq \varphi(U(R_+, \circ)) = U(R_+, \cdot)$. (More precisely, $J(S)$ is the largest of all the ideals I of a ring S such that $\varphi(I) \subseteq U(S, \cdot)$.) This will be useful in the next subsection, devoted to braces.

5.4. Braces. Following Drinfeld [**11**], a *set-theoretic solution of the Yang-Baxter equation* is a pair (X, r), where X is a set and $r: X \times X \to X \times X$ is a bijection such that

$$(r \times \mathrm{id})(\mathrm{id} \times r)(r \times \mathrm{id}) = (\mathrm{id} \times r)(r \times \mathrm{id})(\mathrm{id} \times r).$$

Rump [**29**] proved that there is a connection between radical rings and set-theoretic solutions of the Yang-Baxter equation. A *radical ring* is a rng R for which $R = J(R)$. Hence a rng R is a radical ring if and only if the monoid $(R, \circ)$ is a group, and for such a ring R, the mapping

$$r: R \times R \to R \times R, \quad r(x, y) = (xy + y, (xy + y)' \circ x \circ y),$$

where z' denotes the inverse of z with respect to the operation $\circ$, is a non- degenerate solution of the Yang-Baxter equation such that $r^2 = \mathrm{id}_{R \times R}$.

Rump [**29**] generalized this to braces, and braces were generalized in 2017 to skew braces by L. Guarnieri and L. Vendramin [**17**]. A *(left) skew brace* is a triple $(A, \circ, *)$, where $(A, \circ)$ and $(A, *)$ are groups (not necessarily abelian) such that

$$a \circ (b * c) = (a \circ b) * a^{-1} * (a \circ c)$$

for every $a, b, c \in A$. Here a^{-1} denotes the inverse of a in the group $(A, *)$. The inverse of a in the group $(A, \circ)$ will be denoted by a'. Notice the similarity between this definition and the definition of a left near-ring with identity $(N, *, \cdot)$, where $(N, *)$ is a group (not necessarily abelian), $(N, \cdot)$ is a monoid, and $a(b * c) = ab * ac$. From the next Lemma we will see the close relation between left skew braces and G-groups.

For every skew brace $(A, \circ, *)$, the mapping

$$r: A \times A \to A \times A, \quad r(x, y) = (x^{-1} * (x \circ y), (x^{-1} * (x \circ y))' \circ x \circ y),$$

is a non-degenerate set-theoretic solution of the Yang-Baxter equation ([**17**, Theorem 3.1] and [**21**, p. 96]).

LEMMA 5.1 ([**6**]). *Let A be a skew brace. Then $\lambda: (A, \circ) \to \mathrm{Aut}(A, *)$, given by $\lambda: a \mapsto \lambda_a$, where $\lambda_a(b) = a^{-1} * (a \circ b)$, is a well-defined group homomorphism.*

In other words, for any skew brace $(A, \circ, *)$, we have that $(A, *)$ is an $(A, \circ)$-group with respect to the action $\lambda: (A, \circ) \to \mathrm{Aut}(A, *)$ described in the statement of Lemma 5.1. Conversely, suppose that a set A has two group structures $(A, \circ)$ and $(A, *)$ and that $(A, *)$ is an $(A, \circ)$-group with respect to the action $\lambda: (A, \circ) \to \mathrm{Aut}(A, *)$, defined by $\lambda: a \mapsto \lambda_a$, where $\lambda_a(b) = a^{-1} * (a \circ b)$. Then the fact that each λ_a is an automorphism yields that $\lambda_a(b * c) = \lambda_a(b) * \lambda_a(c)$, i.e., $a^{-1} * (a \circ (b * c)) = a^{-1} * (a \circ b) * a^{-1} * (a \circ c)$, from which $a \circ (b * c) = (a \circ b) * a^{-1} * (a \circ c)$. Hence braces are exactly those particular G-groups (H, λ) for which $G = H$ as sets, and the action λ is defined by $\lambda_a(b) = a^{-1} * (a \circ b)$.

The semidirect product corresponding to such an $(A, \circ)$-group $(A, *)$ is the group $P := (A, *) \ltimes (A, \circ)$, i.e., the cartesian product $P := A \times A$ with group operation defined by

$$(a_1, a_2) \cdot (b_1, b_2) = (a_1 * a_2^{-1} * (a_2 \circ b_1), a_2 \circ b_2). \tag{5.1}$$

Conversely, given two group structures $(A, \circ)$ and $(A, *)$ on the same set A such that $P := A \times A$ with the operation as in (5.1) is a group, then $(A, \circ, *)$ is a left skew brace. This proves that left skew braces are particular groups: they are the groups G for which there is a normal subgroup $A \trianglelefteq G$ and a bijection $\alpha: G/A \to A$ with suitable properties. (The two group structures on A are the original one as a subgroup of G and the one brought from G/A via the bijective 1-cocycle α.)

The original definition of brace given by Rump was that with $(A, *)$ an additive abelian group. Henca a brace is an abelian group $(A, +)$ that is also a group $(A, \circ)$ with a monoid morphism $\lambda': (A, \circ) \to \mathrm{End}(A) \times \mathbb{Z}$ whose image is contained in $\mathrm{End}(A) \times \{1\}$. Notice the similarity with the definition of k-algebra: Given a *commutative* ring k, a k-algebra is a ring A with a ring morphism $\lambda: k \to A$ whose image is contained in the center A. Viewing a left skew brace $(A, \circ, *)$ as the corresponding semidirect product $P := (A, *) \ltimes (A, \circ)$ with the group operation (5.1) suggests us that the correct notion of action of a left skew brace $(A, \circ, *)$ over a group H is that of P-group H. Cf. [**28**].

Notice that the group $U(E \times \mathbb{Z})$ of units of any ring $E \times \mathbb{Z}$ with identity and with an augmentation $\mathbb{Z} \to E \times \mathbb{Z} \to \mathbb{Z}$ has a subgroup $U(E \times \mathbb{Z}) \cap (E \times \{1\})$ of index 2. When E already has an identity, then there is a unique morphism $u: E \times \mathbb{Z} \to E$ of rings with identity that extends the identity morphism $E \to E$. Thus, for a brace A, one has the composite monoid morphism $\lambda := u \circ \lambda': A \to \mathrm{End}(A)$ (cf. Lemma 5.1).

For a skew brace $(A, \circ, *)$, and more generally for any group $(A, *)$, we don't have a ring $\mathrm{End}(A, +)$ anymore, but a nearring $M(A, *)$, consisting of all mappings $A \to A$. Addition is defined by $(f+g)(a) = f(a)*g(a)$, multiplication is composition of mappings, the zero element is the zero map, i.e., the mapping which takes every element of A to the identity element of $(A, *)$, the additive inverse $-f$ of $f \in M(G)$ is $(-f)(a) = (f(a))^{-1}$ (inverse of $f(a)$ in the group $(A, *)$) for all $a \in A$, the identity of the nearring is the identity mapping $\iota_A: A \to A$. For the nearring $M(A, *)$ it is also possible to construct its Dorroh extension as in [**32**], because $M(A, *)$ trivially satisfies condition (BC): the Dorroh extension is the cartesian product $M(A, *) \times \mathbb{Z}$ with addition and multiplication defined by

$$(f, a) + (g, b) = (f + a\iota_A + g - a\iota_A, a + b)$$

$$(f, a)(g, b) = ((f + a\iota_A)(g + b\iota_A) - (ab)\iota_A, ab).$$

As in [**32**], it is easy to verify that $M(A, *) \times \mathbb{Z}$ is a nearring with additive identity $(0, 0)$, multiplicative identity $(0, 1)$, and the additive inverse of (f, a) is $(-a\iota_A -$

$f + a\iota_A, -a)$. Then $\mathbb{Z} \to M(A,*) \times \mathbb{Z} \to \mathbb{Z}$ is an augmented nearring with identity. The kernel of the augmentation $M(A,*) \times \mathbb{Z} \to \mathbb{Z}$, $(f,a) \mapsto a$, is the two-sided ideal $\{ (f,0) \mid f \in M(A,*) \}$ of $M(A,*) \times \mathbb{Z}$, and is a nearring isomorphic to $M(A,*)$. There is a surjective nearring morphism $M(A,*) \times \mathbb{Z} \to M(A,*)$, $(f,a) \mapsto f + a\iota$. We will continue in this direction in a next paper, now we go back to the presentation of a skew brace as a semidirect product P of groups, in order to introduce left skew brace morphisms and ideals of braces.

A *morphism of skew braces* is a mapping $f\colon A \to A'$ such that $f(a \circ b) = f(a) \circ f(b)$ and $f(a * b) = f(a) * f(b)$ for every $a, b \in A$. Every such morphism $f\colon A \to A'$ induces a corresponding group morphism $f \times f\colon P = A \ltimes A \to P' = A' \ltimes A'$, $(f \times f)(a,b) = (f(a), f(b))$. Since $(A, \circ)$ and $(A,*)$ have the same identity, the group morphisms $f\colon (A,\circ) \to (A,\circ)$ and $f\colon (A,*) \to (A,*)$ have the same kernel I, which is a normal subgroup both in $(A,\circ)$ and in $(A,*)$. The kernel of $f \times f\colon P = A \ltimes A \to P' = A' \ltimes A'$ is $I \times I$. The assignements $(A, \circ, *) \mapsto P := (A,*) \ltimes (A,\circ)$ and $f \mapsto f \times f$ define a faithful functor of the category of left skew braces into the category of groups. This allows us to view left skew braces as groups and suggests us the definition of kernel of a left skew brace morphism $f\colon A \to A'$ and of factor skew brace, getting a left skew brace isomorphism $(A/I, *) \ltimes (A/I, \circ) \to (f \times f)(P)$, as follows. The *kernel* of a left skew brace morphism $f\colon A \to A'$ is the inverse image $f^{-1}(1_A)$ of the identity 1_A of $(A,\circ)$, (which coincides with the identity of $(A,*)$).

An *ideal* of a skew brace A is a normal subgroup I both of $(A,\circ)$ and $(A,*)$ such that $I \circ a = I * a$ for every $a \in I$. Notice that this implies that the factor groups of $(A,\circ)$ and $(A,*)$ modulo I coincide as sets. Also notice that $I \circ a = I * a$ for every $a \in A$ if and only if, for every $a, b \in A$, one has $a * b^{-1} \in I$ if and only if $a \circ b' \in I$. Equivalently, if and only if $(x * a) \circ a' \in I$ and $(x \circ a) * a^{-1} \in I$ for every $x \in I$ and every $a \in A$. The semidirect product corresponding to the quotient left skew brace A/I is $(A/I, *) \ltimes (A/I, \circ)$.

As far as the representation of a left skew brace as a G-group is concerned, i.e., as a $(A,\circ)$-group $(A,*)$, notice that if I is an ideal of the left skew brace A, then I is a normal $(A,\circ)$-subgroup of $(A,*)$, so that the group morphism $\lambda\colon (A,\circ) \to \operatorname{Aut}(A,*)$ induces a group morphism $(A,\circ) \to \operatorname{Aut}(A/I,*)$. But now I is contained in the kernel of this morphism $(A,\circ) \to \operatorname{Aut}(A/I,*)$, which therefore induces a group morphism $\overline{\lambda}\colon (A/I,\circ) \to \operatorname{Aut}(A/I,*)$.

For any fixed left skew brace A, there is a canonical one-to-one correspondence between the set of all ideals of the left skew brace A and the set of all equivalence relations on the set A compatible with both the operations $\circ$ and $*$ (i.e., the equivalence relations $\sim$ on the set A such that, for every $a, b, c, d \in A$, $a \sim b$ and $c \sim d$ imply $a \circ c \sim b \circ d$ and $a * c \sim b * d$. Given any ideal I of the left skew brace A, the corresponding equivalence relation on A is $\sim_I$, defined, for every $a, b \in A$, by $a \sim_I b$ if $a^{-1} * b \in I$ (equivalently, if $a' \circ b \in I$; notice that the equivalence classes of A modulo $\sim_I$ are the cosets $a * I = I * a = a \circ I = I \circ a$). Conversely, if $\sim$ is an equivalence relation on A compatible with both the operations $\circ$ and $*$, then the corresponding ideal of A is the equivalence class of 1_A modulo $\sim$.

Any intersection of ideals is an ideal, so that we get a complete lattice $\mathcal{I}(A)$ of ideals for any skew brace A. In particular, every subset X of A generates an ideal, the intersection of the ideals that contain X. It can be "constructively" described as follows. Given a subset X of a skew brace A, consider the increasing sequence X_n, $n \geq 0$, of subsets of A where $X_0 := X$, X_{n+1} is the normal closure of X_n in $(A, *)$ if $n \equiv 0 \pmod 3$, X_{n+1} is the normal closure of X_n in $(A, \circ)$ if $n \equiv 1$ (mod 3, $X_{n+1} := \bigcup_{a \in A} \lambda_a(X_n)$ if $n \equiv 2 \pmod 3$. The ideal of A generated by X is $\bigcup_{n \geq 0} X_n$.

As multiplication in the complete lattice of all ideals of a left skew brace A define as product of two ideals I and J of A the ideal of the skew brace A generated by $\pi_1([I \times I, J \times J]_P) \cup \pi_2([I \times I, J \times J]_P)$, where $[I \times I, J \times J]_P$ is the commutator in P of its normal subgroups $I \times I$ and $J \times J$, and the mappings $\pi_i \colon A \times A \to A$, $i = 1, 2$, are the two canonical projections. With this operation, $\mathcal{I}(A)$ becomes a multiplicative lattice. (Notice: these notations and notions differ from similar previous notations and notions in the literature.) For instance, in our terminology a left skew brace is *abelian* if and only if the two operations $\circ$ and $*$ coincide and are commutative. Similarly, the *socle*

$$\operatorname{soc}(A) := \{\, a \in A \mid ba = 0 \ \text{ for every } b \in A \,\} \tag{4}$$

of a brace A as defined in [**29**] is the *right center* in our terminology. This has been independently observed by Vsevolod Gubarev.)

The notions of prime ideal, solvable, nilpotent, etc., skew braces now become natural. This must be compared with previous notions by Rump (left and right nilpotent braces and radical chains [**29**]), Rowen (2017, see the first paragraph of Section 4 in [**21**]), in [**7**], and [**21**], and will be described in detail in a subsequent paper. But notice that most of these previous notions differ from ours. For instance, the product of two ideals of a left brace A in [**29**] is not an ideal of the left brace, but only a subgroup of $(A, *)$, while in our theory the product of two ideals is always necessarily an ideal. Also, the ideal generated by the product of two ideals in the sense of Rump [**29**] can be different from the product of the two ideals in our sense. For instance, in [**29**, Example 2] one has that $A^{(3)} = \{0, 111\}$ is not an ideal, while in our notation $A^{(3)} = A^2 = A^{(4)}$ is always necessarily an ideal, and A is not right nilpotent and not left nilpotent in our sense, but is left nilpotent in the sense of Rump because $A^{(4)} = 0$.

Acknowledgments

This paper is the result of several conversations and e-mail messages I have exchanged with Professor George Janelidze for the past two years, in particular during the writing of the paper [**13**]. I am extremely grateful to him. I have learnt a lot of things from him.

References

[1] Meltem Altun-Özarslan and Alberto Facchini, *The Krull-Schmidt-Remak-Azumaya Theorem for G-groups*, Rings, modules and codes, Contemp. Math., vol. 727, Amer. Math. Soc., [Providence], RI, [2019] ©2019, pp. 25–38, DOI 10.1090/conm/727/14622. MR3938137

[2] Daniel D. Anderson, *Multiplicative Lattices*, ProQuest LLC, Ann Arbor, MI, 1974. Thesis (Ph.D.)–The University of Chicago. MR2611710

[3] D. D. Anderson, *Abstract commutative ideal theory without chain condition*, Algebra Universalis **6** (1976), no. 2, 131–145, DOI 10.1007/BF02485825. MR419310

[4] María José Arroyo Paniagua and Alberto Facchini, *G-groups and biuniform abelian normal subgroups*, Adv. Group Theory Appl. **2** (2016), 79–111, DOI 10.4399/97888548970146. MR3593016

[5] María José Arroyo Paniagua and Alberto Facchini, *Category of G-groups and its spectral category*, Comm. Algebra **45** (2017), no. 4, 1696–1710, DOI 10.1080/00927872.2016.1222409. MR3576687

[6] David Bachiller, *Extensions, matched products, and simple braces*, J. Pure Appl. Algebra **222** (2018), no. 7, 1670–1691, DOI 10.1016/j.jpaa.2017.07.017. MR3763276

[7] D. Bachiller, F. Cedó, E. Jespers, and J. Okniński, *Asymmetric product of left braces and simplicity; new solutions of the Yang-Baxter equation*, Commun. Contemp. Math. **21** (2019), no. 8, 1850042, 30 pp., DOI 10.1142/S0219199718500426. MR4020748

[8] Francis Borceux, *A survey of semi-abelian categories*, Galois theory, Hopf algebras, and semi-abelian categories, Fields Inst. Commun., vol. 43, Amer. Math. Soc., Providence, RI, 2004, pp. 27–60, DOI 10.1090/fic/043. MR2075580

[9] S. Burris and H. P. Sankappanavar, "A course in universal algebra", Graduate Texts in Mathematics, 78. Springer-Verlag, New York-Berlin, 1981. The "Millennium Edition" is available online at `https://www.math.uwaterloo.ca/~snburris/htdocs/UALG/univ-algebra2012.pdf`

[10] J. L. Dorroh, *Concerning adjunctions to algebras*, Bull. Amer. Math. Soc. **38** (1932), no. 2, 85–88, DOI 10.1090/S0002-9904-1932-05333-2. MR1562332

[11] V. G. Drinfel′d, *On some unsolved problems in quantum group theory*, Quantum groups (Leningrad, 1990), Lecture Notes in Math., vol. 1510, Springer, Berlin, 1992, pp. 1–8, DOI 10.1007/BFb0101175. MR1183474

[12] Samuel Eilenberg and Saunders MacLane, *General theory of natural equivalences*, Trans. Amer. Math. Soc. **58** (1945), 231–294, DOI 10.2307/1990284. MR13131

[13] Alberto Facchini, Carmelo Antonio Finocchiaro, and George Janelidze, *Abstractly constructed prime spectra*, Algebra Universalis **83** (2022), no. 1, Paper No. 8, 38, DOI 10.1007/s00012-021-00764-z. MR4367895

[14] A. Facchini, F. de Giovanni and M. Trombetti, *Spectra of groups*, Algebr Represent Theor (2022). `https://doi.org/10.1007/s10468-022-10138-1`.

[15] Alberto Facchini and Andrea Lucchini, *The Krull-Schmidt theorem holds for finite direct products of biuniform groups*, Algebr. Represent. Theory **21** (2018), no. 2, 309–329, DOI 10.1007/s10468-017-9715-y. MR3780769

[16] Alberto Facchini and Serap Şahinkaya, *Uniqueness of decomposition, factorisations, G-groups and polynomials*, Int. Electron. J. Algebra **24** (2018), 107–128, DOI 10.24330/ieja.440235. MR3828100

[17] L. Guarnieri and L. Vendramin, *Skew braces and the Yang-Baxter equation*, Math. Comp. **86** (2017), no. 307, 2519–2534, DOI 10.1090/mcom/3161. MR3647970

[18] Joachim Hagemann and Christian Herrmann, *A concrete ideal multiplication for algebraic systems and its relation to congruence distributivity*, Arch. Math. (Basel) **32** (1979), no. 3, 234–245, DOI 10.1007/BF01238496. MR541622

[19] J. A. Johnson, *a-adic completions of Noetherian lattice modules*, Fund. Math. **66** (1969/70), 347–373, DOI 10.4064/fm-66-3-347-373. MR263797

[20] E. W. Johnson and J. A. Johnson, *Lattice modules over semi-local Noether lattices*, Fund. Math. **68** (1970), 187–201, DOI 10.4064/fm-68-2-187-201. MR269641

[21] Alexander Konovalov, Agata Smoktunowicz, and Leandro Vendramin, *On skew braces and their ideals*, Exp. Math. **30** (2021), no. 1, 95–104, DOI 10.1080/10586458.2018.1492476. MR4223285

[22] Ralph Freese and Ralph McKenzie, *Commutator theory for congruence modular varieties*, London Mathematical Society Lecture Note Series, vol. 125, Cambridge University Press, Cambridge, 1987. MR909290

[23] Irving Kaplansky, *Fields and rings*, 2nd ed., Chicago Lectures in Mathematics, University of Chicago Press, Chicago, Ill.-London, 1972. MR0349646

[24] Hee Sik Kim and Joseph Neggers, *The semigroups of binary systems and some perspectives*, Bull. Korean Math. Soc. **45** (2008), no. 4, 651–661, DOI 10.4134/BKMS.2008.45.4.651. MR2463141

[25] W. Krull, *Axiomatische Begründung der allgemeinen Idealtheorie*, Sitzber. d. phys.-med. Soc. Erlangen **56** (1924), 47–63.

[26] Sergio R. López-Permouth, Isaac Owusu-Mensah, and Asiyeh Rafieipour, *A monoid structure on the set of all binary operations over a fixed set*, Semigroup Forum **104** (2022), no. 3, 667–688, DOI 10.1007/s00233-022-10280-8. MR4447167

[27] Zachary Mesyan, *The ideals of an ideal extension*, J. Algebra Appl. **9** (2010), no. 3, 407–431, DOI 10.1142/S0219498810003999. MR2659728

[28] Wolfgang Rump, *Modules over braces*, Algebra Discrete Math. **2** (2006), 127–137. MR2320986

[29] Wolfgang Rump, *Braces, radical rings, and the quantum Yang-Baxter equation*, J. Algebra **307** (2007), no. 1, 153–170, DOI 10.1016/j.jalgebra.2006.03.040. MR2278047

[30] Jonathan D. H. Smith, *Mal′cev varieties*, Lecture Notes in Mathematics, Vol. 554, Springer-Verlag, Berlin-New York, 1976. MR0432511

[31] Michio Suzuki, *Group theory. I*, Grundlehren der Mathematischen Wissenschaften [Fundamental Principles of Mathematical Sciences], vol. 247, Springer-Verlag, Berlin-New York, 1982. Translated from the Japanese by the author. MR648772

[32] Stefan Veldsman, *On unital extensions of near-rings and their radicals*, Math. Pannon. **3** (1992), no. 1, 77–81. MR1170539

[33] Morgan Ward, *Residuation in structures over which a multiplication is defined*, Duke Math. J. **3** (1937), no. 4, 627–636, DOI 10.1215/S0012-7094-37-00351-X. MR1546017

[34] Morgan Ward, *Structure residuation*, Ann. of Math. (2) **39** (1938), no. 3, 558–568, DOI 10.2307/1968634. MR1503424

[35] Morgan Ward and R. P. Dilworth, *Residuated lattices*, Trans. Amer. Math. Soc. **45** (1939), no. 3, 335–354, DOI 10.2307/1990008. MR1501995

Dipartimento di Matematica "Tullio Levi-Civita", Università di Padova, 35121 Padova, Italy

Email address: facchini@math.unipd.it

Contemporary Mathematics
Volume **785**, 2023
https://doi.org/10.1090/conm/785/15781

$\mathbb{Z}_2\mathbb{Z}_4$-additive codes as codes over rings

Cristina Fernández-Córdoba and Steve Szabo

Dedicated to Professor Syed Tariq Rizvi for his contribution to Ring and Module Theory

ABSTRACT. In this paper, it is shown that $\mathbb{Z}_2\mathbb{Z}_4$-additive codes with additional structure can be viewed as linear codes over rings. As an example, codes over the finite chain ring of order 8, $R = \frac{\mathbb{Z}_4[x]}{\langle x^2-2,2x\rangle}$, which as additive group is isomorphic to $\mathbb{Z}_2 \times \mathbb{Z}_4$, are shown to be $\mathbb{Z}_2\mathbb{Z}_4$-additive codes. Amongst other results connecting linear codes over R and their $\mathbb{Z}_2\mathbb{Z}_4$-additive images, it is shown that the $\mathbb{Z}_2\mathbb{Z}_4$-additive image of a cyclic code over R is separable, that is, a direct sum of a binary linear code and a linear code over $\mathbb{Z}_4$. The family of chain rings, $\frac{\mathbb{Z}_4[x]}{\langle x^s-2,2x^t\rangle}$, where $1 \leq t < s$, and the finite commutative local Frobenius non-chain ring $\frac{\mathbb{Z}_4[x,y]}{\langle x^2-2,xy-2,y^2,2x,2y\rangle}$ are also considered as alphabets for the study of $\mathbb{Z}_2\mathbb{Z}_4$-additive codes.

1. Introduction

Codes over rings is now a well-studied field. It gained prominence mainly due to the groundbreaking paper [**6**] where it was shown that the Kerdock and Preparata nonlinear binary codes can be viewed as linear codes over $\mathbb{Z}_4$. In the years since that paper, many finite rings such as $\mathbb{Z}_{p^k}$ and the more general family $GR(p^k, r)$ have been used as alphabets, where $GR(p^k, r)$ is the *Galois ring of characteristic* p^n *and degree* r which is characterized as a ring of the form $\mathbb{Z}_{p^k}[x]/\langle f(x)\rangle$ where $f(x)$ is a basic irreducible in $\mathbb{Z}_{p^k}[x]$ of degree r. Note that $\mathbb{Z}_{p^k}$ and $GR(p^k, r)$ are chain rings. Codes over finite chain rings in general have also been well-studied since they have most of the ingredients that codes over fields have.

In 1999, Wood showed in [**24**] that for the most part, the most general class of rings that can be used for alphabets are Frobenius rings. From the details in [**24**], it can be seen that a finite commutative Frobenius ring can be characterized as a direct sum of local rings with simple socles. The *socle* of a commutative ring is the sum of its simple ideals. After finite commutative chain rings, the most logical choice of ring alphabet would be a finite commutative local Frobenius ring. In [**14**], it was shown that there are 7 finite commutative local Frobenius non-chain rings of order 16 which are the smallest such rings. A few of these rings have now also been given at least initial studies as suitable alphabets for codes (see for instance [**11**, **15**, **25**]). Incidentally, $\frac{\mathbb{Z}_4[x,y]}{\langle x^2-2,xy-2,y^2,2x,2y\rangle}$ is one such ring that has not yet been studied but will be used herein as we study $\mathbb{Z}_2\mathbb{Z}_4$-additive codes in Section 5.

This work has been partially supported by the Spanish MINECO under Grant PID2019-104664GB-I00 (AEI / 10.13039/501100011033) and the Catalan AGAUR grant 2021 SGR 00643.

Mixed alphabet codes over rings are a generalization of linear codes over rings. The coordinates of these codes can be partitioned into blocks, each one corresponding to coordinates over the same ring. The first mixed alphabet codes were subgroups of $\mathbb{Z}_2^\alpha \times \mathbb{Z}_4^\beta \times \mathbb{Q}_8^\sigma$, being $\mathbb{Q}_8$ the non-Abelian quaternion group on eight elements. They were introduced by Rifà et al. in 1997 in [**20, 21**] when considering the algebraic structure of translation-invariant propelinear codes. Moreover, Abelian translations-invariant codes corresponding to additive codes were defined by Delsarte in 1973 in terms of association schemes [**9, 10**] when the underlying association scheme is a binary Hamming scheme. Therefore, codes as a subgroups of $\mathbb{Z}_2^\alpha \times \mathbb{Z}_4^\beta$ are the only additive codes in a binary Hamming scheme. These codes are called $\mathbb{Z}_2\mathbb{Z}_4$-*additive codes* and a complete description is given in [**3**] and [**4**].

When a ring R is isomorphic to $\mathbb{Z}_2^{\alpha'} \times \mathbb{Z}_4^{\beta'}$, for some $\alpha', \beta' \geq 1$, we say that R has $\mathbb{Z}_2\mathbb{Z}_4$-*additive structure*. In that case, a linear code over R is isomorphic to a $\mathbb{Z}_2\mathbb{Z}_4$-additive code. In this paper we consider some rings having $\mathbb{Z}_2\mathbb{Z}_4$-additive structure. First, we determine which rings of order 8 have $\mathbb{Z}_2\mathbb{Z}_4$-additive structure and we consider linear codes over those rings. We study the relationship between linear properties over R^n and properties in $\mathbb{Z}_2^\alpha \times \mathbb{Z}_4^\beta$. We see how they are related in terms of their generator matrices, their duality, and their generator polynomials for the case when they are cyclic. Then we consider the $\mathbb{Z}_2\mathbb{Z}_4$-additive codes as codes over other rings.

We have to mention that we consider vectors, matrices and polynomials whose entries are in $\{0, 1\}$ as elements over $\mathbb{Z}_2$ but also as elements over $\mathbb{Z}_4$ depending on the ring base we are considering in each case; it is given by the context. For an element $a \in \mathbb{Z}_4$, we denote by $\tilde{a}$ the element a modulo 2.

This paper is organized as follows. In Section 2 we give the basic definitions of $\mathbb{Z}_2\mathbb{Z}_4$-additive codes. In Section 3 we show that the only ring of order 8 having $\mathbb{Z}_2\mathbb{Z}_4$-additive structure, apart from $\mathbb{Z}_2 \times \mathbb{Z}_4$, is $\frac{\mathbb{Z}_4[x]}{\langle x^2-2,2x\rangle}$ and we give some results on codes over this ring. We see the relationship of a code over this ring with the corresponding code with $\mathbb{Z}_2\mathbb{Z}_4$-additive structure in Section 4. We study the relationship of the generator matrices, the linearity, the rank and kernel, the duality and the cyclicity. Finally, in Section 5, we give some examples of other rings with $\mathbb{Z}_2\mathbb{Z}_4$-additive structure. First, an infinite family of chain rings and second, an example of a Frobenius ring that is a non-chain ring.

2. Preliminaries

A $\mathbb{Z}_2\mathbb{Z}_4$-*additive code* $\mathcal{C}$ is a subgroup of $\mathbb{Z}_2^\alpha \times \mathbb{Z}_4^\beta$ for $\alpha, \beta \in \mathbb{N}$. As a subgroup, $\mathcal{C}$ is isomorphic to an Abelian structure $\mathbb{Z}_2^\gamma \times \mathbb{Z}_4^\delta$ for some non-negative integers γ and δ. We have that the number of codewords is $|\mathcal{C}| = 2^\gamma 4^\delta$. For complete details of $\mathbb{Z}_2\mathbb{Z}_4$-additive codes see [**3**] and [**4**].

Let X be the set of $\mathbb{Z}_2$ coordinate positions and Y the set of $\mathbb{Z}_4$ coordinate positions, so $|X| = \alpha$ and $|Y| = \beta$. Unless otherwise stated, the set X corresponds to the first α coordinates and Y corresponds to the last β coordinates. Call $\mathcal{C}_X$ and $\mathcal{C}_Y$ the code $\mathcal{C}$ restricted to the X and Y coordinates, respectively. Note that for a $\mathbb{Z}_2\mathbb{Z}_4$-additive code $\mathcal{C}$, we have that $\mathcal{C} \subseteq \mathcal{C}_X \times \mathcal{C}_Y$. When there is an equality, $\mathcal{C} = \mathcal{C}_X \times \mathcal{C}_Y$, we say that the code is *separable*.

We denote by $\mathcal{C}_b$ the subcode of $\mathcal{C}$ which contains all codewords of order at most two. Note that $|\mathcal{C}_b| = 2^{\gamma+\delta}$. Let κ be the dimension of the binary linear code $(\mathcal{C}_b)_X$. We say that $\mathcal{C}$ is of type $(\alpha, \beta; \gamma, \delta; \kappa)$. From [**12**], we have that there exists a $\mathbb{Z}_2\mathbb{Z}_4$-additive code $\mathcal{C}$ of type $(\alpha, \beta; \gamma, \delta; \kappa)$ as long as

$$\alpha, \beta, \gamma, \delta, \kappa \geq 0, \quad \alpha + \beta > 0,$$
$$0 \leq \delta + \gamma \leq \beta + \kappa \quad \text{and} \quad \kappa \leq \min\{\alpha, \gamma\}.$$

Let ϕ be the usual quaternary Gray map from $\mathbb{Z}_4$ to $\mathbb{Z}_2^2$ defined in [**6**] by $\phi(0) = (0,0), \phi(1) = (0,1), \phi(2) = (1,1), \phi(3) = (1,0)$. The Gray map is extended to $\mathbb{Z}_2^\alpha \times \mathbb{Z}_4^\beta$ as $\Phi : \mathbb{Z}_2^\alpha \times \mathbb{Z}_4^\beta \to \mathbb{Z}_2^{\alpha+2\beta}$, $\Phi(\mathbf{u}) = (u_1, \dots, u_\alpha \mid \phi(u'_1), \dots, \phi(u'_\beta))$, for $\mathbf{u} = (u_1, \dots, u_\alpha \mid u'_1, \dots, u'_\beta) \in \mathbb{Z}_2^\alpha \times \mathbb{Z}_4^\beta$.

The *standard inner product* of vectors in $\mathbb{Z}_2^\alpha \times \mathbb{Z}_4^\beta$, defined in [**3**], can be written as

$$\langle \mathbf{u}, \mathbf{v} \rangle_{\mathbb{Z}_2\mathbb{Z}_4} = 2\left(\sum_{i=0}^{\alpha-1} u_i v_i\right) + \sum_{j=0}^{\beta-1} u'_j v'_j \in \mathbb{Z}_4, \tag{2.1}$$

where the computations are made taking the zeros and ones in the α binary coordinates as quaternary zeros and ones, respectively. The *dual code* of $\mathcal{C}$, is defined in the standard way by

$$\mathcal{C}^\perp = \{\mathbf{v} \in \mathbb{Z}_2^\alpha \times \mathbb{Z}_4^\beta \mid \langle \mathbf{u}, \mathbf{v} \rangle_{\mathbb{Z}_2\mathbb{Z}_4} = 0, \text{ for all } \mathbf{u} \in \mathcal{C}\}.$$

Two $\mathbb{Z}_2\mathbb{Z}_4$-additive codes $\mathcal{C}_1$ and $\mathcal{C}_2$ of the same type $(\alpha, \beta; \gamma, \delta; \kappa)$ are said to be *permutation equivalent* if one can be obtained from the other by permuting the coordinates. Let $\mathcal{C}$ be a $\mathbb{Z}_2\mathbb{Z}_4$-additive code of type $(\alpha, \beta; \gamma, \delta; \kappa)$. Then, by [**3**], the code $\mathcal{C}$ is permutation equivalent to a $\mathbb{Z}_2\mathbb{Z}_4$-additive code with a generator matrix in standard form

$$\mathcal{G} = \left(\begin{array}{cc|ccc} I_\kappa & T_b & 2T_2 & \mathbf{0} & \mathbf{0} \\ \mathbf{0} & \mathbf{0} & 2T_1 & 2I_{\gamma-\kappa} & \mathbf{0} \\ \mathbf{0} & S_b & S_q & R & I_\delta \end{array}\right),$$

where T_b and S_b are matrices over $\mathbb{Z}_2$; T_1, T_2 and R are matrices over $\mathbb{Z}_4$ with all entries in $\{0,1\} \subset \mathbb{Z}_4$; and S_q is a matrix over $\mathbb{Z}_4$. Note that $\mathcal{C}$ is separable if and only if T_2 and S_b are zero matrices.

Two structural properties of binary non-linear codes are the rank and the dimension of the kernel. The *rank* of a binary code C, denoted by $\text{rank}(C)$, is simply the dimension of the linear span of the codewords of C. The *kernel* of a binary code C, denoted by $K(C)$ and first defined in [**2**], is the set of vectors that leave C invariant under translation. If C contains the all-zero vector, then the kernel is a binary linear subcode of C and we can consider its dimension. The dimension of the kernel of C is denoted by $\ker(C)$. These two invariants, $\text{rank}(C)$ and $\ker(C)$, help in the classification of binary codes but they do not always give a full classification, since two binary codes which are not permutation equivalent could have the same rank and dimension of the kernel. For example, when considering binary Hadamard codes of length 64, in [**18**] authors show that there are 65 pairs of rank and dimension of the kernel for these codes whereas in the Magma database there are 395 of such non-equivalent codes.

In order to study the linearity of $\mathbb{Z}_2\mathbb{Z}_4$-linear codes, the rank and the kernel, we consider the following statement. Let $\mathbf{u} * \mathbf{v}$ denote the component-wise product for any $\mathbf{u}, \mathbf{v} \in \mathbb{Z}_2^\alpha \times \mathbb{Z}_4^\beta$.

PROPOSITION 2.1 ([**12**]). *Let* $\mathbf{u}, \mathbf{v} \in \mathbb{Z}_2^\alpha \times \mathbb{Z}_4^\beta$. *Then,*

$$\Phi(\mathbf{u} + \mathbf{v}) = \Phi(\mathbf{u}) + \Phi(\mathbf{v}) + \Phi(2\mathbf{u} * \mathbf{v}).$$

COROLLARY 2.2 ([**12**]). *Let $\mathcal{C}$ be a $\mathbb{Z}_2\mathbb{Z}_4$-additive code. Then $\Phi(\mathcal{C})$ is linear if and only if $2\mathbf{u} * \mathbf{u} \in \mathcal{C}$ for all $\mathbf{u}, \mathbf{v} \in \mathcal{C}$.*

Let $\mathcal{C}$ be a $\mathbb{Z}_2\mathbb{Z}_4$-additive code. We define the codes

$$\mathcal{R}(\mathcal{C}) = \Phi^{-1}(\langle \Phi(\mathcal{C}) \rangle),$$
$$\mathcal{K}(\mathcal{C}) = \Phi^{-1}(K(C)).$$

LEMMA 2.3 ([**12**]). *If $\mathcal{C}$ is a $\mathbb{Z}_2\mathbb{Z}_4$-additive code, then $\mathcal{R}(\mathcal{C})$ and $\mathcal{K}(\mathcal{C})$ are $\mathbb{Z}_2\mathbb{Z}_4$-additive codes whose image under the extended Gray map are binary linear codes.*

PROPOSITION 2.4 ([**12**]). *Let $\mathcal{C}$ be a $\mathbb{Z}_2\mathbb{Z}_4$-additive code of type $(\alpha, \beta; \gamma, \delta; \kappa)$, and let $C = \Phi(\mathcal{C})$. Let $\mathcal{G}$ be a generator matrix of $\mathcal{C}$, and let $\{\mathbf{u}_i\}_{i=1}^{\gamma}$ be the rows of order two and $\{\mathbf{v}_j\}_{j=1}^{\delta}$ the rows of order four in $\mathcal{G}$. Then,*

(1) $\mathcal{R}(\mathcal{C}) = \langle \mathcal{C}, \{2\mathbf{v}_j * \mathbf{v}_k\}_{1 \le j < k \le \delta} \rangle$.
(2) $\mathcal{K}(\mathcal{C}) = \{\mathbf{u} \in \mathcal{C} \mid 2\mathbf{u} * \mathbf{v}_j \in \mathcal{C}, \forall j \in \{1, \ldots, \delta\}\}$.

COROLLARY 2.5 ([**12**]). *Let $\mathcal{C}$ be a $\mathbb{Z}_2\mathbb{Z}_4$-additive code. Then, $\mathcal{C}_b \subseteq \mathcal{K}(\mathcal{C})$.*

Finally, we give some definitions and remarks on $\mathbb{Z}_2\mathbb{Z}_4$-additive cyclic codes. They were defined in [**1**], and later their duality was studied in [**5**].

Let $\mathbf{u} = (u \mid u') \in \mathbb{Z}_2^{\alpha} \times \mathbb{Z}_4^{\beta}$ and i be an integer. We denote by $u^{(i)}$ (resp. $u'^{(i)}$) the ith cyclic shift of u (resp. u'). Then, we denote by

$$\mathbf{u}^{(i)} = (u^{(i)} \mid u'^{(i)}) = (u_{1+i}, u_{2+i}, \ldots, u_{\alpha+i} \mid u'_{1+i}, u'_{2+i}, \ldots, u'_{\beta+i})$$

the ith double cyclic shift of $\mathbf{u}$, where the subscripts are taken modulo α and β, respectively. A $\mathbb{Z}_2\mathbb{Z}_4$-additive code $\mathcal{C}$ is *cyclic* if for any codeword $\mathbf{u} \in \mathcal{C}$ we have $\mathbf{u}^{(1)} \in \mathcal{C}$.

From [**13**] we have that binary cyclic codes of length n can be seen as ideals in $\mathbb{Z}_2[y]/(y^n - 1)$, and they are principal ideals of the form $\langle b(y) \rangle$, where $b|(y^n - 1)$. Moreover, quaternary cyclic codes of length n can also be seen as ideals in $\mathbb{Z}_4[y]/(y^n - 1)$ and, from [**7**], if n is odd then these ideals are principal of the form $\langle f(y)h(y) + 2f(y) \rangle$, where $f(y)g(y)h(y) = y^n - 1$, and $f(y)$, $g(y)$ and $h(y)$ are pairwise coprime. From [**1**], we have that $\mathbb{Z}_2\mathbb{Z}_4$-cyclic codes of type $(\alpha, \beta; \gamma, \delta; \kappa)$ can be seen as ideals in the ring $R_{\alpha,\beta} = \mathbb{Z}_2[y]/(y^{\alpha}-1) \times \mathbb{Z}_4[y]/(y^{\beta}-1)$, with the operation $\star : \mathbb{Z}_4[y] \times R_{\alpha,\beta} \to R_{\alpha,\beta}$ defined as $\lambda(y) \star (p(y) \mid q(y)) = (\tilde{\lambda}(y)p(y) \mid \lambda(y)q(y))$.

THEOREM 2.6 ([**1**]). *Let $\mathcal{C}$ be a $\mathbb{Z}_2\mathbb{Z}_4$-additive cyclic code of type $(\alpha, \beta; \gamma, \delta; \kappa)$ with β odd. Then, we can define $\mathcal{C}$ as*

$$\mathcal{C} = \langle (b(y) \mid 0), (\ell(y) \mid f(y)h(y) + 2f(y)) \rangle, \tag{2.2}$$

where $f(y)h(y)g(y) = y^{\beta} - 1$ in $\mathbb{Z}_4[y]$, $b(y), \ell(y) \in \mathbb{Z}_2[y]/(y^{\alpha} - 1)$ with $b(y)|(y^{\alpha} - 1)$, $\deg(\ell(y)) < \deg(b(y))$, and $b(y)$ divides $\frac{y^{\beta}-1}{f(y)}\ell(y) \pmod 2$.

3. Codes over Rings of order 8

DEFINITION 3.1. Let R be a commutative ring and $C \subset R^n$. Then C is a *code of length n over R*. If C is an additive subgroup, then C is an *additive* code. If C is an R-submodule then C is a *linear* code.

The goal of this paper is to show that some $\mathbb{Z}_2\mathbb{Z}_4$-additive codes can be viewed as codes over rings. To that end we identify all rings of order 8 which are Frobenius and also have $\mathbb{Z}_2\mathbb{Z}_4$-additive structure.

PROPOSITION 3.2 ([**19**], Theorem 14). *Up to isomorphism there are 11 rings with identity of order 8. The complete list is as follows:*

(1) $\mathbb{F}_8$
(2) $\mathbb{F}_4 \oplus \mathbb{F}_2$
(3) $\mathbb{F}_2 \oplus \mathbb{F}_2 \oplus \mathbb{F}_2$
(4) $\mathbb{F}_2 \oplus \frac{\mathbb{F}_2[x]}{\langle x^2 \rangle}$
(5) $\begin{bmatrix} \mathbb{F}_2 & \mathbb{F}_2 \\ 0 & \mathbb{F}_2 \end{bmatrix}$
(6) $\frac{\mathbb{F}_2[u,v]}{\langle u^2, v^2, uv \rangle}$
(7) $\frac{\mathbb{F}_2[x]}{\langle x^3 \rangle}$
(8) $\mathbb{F}_2 \oplus \mathbb{Z}_4$
(9) $\frac{\mathbb{Z}_4[x]}{\langle x^2-2, 2x \rangle}$
(10) $\frac{\mathbb{Z}_4[x]}{\langle x^2, 2x \rangle}$
(11) $\mathbb{Z}_8$

COROLLARY 3.3. *Up to isomorphism the only Frobenius rings with $\mathbb{Z}_2\mathbb{Z}_4$-additive structure are $\mathbb{F}_2 \oplus \mathbb{Z}_4$ and $\frac{\mathbb{Z}_4[x]}{\langle x^2-2,2x \rangle}$.*

PROOF. The only 3 rings from Proposition 3.2 with the correct additive structure are $\mathbb{F}_2 \oplus \mathbb{Z}_4$, $\frac{\mathbb{Z}_4[x]}{\langle x^2-2,2x \rangle}$ and $\frac{\mathbb{Z}_4[x]}{\langle x^2,2x \rangle}$. The ring $\mathbb{F}_2 \oplus \mathbb{Z}_4$ is a direct sum of chain rings, so Frobenius. The ring $\frac{\mathbb{Z}_4[x]}{\langle x^2,2x \rangle}$ is local with maximal ideal $\langle x, 2 \rangle$. Also, its socle is $\langle x \rangle \oplus \langle 2 \rangle$. From [**24**], we know that a finite local commutative Frobenius ring has a simple socle. So, it is not Frobenius. Now, $\frac{\mathbb{Z}_4[x]}{\langle x^2-2,2x \rangle}$ is a chain ring since its maximal ideal is principally generated by x, so it is Frobenius. □

3.1. Codes over $\frac{\mathbb{Z}_4[x]}{\langle x^2-2,2x \rangle}$. Let $R = \frac{\mathbb{Z}_4[x]}{\langle x^2-2,2x \rangle}$. In the following sections, we will consider codes over R and show that they have images that are $\mathbb{Z}_2\mathbb{Z}_4$-additive codes. Furthermore, cyclic codes over R are cyclic $\mathbb{Z}_2\mathbb{Z}_4$-additive codes. Finally, under the standard inner product on R^n, the image of the dual of a code over R is the dual of the image of the code for the standard inner product defined for $\mathbb{Z}_2\mathbb{Z}_4$-additive codes.

The following is the ideal lattice of R.

$$\langle 1 \rangle$$
$$|$$
$$\langle x \rangle$$
$$|$$
$$\langle x^2 \rangle = \langle 2 \rangle$$
$$|$$
$$\langle 0 \rangle$$

Notice that multiplication by x "moves" elements down the chain. As an additive group, $R = \langle 1, x \rangle$. So, any element can be represented as $\mathbf{a} = a' + ax$ where $a' \in \{0,1,2,3\}$ and $a \in \{0,1\}$.

Let $\mathfrak{C}$ be a linear code over R. Define $\mathfrak{C}_b = \{\mathbf{a} \in \mathfrak{C} \mid 2\mathbf{a} = 0\}$, $\mathfrak{C}_X = \{ax \mid a' + ax \in \mathfrak{C}\}$ and $\mathfrak{C}_Y = \mathfrak{C}/\langle x \rangle$. Note that $\mathfrak{C}_b, \mathfrak{C}_X$ and $\mathfrak{C}_Y$ are isomorphic to binary codes.

LEMMA 3.4. *Let $\mathfrak{C}$ be a linear code over R and $C = \rho(\mathfrak{C})$. Then,*

(1) $2\mathfrak{C} \subseteq \mathfrak{C}_b$,
(2) $x\mathfrak{C} \subseteq \mathfrak{C}_b$.

PROOF. Clearly, $2\mathfrak{C} \in \mathfrak{C}_b$. Let $\mathbf{a} = a' + ax \in \mathfrak{C}$. Since $x\mathbf{a} = 2a + \tilde{a'}x$, we have that $x\mathbf{a} \in \mathfrak{C}_b$. Therefore $x\mathfrak{C} \subset \mathfrak{C}_b$. □

Now we can give a generator matrix of a linear code $\mathfrak{C}$ over R. By Proposition 3.2 and Theorem 3.5 in [**17**], any linear code over R is permutation equivalent to a code with a generator matrix in the form

$$G = \begin{pmatrix} I_{k_0} & G_{0,1} & G_{0,2,1} + xG_{0,2,2} & G_{0,3,1} + xG_{0,3,2} \\ \mathbf{0} & xI_{k_1} & xG_{1,2} & xG_{1,3} \\ \mathbf{0} & \mathbf{0} & 2I_{k_2} & 2G_{2,3} \end{pmatrix}, \tag{3.1}$$

where all the entries in the label submatrices are from $\{0, 1\}$ with the exception of $G_{0,3,1}$ whose entries are from $\{0, 1, 2, 3\}$.

If $\mathfrak{C}$ is a linear code over R of length n with generator matrix as in (3.1), then we say the $\mathfrak{C}$ is of *type* $(n; k_0, k_1, k_2)$.

PROPOSITION 3.5. *Let $\mathfrak{C}$ be a linear code over R of type $(n; k_0, k_1, k_2)$. Then, $|\mathfrak{C}| = 2^{k_0+2k_1+k_2}4^{k_0}$, and $|\mathfrak{C}_b| = 2^{2k_0+2k_1+k_2}$.*

PROOF. Let $\mathfrak{C}$ be a linear code over R of type $(n; k_0, k_1, k_2)$ and generator matrix G as in (3.1). Let $\{\mathbf{v}_i\}_{i=1}^{k_0}$ be the first k_0 vectors of G, $\{\mathbf{w}_j\}_{j=1}^{k_1}$ the next k_1 codewords of G and $\{\mathbf{u}_\ell\}_{\ell=1}^{k_2}$ the last codewords of G. Any codeword $\mathbf{a} \in \mathfrak{C}$ can be written as

$$\mathbf{a} = \sum_{i=0}^{k_0}(\lambda_i + \omega_i x)\mathbf{v}_i + \sum_{j=1}^{k_1}(\mu_j + \tau_i x)\mathbf{w}_j + \sum_{\ell=0}^{k_2} \nu_\ell \mathbf{u}_\ell,$$

where $\lambda_i \in \{0, 1, 2, 3\}$, $\omega_i, \mu_j, \tau_j, \nu_\ell \in \{0, 1\}$. Therefore, we have that $|\mathfrak{C}| = 2^{k_0+2k_1+k_2}4^{k_0}$.

Multiplying any of the first k_0 rows by x or 2 or multiplying any of the next k_1 rows by x creates a codeword of order 2. Any of the last $k_1 + k_2$ rows are also of order at most 2. So, there are $2^{2k_0+2k_1+k_2}$ codewords of order 2; that is, $|\mathfrak{C}_b| = 2^{2k_0+2k_1+k_2}$. □

4. Codes over $\frac{\mathbb{Z}_4[x]}{\langle x^2-2,2x\rangle}$ as $\mathbb{Z}_2\mathbb{Z}_4$-additive codes

Recall that any element can be represented as $\mathbf{a} = a' + ax$ where $a' \in \{0, 1, 2, 3\}$ and $a \in \{0, 1\}$. With this we define the bijective map

$$\rho : R \to \mathbb{Z}_2 \times \mathbb{Z}_4;\ \mathbf{a} = a' + ax \mapsto (a \mid a').$$

We also denote by ρ the coordinate-wise extended map to elements in R^n as

$$\rho : R^n \to \mathbb{Z}_2^n \times \mathbb{Z}_4^n;\ (\mathbf{a}_1, \dots, \mathbf{a}_n) \mapsto (a_1, \dots, a_n \mid a_1', \dots, a_n'),$$

where $\mathbf{a}_i = a_i' + a_i x$ for $1 \le i \le n$.

EXAMPLE 4.1. Let $(1, x, 2) \in R^3$. We have that $x(1, x, 2) = (x, 2, 0)$ and $x^2(1, x, 2) = (2, 0, 0)$. Note that $\rho((1, x, 2)) = (010 \mid 102)$, $\rho(x(1, x, 2)) = \rho(x, 2, 0) = (100 \mid 020)$, and $\rho(x^2(1, x, 2)) = \rho(2, 0, 0) = (000 \mid 200)$.

We also define the map $\varphi : R^n \to \mathbb{Z}_2^{3n}$ given by $\varphi(\mathbf{a}) = \Phi(\rho(\mathbf{a}))$ for $\mathbf{a} \in R^n$. Note that if $\mathbf{a} = (\mathbf{a}_1, \dots \mathbf{a}_n)$, where $\mathbf{a}_i = a_i' + a_i x \in R$ for all i, then $\varphi(\mathbf{a}) = (a_1, \dots, a_n, \phi(a_1'), \dots, \phi(a_n'))$.

Let $\mathfrak{C}$ be a linear code over R and $\mathcal{C} = \rho(\mathfrak{C})$. Recall that $\mathfrak{C}_X = \{ax \mid a' + ax \in \mathfrak{C}\}$ and $\mathfrak{C}_Y = \mathfrak{C}/\langle x \rangle$. From the definition of $\mathcal{C}_X$ and $\mathcal{C}_Y$, we have that $\mathcal{C}_X = \rho(\mathfrak{C}_X)$ and $\mathcal{C}_Y = \rho(\mathfrak{C}_Y) = \mathfrak{C}_Y$.

LEMMA 4.2. *Let $\mathfrak{C}$ be a linear code over R and let $\mathcal{C} = \rho(\mathfrak{C})$. For all $(a \mid a') \in \mathcal{C}$, $(\tilde{a'} \mid 2a) \in \mathcal{C}$.*

PROOF. Let $(a \mid a') \in \mathcal{C}$. Then, $a' + ax \in \mathfrak{C}$ and $x\mathbf{a} = x(a' + ax) = 2a + \tilde{a'}x \in \mathfrak{C}$. Therefore $(\tilde{a'} \mid 2a) \in \mathcal{C}$. □

LEMMA 4.3. *Let $\mathfrak{C}$ be a linear code over R and $\mathcal{C} = \rho(\mathfrak{C})$. Then, $\rho(\mathfrak{C}_b) = \mathcal{C}_b$.*

PROOF. By definition, if $\mathbf{a} = a' + ax \in \mathfrak{C}_b$, then $a' \in \{0, 2\}$. Since $\rho(\mathbf{a}) = (a \mid a')$, we have that $\mathbf{a} \in \mathfrak{C}_b$ if and only if $(a \mid a') \in \mathcal{C}_b$. □

PROPOSITION 4.4. *Let $\mathfrak{C}$ be a linear code over R of type $(n; k_0, k_1, k_2)$. Then $\mathcal{C} = \rho(\mathfrak{C})$ is of type $(n, n; \gamma, \delta; \kappa)$ where $\gamma = k_0 + 2k_1 + k_2$, $\delta = k_0$ and $\kappa = k_0 + k_1$.*

PROOF. Let $\mathfrak{C}$ be a linear code over R of type $(n; k_0, k_1, k_2)$ and let $\mathcal{C} = \rho(\mathfrak{C})$ be of type $(\alpha, \beta; \gamma, \delta; \kappa)$.

Clearly, $\alpha = \beta = n$. By definition, $\mathcal{C}$ has $2^\gamma 4^\delta$ codewords and $2^{\gamma+2\delta}$ codewords of order at most 2. We have that $\mathcal{C} = \rho(\mathfrak{C})$ and, by Lemma 4.3, $\mathcal{C}_b = \rho(\mathfrak{C}_b)$ and hence $|\mathcal{C}| = |\mathfrak{C}|$, and $|\mathcal{C}_b| = |\mathfrak{C}_b|$. Then, from Proposition 3.5, we have that $2^\gamma 4^\delta = 2^{k_0+2k_1+k_2} 4^{k_0}$, and $2^{\gamma+2\delta} = 2^{2k_0+2k_1+k_2}$. Therefore, $\delta = k_0$ and $\gamma = k_0 + 2k_1 + k_2$.

Note that the generator matrix of $(\mathfrak{C}_b)_X$ is

$$G = \begin{pmatrix} xI_{k_0} & xG_{0,1} & xG_{0,2} & xG_{0,3} \\ \mathbf{0} & xI_{k_1} & xG_{1,2} & xG_{1,3} \end{pmatrix}.$$

Therefore, since κ is the dimension of $\mathcal{C}_b$ and $\mathcal{C}_b = \rho(\mathfrak{C}_b)$ by Lemma 4.3, we have that $\kappa = k_0 + k_1$. □

LEMMA 4.5. *Let $\mathfrak{C}$ be a linear code over R. If $\{\mathbf{a}_i\}_{i=1}^r$ are generator vectors of $\mathfrak{C}$, then $\{\rho(\mathbf{a}_i), \rho(x\mathbf{a}_i)\}_{i=1}^r$ are generator vectors of $\mathcal{C} = \rho(\mathfrak{C})$.*

PROOF. Let $\mathbf{a} \in \mathcal{C}$. We have that $\rho^{-1}(\mathbf{a}) = \sum_{i=0}^r (\lambda_i + \omega_i x)\mathbf{a}_i = \sum_{i=0}^r \lambda_i \mathbf{a}_i + \sum_{i=0}^r \omega_i x\mathbf{a}_i$, where $\lambda_i \in \{0,1,2,3\}, \omega_i \in \{0,1\}$. Therefore, $\mathbf{a} = \sum_{i=0}^r \lambda_i \rho(\mathbf{a}_i) + \sum_{i=0}^r \omega_i \rho(x\mathbf{a}_i)$, and $\mathbf{a}$ is generated by $\{\rho(\mathbf{a}_i), \rho(x\mathbf{a}_i)\}_{i=1}^r$. □

PROPOSITION 4.6. *Let $\mathfrak{C}$ be a linear code over R of type $(n; k_0, k_1, k_2)$ with generator matrix as in (3.1). Then, the matrix*

(4.1)

$$\mathcal{G} = \left(\begin{array}{ccc|ccc} I_{k_0+k_1} & \tilde{A} & \tilde{B} & \mathbf{0} & \mathbf{0} & \begin{pmatrix} 2(G_{0,3,2} - G_{0,2,2}G_{2,3}) \\ \mathbf{0} \end{pmatrix} \\ \mathbf{0} & \mathbf{0} & G_{0,3,2} & I_{k_0} & (G_{0,1}\ G_{0,2}) & G_{0,3,1} \\ \mathbf{0} & \mathbf{0} & \mathbf{0} & \mathbf{0} & 2I_{k_1+k_2} & \begin{pmatrix} 2(G_{1,3} - G_{1,2}G_{2,3}) \\ 2G_{2,3} \end{pmatrix} \end{array} \right),$$

is a generator matrix of $\mathcal{C} = \rho(\mathfrak{C})$, where

$$A = \begin{pmatrix} G_{0,2,1} - G_{0,1}G_{1,2} \\ G_{1,2} \end{pmatrix} \text{ and } B = \begin{pmatrix} \tilde{G}_{0,3,1} - G_{0,1}G_{1,3} \\ G_{1,3} \end{pmatrix}.$$

PROOF. Let $\mathfrak{C}$ be a linear code over R of type $(n; k_0, k_1, k_2)$ and generator matrix G as in (3.1). Let $\{\mathbf{v}_i\}_{i=1}^{k_0}$ be the first k_0 vectors of G, $\{\mathbf{w}_j\}_{j=1}^{k_1}$ the next k_1

codewords of G and $\{\mathbf{u}_\ell\}_{\ell=1}^{k_2}$ the last codewords of G. Therefore, $\{\{\mathbf{v}_i\}_{i=1}^{k_0}, \{\mathbf{w}_j\}_{j=1}^{k_1}, \{\mathbf{u}_\ell\}_{\ell=1}^{k_2}\}$ are the generator vectors of $\mathfrak{C}$. By 4.5, we have that

$$S = \{\{\rho(\mathbf{v}_i), \rho(x\mathbf{v}_i)\}_{i=1}^{k_0}, \{\rho(\mathbf{w}_j), \rho(x\mathbf{w}_j)\}_{j=1}^{k_1}, \{\rho(\mathbf{u}_\ell)\}_{\ell=1}^{k_2}\}$$

are the generator vectors of $\mathcal{C}$. Therefore, if we consider the elements in S as row vectors in a appropriate order and consider a few more obvious eliminations, we obtain the matrix

$$\left(\begin{array}{cccc|cccc} \mathbf{0} & \mathbf{0} & G_{0,2,2} & G_{0,3,2} & I_{k_0} & G_{0,1} & G_{0,2,1} & G_{0,3,1} \\ I_{k_0} & \tilde{G}_{0,1} & G_{0,2,1} & G_{0,3,1} & \mathbf{0} & \mathbf{0} & 2G_{0,2,1} & 2G_{0,3,2} \\ \mathbf{0} & I_{k_1} & G_{1,2} & G_{1,3} & \mathbf{0} & \mathbf{0} & \mathbf{0} & \mathbf{0} \\ \mathbf{0} & \mathbf{0} & \mathbf{0} & \mathbf{0} & \mathbf{0} & 2I_{k_1} & 2G_{1,2} & 2G_{1,3} \\ \mathbf{0} & \mathbf{0} & \mathbf{0} & \mathbf{0} & \mathbf{0} & \mathbf{0} & 2I_{k_2} & 2G_{2,3} \end{array}\right).$$

Applying transformations to the last matrix to obtain a generator matrix in standard form, we obtain $\mathcal{G}$. □

Let $\mathfrak{C}$ be a linear code over R and $\mathcal{C} = \rho(\mathfrak{C})$. We say that $\mathfrak{C}$ is separable if $\mathcal{C}$ is separable. The following lemma follows from the definition of a separable $\mathbb{Z}_2\mathbb{Z}_4$-additive code and the function ρ.

LEMMA 4.7. *Let $\mathfrak{C}$ be a linear code over R with generator matrix G as in* (3.1). *Then, $\mathfrak{C}$ is separable if and only if $G_{0,3,2}$ and $2(G_{0,2,2}G_{2,3})$ are the zero matrices.*

PROOF. Let $\mathcal{C} = \rho(\mathfrak{C})$. We have that $\mathfrak{C}$ is separable is and only if $\mathcal{C}$ is separable. By 4.6, from the matrix G of $\mathfrak{C}$ we can obtain the generator matrix $\mathcal{G}$ given in (4.1). Then, $\mathcal{C}$ is separable if and only if $G_{0,3,2} = \mathbf{0}$ and $2(G_{0,2,2}G_{2,3}) = \mathbf{0}$. □

Finally, if we consider a $\mathbb{Z}_2\mathbb{Z}_4$-additive code $\mathcal{C}$, we characterize when it is the image under ρ of a linear code over R. Also, in the case it is not the image of a linear code over R, we give the minimum linear code over R containing $\rho^{-1}(\mathcal{C})$.

THEOREM 4.8. *Let $\mathcal{C}$ be a $\mathbb{Z}_2\mathbb{Z}_4$-additive code of type $(\alpha,\alpha;\gamma,\delta;\kappa)$. Let $\mathcal{G}$ be a generator matrix of $\mathcal{C}$ whose row vectors are $\{\mathbf{w}_i = (w_i \mid w_i')\}_{i=1}^{\gamma+\delta}$. Then, $\rho^{-1}(\mathcal{C})$ is a linear code over R if and only if for all $i \in \{1,\dots,\gamma+\delta\}$, $(\tilde{w}_i' \mid 2w_i) \in \mathcal{C}$.*

PROOF. Let $\mathfrak{C} = \rho^{-1}(\mathcal{C})$. Assume $\mathfrak{C}$ is linear over R and let $\tilde{w}_i'+w_ix = \rho^{-1}(\mathbf{w}_i)$, for $i \in \{1,\dots,\gamma+\delta\}$. Then, $x(\tilde{w}_i'+w_ix) = 2w_j+\tilde{w}_i'x \in \mathfrak{C}$ and hence $\rho(2w_j+\tilde{w}_i'x) = (\tilde{w}_i' \mid 2w_i) \in \mathcal{C}$.

Assume now that for all $i \in \{1,\dots,\gamma+\delta\}$, $(\tilde{w}_i' \mid 2w_i) \in \mathcal{C}$. We only have to check that for all $\mathbf{a} \in \mathfrak{C}$, we have $x\mathbf{a} \in \mathfrak{C}$ or, equivalently, $\rho(x\mathbf{a}) \in \mathcal{C}$. Let $\mathbf{a} = a' + ax \in \mathfrak{C}$. We have that $\rho(\mathbf{a}) = (a \mid a') \in \mathcal{C}$ and hence $\rho(\mathbf{a}) = \sum_{i=1}^{\gamma+\delta}\lambda_i\mathbf{w}_i = (\sum_{i=1}^{\gamma+\delta}\lambda_iw_i \mid \sum_{i=1}^{\gamma+\delta}\lambda_iw_i') = (a \mid a')$, for $\lambda_i \in \mathbb{Z}_4$ for $i \in \{1,\dots,\gamma+\delta\}$. Since $\rho(x\mathbf{a}) = (\tilde{a'} \mid 2a)$, we have that $\rho(x\mathbf{a}) = (\sum_{i=1}^{\gamma+\delta}\tilde{\lambda}_i\tilde{w}_i' \mid 2\sum_{i=1}^{\gamma+\delta}\lambda_iw_i) = \sum_{i=0}^{\gamma+\delta}\lambda_i(\tilde{w}_i' \mid 2u_i) \in \mathcal{C}$. □

COROLLARY 4.9. *Let $\mathcal{C}$ be a $\mathbb{Z}_2\mathbb{Z}_4$-additive code of type $(\alpha,\alpha;\gamma,\delta;\kappa)$. Let G_S be a generator matrix of $\mathcal{C}$ in standard form. Let $\{\mathbf{u}_i = (u_i \mid u_i')\}_{i=1}^{\gamma}$ and $\{\mathbf{v}_j = (v_j \mid v_j')\}_{j=1}^{\delta}$ be the row vectors of order two and four in G_S, respectively. Then,*

$$\rho^{-1}(\langle\mathcal{C}, \{(\mathbf{0} \mid 2u_i)\}_{i=1}^{\gamma}, \{(\tilde{v}_j' \mid 2v_j')\}_{j=1}^{\delta}\rangle)$$

is the minimum linear code over R containing $\rho^{-1}(\mathcal{C})$.

4.1. Linearity, rank and kernel. Let $\mathfrak{C}$ be a linear code over R. We have that $\mathcal{C} = \rho(\mathfrak{C})$ is a $\mathbb{Z}_2\mathbb{Z}_4$-additive code and $C = \Phi(\mathcal{C}) = \varphi(\mathfrak{C})$ is a binary, not necessarily linear, code.

EXAMPLE 4.10. Let $\mathfrak{C}$ be the linear code over R generated by $\langle(2+x,2,0,0)$, $(1,x,1,0)$, $(1,x,x,1+x)\rangle$. We see that $C = \varphi(\mathfrak{C})$ is not linear. Let $\mathbf{u} = (1,x,1,0)$ and $\mathbf{v} = (1,x,x,1+x)$. We have that

$$\begin{aligned}\varphi(\mathbf{u}) + \varphi(\mathbf{v}) &= (0,1,0,0,0,1,0,0,0,1,0,0) + (0,1,1,1,0,1,0,0,0,0,0,1)\\ &= (0,0,1,1,0,0,0,0,0,1,0,1).\end{aligned}$$

Note that $\varphi^{-1}(0,0,1,1,0,0,0,0,0,1,0,1) = (0,0,1+x,1+x) \notin R$. Therefore, $\varphi(\mathbf{u}) + \varphi(\mathbf{v}) \notin C$ and C is not linear.

By definition, we have that $\varphi(\mathfrak{C})$ is linear if and only if $\Phi(\mathcal{C})$ is linear. The linearity of the image of $\mathbb{Z}_2\mathbb{Z}_4$-additive codes under the extended Gray map is studied in [**12**]. A necessary and sufficient condition for a $\mathbb{Z}_2\mathbb{Z}_4$-additive linear code $\mathcal{C}$ such that $\Phi(\mathcal{C})$ is given in Corollary 2.2 and is based on the operation $*$. In order to give necessary and sufficient conditions for the image of a linear code over R, $\mathfrak{C}$, under φ to be linear depending only on the codewords of $\mathfrak{C}$, we define the following operation. For $\mathbf{u} = (\mathbf{u}_1,\dots,\mathbf{u}_n), \mathbf{v} = (\mathbf{v}_1,\dots,\mathbf{v}_n) \in \mathcal{R}^n$, we define the operation $\mathbf{u}\star\mathbf{v}$ as the component-wise operation, where, for $\mathbf{u}_i = u_i' + u_i x, \mathbf{v}_i = v_i' + v_i x \in R$, $2\mathbf{u}_i \star \mathbf{v}_i = u_i' \cdot v_i'$. Since $\rho(\mathbf{u}\star\mathbf{v}) = \rho(\mathbf{u}) * \rho(\mathbf{u})$, as a corollary of Corollary 2.2, we obtain the following statement.

COROLLARY 4.11. *Let $\mathfrak{C}$ be a linear code over R and $C = \varphi(\mathfrak{C})$. Then, C is linear if and only if $\mathbf{u}\star\mathbf{v} \in \mathfrak{C}$ for all $\mathbf{u},\mathbf{v} \in \mathfrak{C}$.*

Let $\mathfrak{C}$ be a linear code over R, $\mathcal{C} = \rho(\mathfrak{C})$ and $C = \varphi(\mathfrak{C}) = \Phi(\mathcal{C})$. Let $K(C)$ be the kernel of C. We define

$$\begin{aligned}\mathfrak{R}(\mathfrak{C}) &= \varphi^{-1}(\langle\varphi(\mathfrak{C})\rangle),\\ \mathfrak{K}(\mathfrak{C}) &= \varphi^{-1}(K(\varphi(\mathfrak{C}))).\end{aligned}$$

By Lemma 2.3, we have that $\mathcal{R}(\mathcal{C}) = \Phi^{-1}(\langle\Phi(\mathcal{C})\rangle)$ and $\mathcal{K}(\mathcal{C}) = \Phi^{-1}(K(\Phi(\mathcal{C}))$ are $\mathbb{Z}_2\mathbb{Z}_4$-additive codes. We prove that a similar property is given for $\mathfrak{R}(\mathfrak{C})$ and $\mathfrak{K}(\mathfrak{C})$.

LEMMA 4.12. *Let $\mathfrak{C}$ be a linear code over R and $\mathcal{C} = \rho(\mathcal{C})$. Then,*

$$\mathfrak{R}(\mathfrak{C}) = \rho^{-1}(\mathcal{R}(\mathcal{C})).$$

PROOF. By definition, $\mathcal{R}(\mathcal{C}) = \Phi^{-1}(\langle\Phi(\mathcal{C})\rangle) = \Phi^{-1}(\langle\varphi(\mathfrak{C})\rangle)$. Since $\varphi = \rho\circ\Phi$, $\rho^{-1}(\mathcal{R}(\mathcal{C})) = \rho^{-1}(\Phi^{-1}(\langle\varphi(\mathfrak{C})\rangle)) = \varphi^{-1}(\langle\varphi(\mathfrak{C})\rangle) = \mathfrak{R}(\mathfrak{C})$ □

PROPOSITION 4.13. *Let $\mathfrak{C}$ be a linear code over R. Then, $\mathfrak{R}(\mathfrak{C})$ is a linear code over R.*

PROOF. By Lemma 4.12, $\mathfrak{R}(\mathfrak{C}) = \rho^{-1}(\mathcal{R}(\mathcal{C}))$. Let $\mathcal{G}$ be a generator matrix of $\mathcal{C}$, and let $\{\mathbf{u}_i = (u_u \mid u_i')\}_{i=1}^{\gamma}$ be the rows of order two and $\{\mathbf{v}_j = (v_j \mid v_j')\}_{j=1}^{\delta}$ the rows of order four in $\mathcal{G}$. By Proposition 2.4, $\mathcal{R}(\mathcal{C}) = \langle\mathcal{C}, \{2\mathbf{v}_j * \mathbf{v}_k\}_{1\le j<k\le\delta}\rangle$, and hence $\rho^{-1}(\mathcal{R}(\mathcal{C})) = \langle\mathfrak{C}, \{\rho^{-1}(2\mathbf{v}_j * \mathbf{v}_k)\}_{1\le j<k\le\delta}\rangle$. Clearly $\rho^{-1}(\mathcal{R}(\mathcal{C}))$ is additive. Note that for $j,k \in \{1,\dots,\delta\}$, $2\mathbf{v}_j * \mathbf{v}_k = (\mathbf{0} \mid 2v_j' * v_k')$ and hence $\rho^{-1}(2\mathbf{v}_j * \mathbf{v}_k) = 2v_j' * v_k'$. Since $x(2v_j' * v_k') = 0 \in \mathfrak{C}$, we have that $\rho^{-1}(\mathcal{R}(\mathcal{C}))$ is a linear code over R. □

Let $\mathfrak{C}$ be a lincar code over R, $\mathcal{C} = \rho(\mathfrak{C})$ and $C = \varphi(\mathfrak{C}) = \Phi(\mathcal{C})$. Note that, by definition, $\mathcal{K}(\mathcal{C}) = \Phi^{-1}(K(C))$, and we have that

$$\mathfrak{K}(\mathfrak{C}) = \varphi^{-1}(K(C)) = \rho^{-1}(\mathcal{K}(\mathcal{C})). \tag{4.2}$$

By Lemma 2.3, $\mathcal{K}(\mathcal{C})$ is a $\mathbb{Z}_2\mathbb{Z}_4$-additive code, we see that $\mathfrak{K}(\mathfrak{C})$ is a linear code over R.

LEMMA 4.14. *Let $\mathfrak{C}$ be a linear code over R. Then, $\mathfrak{C}_b \subseteq \mathfrak{K}(\mathfrak{C})$.*

PROOF. The statement is given due to the fact that $\mathfrak{C}_b = \rho^{-1}(\mathcal{C}_b)$ by Lemma 4.3 and $\mathcal{C}_b \subseteq \mathcal{K}(\mathcal{C})$. □

PROPOSITION 4.15. *Let $\mathfrak{C}$ be a linear code over R. Then, $\mathfrak{K}(\mathfrak{C})$ is a linear code over R.*

PROOF. Let $\mathfrak{C}$ be a linear code over R, $\mathcal{C} = \rho(\mathfrak{C})$ the corresponding $\mathbb{Z}_2\mathbb{Z}_4$-additive code, and $C = \varphi(\mathfrak{C})$.

By (4.2), $\mathfrak{K}(\mathfrak{C}) = \rho^{-1}(\mathcal{K}(\mathcal{C}))$. Clearly, $\mathfrak{K}(\mathfrak{C})$ is additive and we only have to check that $x\mathfrak{K}(\mathfrak{C}) \subseteq \mathfrak{K}(\mathfrak{C})$. Let $\mathbf{u} \in \mathfrak{K}(\mathfrak{C}) \subseteq \mathfrak{C}$. By Lemma 3.4, $x\mathbf{u} \in \mathfrak{C}_b$. Since $\mathfrak{C}_b \subseteq \mathfrak{R}(\mathfrak{C})$ by Lemma 4.14, $x\mathbf{u} \in \mathfrak{R}(\mathfrak{C})$ and $\mathfrak{R}(\mathfrak{C})$ is linear. □

4.2. Duality. Let $\mathbf{a} = (\mathbf{a}_1, \ldots \mathbf{a}_n), \mathbf{b} = (\mathbf{b}_1, \ldots, \mathbf{b}_n) \in R^n$. The inner product of $\mathbf{a}$, $\mathbf{b}$, denoted be $\langle \mathbf{a}, \mathbf{b}\rangle_R$ is the usual inner product of elements over a ring,

$$\langle \mathbf{a}, \mathbf{b}\rangle_R = \sum_{i=1}^{n} \mathbf{a}_i \mathbf{b}_i.$$

Recall the standard inner product $\langle \mathbf{u}, \mathbf{v}\rangle_{\mathbb{Z}_2\mathbb{Z}_4}$ of vectors $\mathbf{u}, \mathbf{v} \in \mathbb{Z}_2^\alpha \times \mathbb{Z}_4^\beta$ given in (2.1).

LEMMA 4.16. *Let $\mathbf{a}, \mathbf{b} \in R^n$. Then,*

$$\langle \mathbf{a}, \mathbf{b}\rangle_R = \langle \rho(\mathbf{a}), \rho(\mathbf{b})\rangle_{\mathbb{Z}_2\mathbb{Z}_4} + \lambda x,$$

for $\lambda \in \{0, 1\}$.

PROOF. Let $\mathbf{a}_i$ and $\mathbf{b}_i$ be the i-th component of $\mathbf{a}$ and $\mathbf{b}$ respectively. Then

$$\begin{aligned}
\langle \mathbf{a}, \mathbf{b}\rangle_R &= \sum_{i=1}^{n} \mathbf{a}_i \mathbf{b}_i \\
&= \sum_{i=1}^{n} (a_i' + a_i x)(b_i' + b_i x) \\
&= \sum_{i=1}^{n} (a_i' b_i' + (a_i' b_i + a_i b_i')x + 2a_i b_i) \\
&= \sum_{i=1}^{n} (2a_i b_i + a_i' b_i') + (a_i' b_i + a_i b_i')x \\
&= \langle \rho(\mathbf{a}), \rho(\mathbf{b})\rangle_{\mathbb{Z}_2\mathbb{Z}_4} + \sum_{i=1}^{n} (a_i' b_i + a_i b_i')x
\end{aligned}$$

□

COROLLARY 4.17. *Let* $\mathbf{a}, \mathbf{b} \in R^n$. *If* $\langle \mathbf{a}, \mathbf{b} \rangle_R = 0$ *then* $\langle \rho(\mathbf{a}), \rho(\mathbf{b}) \rangle_{\mathbb{Z}_2\mathbb{Z}_4} = 0$.

THEOREM 4.18. *Let* $\mathfrak{C}$ *be a linear code over* R *of length* n. *Then,*

$$\rho(\mathfrak{C}^\perp) = \rho(\mathfrak{C})^\perp.$$

PROOF. By Corollary 4.17, $\rho(\mathfrak{C}^\perp) \subset \rho(\mathfrak{C})^\perp$. Also, $|\mathfrak{C}| = |\rho(\mathfrak{C})|$. Lemma 1 in [**3**] shows that $|\rho(\mathfrak{C})||\rho(\mathfrak{C})^\perp| = 4^n 2^n = 8^n$. From [**23**], since R is a Frobenius ring, $|\mathfrak{C}||\mathfrak{C}^\perp| = 8^n$. So,

$$|\mathfrak{C}^\perp| = \frac{|8|^n}{|\mathfrak{C}|} = \frac{|8|^n}{|\rho(\mathfrak{C})|} = |\rho(\mathfrak{C})^\perp|.$$

Hence, $\rho(\mathfrak{C}^\perp) = \rho(\mathfrak{C})^\perp$. □

From the previous theorem, we have that the following diagram commutes:

$$\begin{array}{ccc} \mathfrak{C} & \xrightarrow{\rho} & C = \rho(\mathfrak{C}) \\ \perp\big\downarrow & & \perp\big\downarrow \\ \mathfrak{C}^\perp & \xrightarrow{\rho} & C^\perp = \rho(\mathfrak{C}^\perp). \end{array} \tag{4.3}$$

4.3. Cyclicity. Let $\mathbf{u} \in R^n$, and $i \geq 0$. We denote by $\mathbf{u}^{(i)}$ the ith cyclic shift of $\mathbf{u}$. A code $\mathfrak{C}$ over R is *cyclic* if for any $\mathbf{u} \in \mathfrak{C}$, $\mathbf{u}^{(1)} \in \mathfrak{C}$. It is well known that a cyclic code over R of length n can be viewed as an ideal in the ambient ring $\frac{R[y]}{\langle y^n - 1 \rangle}$.

EXAMPLE 4.19. Let $\mathfrak{C}$ be code over R generated by

$$G = \begin{pmatrix} 1 & 0 & 0 & 0 & 1 \\ 0 & 1 & 0 & 0 & 1 \\ 0 & 0 & 1 & 0 & 1 \\ 0 & 0 & 0 & 1 & 1 \\ 0 & 0 & 0 & 0 & x \end{pmatrix}.$$

Note that for any row vector $\mathbf{u}$ in G, we have $\mathbf{u}^{(1)} \in \mathfrak{C}$. Therefore, it is easy to see that $\mathfrak{C}$ is a cyclic code over R.

If we consider the polynomial ring $\mathcal{R} = R[y]/(y^5 - 1)$, and the code $\mathfrak{C}$ as an ideal of $\mathcal{R}$, then it is easy to check that $\mathfrak{C} = \langle y+1, x \rangle$. Note, $y - 1 + x = y + 1 + x^2 + x = 1(y+1) + (x+1)x$, $y + 1 = (y - 1 + x)((x-1)y^4 + y^3 + (1-x)y^2 + 3y + x)$ and $x = (y - 1 + x)(y^4 + (1-x)y^3 + 3y^2 + (x-1)y + 1)$. So, $\mathfrak{C} = \langle y+1, x \rangle = \langle y - 1 + x \rangle$.

Since R is a finite chain ring, in [**17**] it is prove that the ring $R[y]/\langle y^n - 1 \rangle$ is principal. Therefore, we obtain the following statement.

PROPOSITION 4.20 ([**17**]). *Let* $\mathfrak{C}$ *be a cyclic code over* R *of odd length* n. *Then, there exist monic polynomials* $f_0(y), f_1(y)$ *and* $f_2(y) \in \mathbb{Z}_2[y]$ *such that* $deg(f_0(y)) < deg(f_1(y)) < deg(f_2(y))$, $f_2(y)|f_1(y)|f_0(y)|y^n - 1$ *and*

$$\mathfrak{C} = \langle f_0(y), x f_1(y), 2 f_2(y) \rangle = \langle f_0(y) + x f_1(y) + 2 f_2(y) \rangle.$$

PROPOSITION 4.21. *Let* $\mathfrak{C}$ *be a cyclic code over* R. *Then, the* $\mathbb{Z}_2\mathbb{Z}_4$*-additive code* $\rho(\mathfrak{C})$ *is cyclic.*

PROOF. Let $\mathfrak{C}$ be a cyclic code over R and $C = \rho(\mathfrak{C})$. Let $\mathbf{u} = (u' \mid u) \in C$. We have that $\rho^{-1}(\mathbf{u}) = u' + ux \in \mathfrak{C}$. Since $\mathfrak{C}$ is cyclic, for $i \geq 0$, $(\rho^{-1}(\mathbf{u}))^{(i)} = (u')^{(i)} + u^{(i)} x \in \mathfrak{C}$. Therefore, $\rho((\rho^{-1}(\mathbf{u}))^{(i)}) = (u^{(i)} \mid (u')^{(i)}) \in C$ and C is cyclic. □

EXAMPLE 4.22. Let $\mathfrak{C}$ be the cyclic code over R given in Example 4.19. The $\mathbb{Z}_2\mathbb{Z}_4$-additive code $\mathcal{C} = \rho(\mathfrak{C})$ is generated by

$$G = \left(\begin{array}{ccccc|ccccc} 1 & 0 & 0 & 0 & 0 & 0 & 0 & 0 & 0 & 0 \\ 0 & 1 & 0 & 0 & 0 & 0 & 0 & 0 & 0 & 0 \\ 0 & 0 & 1 & 0 & 0 & 0 & 0 & 0 & 0 & 0 \\ 0 & 0 & 0 & 1 & 0 & 0 & 0 & 0 & 0 & 0 \\ 0 & 0 & 0 & 0 & 1 & 0 & 0 & 0 & 0 & 0 \\ 0 & 0 & 0 & 0 & 0 & 1 & 0 & 0 & 0 & 1 \\ 0 & 0 & 0 & 0 & 0 & 0 & 1 & 0 & 0 & 1 \\ 0 & 0 & 0 & 0 & 0 & 0 & 0 & 1 & 0 & 1 \\ 0 & 0 & 0 & 0 & 0 & 0 & 0 & 0 & 1 & 1 \\ 0 & 0 & 0 & 0 & 0 & 0 & 0 & 0 & 0 & 2 \end{array}\right)$$

and it is also cyclic. Note that $\mathcal{C} = \langle (1 \mid 0), (0 \mid y - 1)\rangle$; that is, $f(y) = 1, h(y) = y + 1, b(y) = 1, \ell(y) = 0$.

Note that in the last example, we have that the code $\mathcal{C}$ is separable, and so it is $\mathfrak{C}$ by definition. This fact is true in general for all cyclic code over R with odd length as it is shown in the following proposition.

PROPOSITION 4.23. *Let $\mathfrak{C}$ be a cyclic code over R of odd length n. Then, $\mathfrak{C}$ is separable.*

PROOF. From Proposition 4.21, the $\mathbb{Z}_2\mathbb{Z}_4$-additive code $\mathcal{C} = \rho(\mathfrak{C})$ is also cyclic. We have that, by Proposition 4.20,

$$\mathfrak{C} = \langle f_0(y), xf_1(y), 2f_2(y)\rangle,$$

for $f_0(y), f_1(y), f_2(y) | y^n - 1$. Therefore,

$$\begin{aligned} \mathcal{C} &= \langle \rho(f_0(y)), \rho(xf_1(y)), \rho(2f_2(y))\rangle \\ &= \langle (0 \mid f_0(y)), (f_1(y) \mid 0)), (0, 2f_2(y))\rangle \\ &= \langle (f_1(y) \mid 0), (0, f_0(y) + 2f_2(y))\rangle. \end{aligned}$$

Hence, the code $\mathcal{C}$ is separable and so it is $\mathfrak{C}$. □

As a corollary of Theorem 2.6, Proposition 4.21 and Proposition 4.23 we have the following statement.

COROLLARY 4.24. *Let $\mathfrak{C}$ be a cyclic code over R of odd length n, $\mathcal{C} = \rho(\mathfrak{C})$. Then,*

- $\mathfrak{C} = \langle f(y)h(y) + xp(y)f(y) + 2f(y)\rangle$,
- $\mathcal{C} = \langle (p(y)f(y) \mid 0), (0, f(y)h(y) + 2f(y))\rangle$,

where $f(y)h(y)g(y) = y^n - 1$, for $f(y), h(y)$ and $g(y)$ pairwise coprime polynomials in $\mathbb{Z}_2[y]$ and $p(y)|h(y)$.

EXAMPLE 4.25. Let $\mathfrak{C}$ be the cyclic code over R given in Example 4.19, and $\mathcal{C} = \rho(\mathfrak{C})$. From Examples 4.19 and 4.22, we have that $\mathfrak{C} = \langle y - 1 + x\rangle$ and $\mathcal{C} = \langle (1 \mid 0), (0 \mid y - 1)\rangle$. From Corollary 4.24, $\mathfrak{C} = \langle f(y)h(y) + xp(y)f(y) + 2f(y)\rangle$, where $f(y) = p(y) = 1$ and $h(y) = y - 1$.

5. Other rings with $\mathbb{Z}_2\mathbb{Z}_4$-additive structure

In previous sections a chain ring of order 8 was studied. In this section, we find a family of finite commutative chain rings with $\mathbb{Z}_2\mathbb{Z}_4$-additive structure as well as consider a local Frobenius non-chain ring of order 16 with $\mathbb{Z}_2\mathbb{Z}_4$-additive structure.

5.1. Infinite family of finite commutative chain ring.

LEMMA 5.1 (Theorem XVII.5, [**16**]). *Let R be a finite commutative chain ring. Suppose the maximal ideal J of R has nilpotency e, the characteristic of R is p^k and $r = [R/J : Z_p]$. Then there exist integers t and s such that*

$$R \cong \frac{GR(p^k, r)[x]}{\langle g(x), p^{k-1}x^t\rangle},$$

where $1 \leq t = e - (k-1)s \leq s$ and $g(x)$ is a monic Eisenstein polynomial of degree s over $GR(p^k, r)$. Furthermore, when $t = s$, $p^{k-1}x^t = p^{k-1}g(x)$, so the $p^{k-1}x^t$ term can be omitted.

By Lemma 5.1, when the characteristic of a finite chain ring R is 4,

$$R \cong \frac{GR(4, r)[x]}{\langle g(x), 2x^t\rangle}$$

where $g(x)$ is a monic Eisenstein polynomial of degree s over $GR(4, r)$. If $t = s$, as an additive group, R is isomorphic to $\mathbb{Z}_4^s$. To guarantee that R has $\mathbb{Z}_2\mathbb{Z}_4$-additive structure, we exclude the cases where $t = s$. Also, by Theorem 1 and Lemma 5 in [**8**], when s is odd, we may assume $g(x) = x^s - 2$. Although having $r > 1$ will produce rings with $\mathbb{Z}_2\mathbb{Z}_4$-additive structure, we will keep it simple and only consider cases when $r = 1$ and also assume $g(x) = x^s - 2$. Specifically, let

$$R = \frac{\mathbb{Z}_4[x]}{\langle x^s - 2, 2x^t\rangle},$$

where $1 \leq t < s$.

The following is the ideal lattice of R.

$$\begin{array}{c}
\langle 1 \rangle \\
| \\
\langle x \rangle \\
| \\
\langle x^2 \rangle \\
| \\
\vdots \\
| \\
\langle x^s \rangle = \langle 2 \rangle \\
| \\
\langle x^{s+1} \rangle = \langle 2x \rangle \\
| \\
\vdots \\
| \\
\langle x^{s+t-1} \rangle = \langle 2x^{t-1} \rangle \\
| \\
\langle x^{s+t} \rangle = \langle 2x^t \rangle = \langle 0 \rangle
\end{array}$$

Notice that multiplication by x "moves" elements down the chain. It is clear that the additive structure is $\mathbb{Z}_2^{s-t} \times \mathbb{Z}_4^t$. Note that any element $\mathbf{a} \in R$ can be written as $\mathbf{a} = a_1 x^{s-1} + \cdots + a_{s-t}x^t + \cdots + a_{s-1}x + a_s$, for $a_1, \ldots, a_{s-t} \in \{0,1\}$ and $a_{s-t+1}, \ldots, a_s \in \{0,1,2,3\}$. We define the map

$$\begin{aligned}
\rho : R &\to \mathbb{Z}_2^{s-t} \times \mathbb{Z}_4^t \\
\mathbf{a} &\mapsto (a_1, \ldots, a_{s-t} \mid a_{s-t+1}, \ldots, a_s).
\end{aligned}$$

We denote also by ρ the coordinate extended map from R^n to $\mathbb{Z}_2^{(s-t)n} \times \mathbb{Z}_4^{tn}$ as follows. Let $\mathbf{a} \in R^n$. We have that

$$\begin{aligned}
\mathbf{a} &= (a_{1,1}x^{s-1} + \cdots + a_{1,s-1}x + a_{1,s}, \ldots, a_{n,1}x^{s-1} + \cdots + a_{n,s-1}x + a_{1,s}) \\
&= \mathbf{a}_1 x^{s-1} + \ldots \mathbf{a}_{s-t}x^t + \cdots + \mathbf{a}_{s-1}x + \mathbf{a}_s,
\end{aligned}$$

where $\mathbf{a}_i = (a_{1,i}, \ldots, a_{n,i})$. Then,

$$\begin{aligned}
\rho : R^n &\to \mathbb{Z}_2^{(s-t)n} \times \mathbb{Z}_4^{tn} \\
\mathbf{a} &\mapsto (\mathbf{a}_1, \ldots, \mathbf{a}_{s-t} \mid \mathbf{a}_{s-t+1}, \ldots, \mathbf{a}_s).
\end{aligned}$$

Note that $\mathbf{a}_1, \ldots, \mathbf{a}_{s-t} \in \{0,1\}^n$ and $\mathbf{a}_{s-t+1}, \ldots, \mathbf{a}_s \in \{0,1,2,3\}^n$.

EXAMPLE 5.2. Let $t = 2$ and $s = 4$ so that $R = \frac{\mathbb{Z}_4[x]}{\langle x^4-2,2x^2\rangle}$. Let $n = 3$ and $\mathbf{a} = (x^3+3, x^2+2x+1, x^3+x) \in R^3$. We can write $\mathbf{a} = (1,0,1)x^3 + (0,1,0)x^2 + (0,2,1)x + (3,1,0)$. Therefore,

$$\rho(\mathbf{a}) = (1,0,1,0,1,0 \mid 0,2,1,3,1,0).$$

By considering the ρ map, we have the following proposition.

PROPOSITION 5.3. *Let $R = \frac{\mathbb{Z}_4[x]}{\langle x^s-2,2x^t\rangle}$ with $1 \leq t < s$. Let $\mathfrak{C}$ be a linear code over R of length n. Then, $\mathfrak{C}$ is isomorphic to a $\mathbb{Z}_2\mathbb{Z}_4$-additive code $\mathcal{C} \subseteq \mathbb{Z}_2^{(s-t)n} \times \mathbb{Z}_4^{tn}$.*

In the next corollary we show how multiplication by x effects the image a codeword.

COROLLARY 5.4. *Let $\mathfrak{C}$ be a linear code over $\frac{\mathbb{Z}_4[x]}{\langle x^s-2,2x^t\rangle}$ of length n and let $\mathcal{C} = \rho(\mathfrak{C})$. Then, for all $(\mathbf{a}_1,\dots,\mathbf{a}_{s-t} \mid \mathbf{a}_{s-t+1},\dots,\mathbf{a}_s) \in \mathcal{C}$, where $\mathbf{a}_i \in \mathbb{Z}_2^n$ for $1 \leq i \leq s-t$ and $\mathbf{a}_j \in \mathbb{Z}_4^n$ for $s-t+1 \leq j \leq s$, we have*

$$(\mathbf{a}_2,\dots,\mathbf{a}_{s-t},\tilde{\mathbf{a}}_{s-t+1} \mid \mathbf{a}_{s-t+2},\dots,\mathbf{a}_s, 2\mathbf{a}_1) \in \mathcal{C}.$$

PROOF. Let $(\mathbf{a}_1,\dots,\mathbf{a}_{s-t} \mid \mathbf{a}_{s-t+1},\dots,\mathbf{a}_s) \in \mathcal{C}$, and $\mathbf{a} = \rho^{-1}(\mathbf{a}_1,\dots,\mathbf{a}_{s-t} \mid \mathbf{a}_{s-t+1},\dots,\mathbf{a}_s)$. The statement follows from Proposition 5.3 and the fact that for $x\mathbf{a} \in \mathfrak{C}$ and $\rho(x\mathbf{a}) = (\mathbf{a}_2,\dots,\mathbf{a}_{s-t},\tilde{\mathbf{a}}_{s-t+1} \mid \mathbf{a}_{s-t+2},\dots,\mathbf{a}_s,2\mathbf{a}_1)$. □

EXAMPLE 5.5. Let $t = 1$ and $s = 3$. Note that the ring $\frac{\mathbb{Z}_4[x]}{\langle x^3-2,2x\rangle}$ has order 16. Let $\mathfrak{C}$ be a linear code of length n over R. By Proposition 5.3 $\mathcal{C} = \rho(\mathfrak{C})$ is a $\mathbb{Z}_2\mathbb{Z}_4$-additive, $\mathcal{C} \subseteq \mathbb{Z}_2^{2n} \times \mathbb{Z}_4^{3n}$.

Any codeword in $\mathfrak{C}$ can be written as $\mathbf{a} = \mathbf{a}_1x^2 + \mathbf{a}_2x + \mathbf{a}_3$, for $\mathbf{a}_1, \mathbf{a}_2 \in \{0,1\}^n$ and $\mathbf{a}_3 \in \{0,1,2,3\}^n$. We have that $\rho(\mathbf{a}) = (\mathbf{a}_1, \mathbf{a}_2 \mid \mathbf{a}_3)$. Note that $\rho(x\mathbf{a}) = \rho(2\mathbf{a}_1 + \mathbf{a}_2x^2 + \tilde{\mathbf{a}_3}x) = (\mathbf{a}_2, \tilde{\mathbf{a}_3} \mid 2\mathbf{a}_1)$.

5.2. Frobenius rings of order 16. We have seen in Example 5.5 that $\frac{\mathbb{Z}_4[x]}{\langle x^3-2,2x\rangle}$, which is of order 16, have $\mathbb{Z}_2\mathbb{Z}_4$-additive structure. In the following proposition we give a list of all local Frobenius rings of order 16 with $\mathbb{Z}_2\mathbb{Z}_4$-additive structure which can be obtained from Corollary 1 in [**14**].

PROPOSITION 5.6. *Up to isomorphism there are 3 local commutative Frobenius rings with $\mathbb{Z}_2\mathbb{Z}_4$-additive structure of order 16. These rings, which are isomorphic to $\mathbb{Z}_2^2 \times \mathbb{Z}_4$, are the following:*

(1) $\frac{\mathbb{Z}_4[x]}{\langle x^3-2,2x\rangle}$,
(2) $\frac{\mathbb{Z}_4[x,y]}{\langle x^2,xy-2,y^2,2x,2y\rangle}$,
(3) $\frac{\mathbb{Z}_4[x,y]}{\langle x^2-2,xy-2,y^2,2x,2y\rangle}$.

Notice the first in the previous list is a chain ring of the type we have been discussing in Section 5.1. The other two are non-chain rings. These are two of the set of smallest local Frobenius non-chain rings. It can be shown that $\frac{\mathbb{Z}_4[x,y]}{\langle x^2-2,xy-2,y^2,2x,2y\rangle} \cong \frac{\mathbb{Z}_4[x,y]}{\langle x^2-2,xy,y^2-2,2x,2y\rangle}$. For the remainder of the section we set $R = \frac{\mathbb{Z}_4[x,y]}{\langle x^2-2,xy,y^2-2,2x,2y\rangle}$. We will consider codes over R and their images as $\mathbb{Z}_2\mathbb{Z}_4$-additive codes. We mainly do so as an example of how the "mixing" of the binary and quaternary parts can be controlled. The following is the ideal lattice of R.

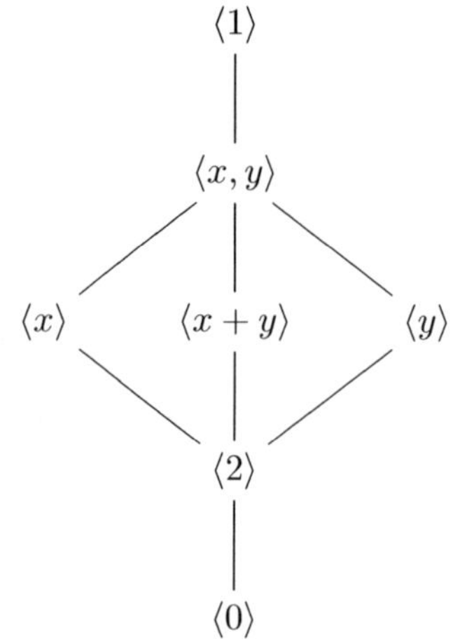

As an additive group, $R = \langle 1, x, y\rangle$. So, any element can be represented as $\mathbf{a} = a' + a_x x + a_y y$ where $a' \in \{0,1,2,3\}$ and $a_x, a_y \in \{0,1\}$. With this we define the map

$$\rho : R \to \mathbb{Z}_2^2 \times \mathbb{Z}_4;\ \mathbf{a} = a' + a_x x + a_y y \mapsto (a_x, a_y \mid a').$$

We also denote ρ the coordinate-wise extended map to elements in R^n. Let $\mathbf{a} \in R^n$. We have that

$$\mathbf{a} = (a_1' + a_{1,x}x + a_{1,y}y, \ldots, a_n' + a_{n,x}x + a_{n,y}y) = \mathbf{a}_1 + \mathbf{a}_x x + \mathbf{a}_y y,$$

where $\mathbf{a}' = (a_1', \ldots, a_n')$, $\mathbf{a}_x = (a_{1,x}, \ldots, a_{n,x})$ and $\mathbf{a}_y = (a_{1,y}, \ldots, a_{n,y})$. Then,

$$\rho : R^n \to \mathbb{Z}_2^{2n} \times \mathbb{Z}_4^n;\ \mathbf{a}' + \mathbf{a}_x x + \mathbf{a}_y y \mapsto (\mathbf{a}_x, \mathbf{a}_y \mid \mathbf{a}'),$$

where $\mathbf{a}' \in \{0,1,2,4\}^n$ and $\mathbf{a}_x, \mathbf{a}_y \in \{0,1\}^n$. From this map, we can obtain the following proposition.

PROPOSITION 5.7. *Let $R = \frac{\mathbb{Z}_4[x,y]}{\langle x^2-2, xy, y^2-2, 2x, 2y\rangle}$. Let $\mathfrak{C}$ be a linear code over R of length n. Then, $\mathfrak{C}$ is isomorphic to a $\mathbb{Z}_2\mathbb{Z}_4$-additive code $\mathcal{C} \subseteq \mathbb{Z}_2^{2n} \times \mathbb{Z}_4^n$.*

In the next corollary we show how multiplication by x and by y effects the image a codeword.

COROLLARY 5.8. *Let $\mathfrak{C}$ be a linear code over $\frac{\mathbb{Z}_4[x,y]}{\langle x^2-2, xy, y^2-2, 2x, 2y\rangle}$ of length n and let $\mathcal{C} = \rho(\mathfrak{C})$. Then, for all $(\mathbf{a}_x, \mathbf{a}_y \mid \mathbf{a}') \in \mathcal{C}$, where $\mathbf{a}_x, \mathbf{a}_y \in \mathbb{Z}_2^n$ and $\mathbf{a}' \in \mathbb{Z}_4^n$, we have*

(1) $(\tilde{a'}, 0 \mid 2a_x) \in \mathcal{C}$, *and*
(2) $(0, \tilde{a'} \mid 2a_y) \in \mathcal{C}$.

PROOF. Let $(\mathbf{a}_x, \mathbf{a}_y \mid \mathbf{a}') \in \mathcal{C}$, and $\mathbf{a} = \rho^{-1}(\mathbf{a}_x, \mathbf{a}_y \mid \mathbf{a}')$. The statement follows from Proposition 5.7 and the fact that $x\mathbf{a} \in \mathfrak{C}$ and $y\mathbf{a} \in \mathcal{C}$. □

Finally we look at the inner product over R and some conditions on the duality.

LEMMA 5.9. *Let $\mathbf{a}, \mathbf{b} \in R^n$. Then,*

$$\langle \mathbf{a}, \mathbf{b}\rangle_R = \langle \rho(\mathbf{a}), \rho(\mathbf{b})\rangle_{\mathbb{Z}_2\mathbb{Z}_4} + \lambda x + \mu y,$$

for $\lambda, \mu \in \{0,1\}$.

PROOF. Let $\mathbf{a}_i$ and $\mathbf{b}_i$ be the i-th component of $\mathbf{a}$ and $\mathbf{b}$ respectively. Then

$$\begin{aligned}
\langle \mathbf{a}, \mathbf{b} \rangle_R &= \sum_{i=1}^{n} \mathbf{a}_i \mathbf{b}_i \\
&= \sum_{i=1}^{n} (a_i' + a_{ix}x + a_{iy}y)(b_i' + b_{ix}x + b_{iy}y) \\
&= \sum_{i=1}^{n} a_i'b_i' + (a_{ix}b_i' + a_i'b_{ix})x + (a_{iy}b_i' + a_i'b_{iy})y + 2(a_{ix}b_{ix} + a_{iy}b_{iy}) \\
&= \sum_{i=1}^{n} (2(a_{ix}b_{ix} + a_{iy}b_{iy}) + a_i'b_i') + (a_{ix}b_i' + a_i'b_{ix})x + (a_{iy}b_i' + a_i'b_{iy})y \\
&= \langle \rho(\mathbf{a}), \rho(\mathbf{b}) \rangle_{\mathbb{Z}_2\mathbb{Z}_4} + \sum_{i=1}^{n} (a_{ix}b_i' + a_i'b_{ix})x + (a_{iy}b_i' + a_i'b_{iy})y
\end{aligned}$$

□

To close, we notice in the previous example, that the duality preservation is dependent on the definition of ρ. In previous sections, and here with this last example, the map ρ was defined in the natural way. But, we could have used the ring $\frac{\mathbb{Z}_4[x,y]}{\langle x^2-2, xy-2, y^2, 2x, 2y \rangle}$ with the generating set $1, x, x+y$ of the ring to define ρ. For some clarity on this idea of duality preserving basis, see [**22**].

Acknowledgment

This paper is a result of discussions during the conference "Noncommutative rings and their Applications, VII", Lens, France, 5-7th July 2021. The authors would like to thank Dr. André Leroy for his ongoing effort to connect the ring theory and coding theory communities through this biennial conference.

References

[1] T. Abualrub, I. Siap, and N. Aydin, *$\mathbb{Z}_2\mathbb{Z}_4$-additive cyclic codes*, IEEE Trans. Inform. Theory **60** (2014), no. 3, 1508–1514, DOI 10.1109/TIT.2014.2299791. MR3168415

[2] H. Bauer, B. Ganter, and F. Hergert, *Algebraic techniques for nonlinear codes*, Combinatorica **3** (1983), no. 1, 21–33, DOI 10.1007/BF02579339. MR716419

[3] J. Borges, C. Fernández-Córdoba, J. Pujol, J. Rifà, and M. Villanueva, *$\mathbb{Z}_2\mathbb{Z}_4$-linear codes: generator matrices and duality*, Des. Codes Cryptogr. **54** (2010), no. 2, 167–179, DOI 10.1007/s10623-009-9316-9. MR2576874

[4] J. Borges, C. Fernández-Córdoba, J. Pujol, J. Rifà, and M. Villanueva, *$\mathbb{Z}_2\mathbb{Z}_4$-Linear Codes.* Springer, 2022. ISBN: 978-3-031-05440-2.

[5] J. Borges, C. Fernández-Córdoba, and R. Ten-Valls, *$\mathbb{Z}_2\mathbb{Z}_4$-additive cyclic codes, generator polynomials, and dual codes*, IEEE Trans. Inform. Theory **62** (2016), no. 11, 6348–6354, DOI 10.1109/TIT.2016.2611528. MR3565120

[6] A. R. Hammons Jr., P. V. Kumar, A. R. Calderbank, N. J. A. Sloane, and P. Solé, *The $\mathbf{Z}_4$-linearity of Kerdock, Preparata, Goethals, and related codes*, IEEE Trans. Inform. Theory **40** (1994), no. 2, 301–319, DOI 10.1109/18.312154. MR1294046

[7] A. R. Calderbank and N. J. A. Sloane, *Modular and p-adic cyclic codes*, Des. Codes Cryptogr. **6** (1995), no. 1, 21–35, DOI 10.1007/BF01390768. MR1334198

[8] W. E. Clark and J. J. Liang, *Enumeration of finite commutative chain rings*, J. Algebra **27** (1973), 445–453, DOI 10.1016/0021-8693(73)90055-0. MR337910

[9] P. Delsarte, *An algebraic approach to the association schemes of coding theory*, Philips Res. Rep. Suppl. **10** (1973), vi+97. MR384310

[10] P. Delsarte and V. I. Levenshtein, *Association schemes and coding theory*, IEEE Trans. Inform. Theory **44** (1998), no. 6, 2477–2504, DOI 10.1109/18.720545. Information theory: 1948–1998. MR1658771

[11] S. T. Dougherty, B. Yildiz, and S. Karadeniz, *Codes over R_k, Gray maps and their binary images*, Finite Fields Appl. **17** (2011), no. 3, 205–219, DOI 10.1016/j.ffa.2010.11.002. MR2793245

[12] C. Fernández-Córdoba, J. Pujol, and M. Villanueva, *$\mathbb{Z}_2\mathbb{Z}_4$-linear codes: rank and kernel*, Des. Codes Cryptogr. **56** (2010), no. 1, 43–59, DOI 10.1007/s10623-009-9340-9. MR2644247

[13] F. J. MacWilliams and N. J. A. Sloane, *The theory of error-correcting codes. I*, North-Holland Mathematical Library, Vol. 16, North-Holland Publishing Co., Amsterdam-New York-Oxford, 1977. MR0465509

[14] E. Martínez-Moro and S. Szabo, *On codes over local Frobenius non-chain rings of order 16*, Noncommutative rings and their applications, Contemp. Math., vol. 634, Amer. Math. Soc., Providence, RI, 2015, pp. 227–241, DOI 10.1090/conm/634/12702. MR3307401

[15] E. Martínez-Moro, S. Szabo, and B. Yildiz, *Linear codes over $\frac{\mathbb{Z}_4[x]}{\langle x^2+2x\rangle}$*, Int. J. Inf. Coding Theory **3** (2015), no. 1, 78–96. MR3341091

[16] B. R. McDonald, *Finite rings with identity*, Pure and Applied Mathematics, Vol. 28, Marcel Dekker, Inc., New York, 1974. MR0354768

[17] G. H. Norton and A. Sălăgean, *On the structure of linear and cyclic codes over a finite chain ring*, Appl. Algebra Engrg. Comm. Comput. **10** (2000), no. 6, 489–506, DOI 10.1007/PL00012382. MR1770877

[18] K. T. Phelps, J. Rifà, and M. Villanueva, *Rank and kernel of binary Hadamard codes*, IEEE Trans. Inform. Theory **51** (2005), no. 11, 3931–3937, DOI 10.1109/TIT.2005.856940. MR2239008

[19] R. Raghavendran, *Finite associative rings*, Compositio Math. **21** (1969), 195–229. MR246905

[20] J. Rifà, J. M. Basart, and L. Huguet, *On completely regular propelinear codes*, Applied algebra, algebraic algorithms and error-correcting codes (Rome, 1988), Lecture Notes in Comput. Sci., vol. 357, Springer, Berlin, 1989, pp. 341–355, DOI 10.1007/3-540-51083-4_71. MR1008511

[21] J. Rifà and J. Pujol, *Translation-invariant propelinear codes*, IEEE Trans. Inform. Theory **43** (1997), no. 2, 590–598, DOI 10.1109/18.556115. MR1447533

[22] S. Szabo and F. Ulmer, *Duality preserving Gray maps for codes over rings*, J. Algebra Appl. **16** (2017), no. 9, 1750161, 23, DOI 10.1142/S0219498817501614. MR3661628

[23] S. Szabo and J. A. Wood, *Properties of dual codes defined by nondegenerate forms*, J. Algebra Comb. Discrete Struct. Appl. **4** (2017), no. 2, 105–113, DOI 10.13069/jacodesmath.284934. MR3601342

[24] J. A. Wood, *Duality for modules over finite rings and applications to coding theory*, Amer. J. Math. **121** (1999), no. 3, 555–575. MR1738408

[25] B. Yildiz and S. Karadeniz, *Linear codes over $\mathbb{F}_2 + u\mathbb{F}_2 + v\mathbb{F}_2 + uv\mathbb{F}_2$*, Des. Codes Cryptogr. **54** (2010), no. 1, 61–81, DOI 10.1007/s10623-009-9309-8. MR2576881

Department of Information and Communications Engineering, Universitat Autònoma de Barcelona, Spain

Email address: `cristina.fernandez@uab.es`

Department of Mathematics and Statistics, Eastern Kentucky University, Richmond, Kentucky 40475

Email address: `steve.szabo@eku.edu`

Contemporary Mathematics
Volume **785**, 2023
https://doi.org/10.1090/conm/785/15782

On Utumi rings and continuous regular Baer rings

Gangyong Lee, Cosmin S. Roman, Nguyen Khanh Tung,
and Xiaoxiang Zhang

Dedicated to Professor S. Tariq Rizvi for his contribution to Ring and Module Theory

Abstract. In 1963, Utumi determined under what condition the maximal one-sided ring of quotients of a ring is, in fact, two-sided. That is, he showed if a ring is nonsingular and cononsingular then its maximal right and left rings of quotients coincide. In this paper, we provide more properties of an Utumi ring. A ring is called (right) Utumi if it is both (right) nonsingular and (right) cononsingular. We show that every right Utumi ring is left nonsingular. Moreover, we prove that, for any one-sided nonsingular ring R, its maximal right and left rings of quotients coincide if and only if R is a cononsingular ring. Thus, for a right nonsingular R, which is not left nonsingular, the maximal right and left rings of quotients of R are different. We also prove that every left cononsingular, right extending ring is directly finite. In particular, every left cononsingular, right nonsingular, right continuous ring is a unit-regular Baer ring. Furthermore, we characterize right and left continuous regular rings in terms of Utumi rings. Examples which delineate the concepts and results are provided.

1. Introduction

As we know that a commutative integral domain can be embedded in a field, called its field of fractions. A construction of a "ring of quotients" from a given ring has been studied widely. So, rings of quotients play an important role in the ring theory (see, e.g., [**2, 4, 5, 9, 11, 12, 20, 23**]). Utumi [**25**] gave another construction of a ring of quotients for any ring with zero left annihilator. While a ring R need not have the classical right ring of quotients in general, it always possesses the maximal right (resp., left) ring of quotients, denoted by $Q(R)$ (resp., $Q^\ell(R)$), which is a useful tool to characterize the ring R. According to the theorem due to Johnson [**11**, Theorem 3], a ring is right nonsingular if and only if its maximal right ring of quotients is (von Neumann) regular. Utumi [**26**, Theorem 3.3] proved that the maximal right and left rings of quotients of a nonsingular ring R coincide, i.e., $Q(R) = Q^\ell(R)$, precisely when R is cononsingular.

Recall that a ring R is *right nonsingular* if there is no nonzero element $s \in R$ such that the right annihilator $\mathbf{r}_R(s)$ is essential in R_R. A ring R is called *right cononsingular* if any right ideal, with zero left annihilator, is essential in R_R. The

2020 *Mathematics Subject Classification.* Primary 16D50, 16E50, 16R20.

Key words and phrases. Utumi ring, maximal right rings of quotients, directly finite, regular Baer ring.

left version of (co)nonsingularity is defined similarly. If R is both right and left (co)nonsingular then it is called a *(co)nonsingular ring.* Hence R is said to be a *right Utumi ring* if it is both right nonsingular and right cononsingular. A *left Utumi ring* is defined similarly. A two-sided Utumi ring shall be referred to as an *Utumi ring.* It turns out that every right Utumi ring is left nonsingular (Proposition 2.4). Moreover, we prove that, for any one-sided nonsingular ring R, $Q(R) = Q^{\ell}(R)$ if and only if R is a cononsingular ring (Theorem 2.6).

Chatters and Khuri [**3**] proved that R is a right nonsingular right extending ring if and only if R is a right cononsingular Baer ring. Recall that a ring R is said to be *right extending* if every right ideal of R is essential in a direct summand of R_R. A ring R is called *Baer* if the right annihilator of any nonempty subset of R is a direct summand of R_R. It is well known that R is a Baer ring if and only if the left annihilator of any nonempty subset of R is a direct summand of ${}_RR$. We prove in Theorem 2.13 that every left cononsingular, right extending ring R is directly finite (i.e., for any $x, y \in R$, $xy = 1$ implies $yx = 1$).

It is well known that every right and left continuous (von Neumann) regular ring is unit-regular. We show that every left cononsingular, right nonsingular, right continuous ring is a unit-regular Baer ring (Theorem 3.1). We also characterize right and left continuous regular rings in terms of Utumi rings (see Theorem 3.6 for more details).

Throughout this paper, R is a ring with unity. The notations $I \leq R$, $I \trianglelefteq R$, or $I \leq^{\text{ess}} R$ mean that I is a right ideal, a right and left ideal, or an essential right ideal of R, respectively. $\text{CFM}_{\mathbb{N}}(F)$ denotes the $\mathbb{N}\times\mathbb{N}$ column finite matrix ring over a field F. We also denote $\mathbf{r}_R(I) = \{r \in R \,|\, Ir = 0\}$ and $\mathbf{l}_R(I) = \{r \in R \,|\, rI = 0\}$ for $I \subseteq R$. For unexplained concepts and notations, we refer the readers to [**1, 2, 4, 5, 12, 17**].

For a ring, it is of interest to consider the relations among the properties of being Baer, extending, continuous, regular, and unit-regular. For instance, the following results have been established.

THEOREM 1.1. *The following statements hold true:*

(i) ([**11**, Theorem 3]) *A ring is right nonsingular if and only if its maximal right ring of quotients is regular.*
(ii) ([**5**, Corollary 13.23]) *Every right and left continuous regular ring is unit-regular.*
(iii) ([**3**, Theorem 2.1]) *A ring R is right nonsingular, right extending if and only if R is a right cononsingular Baer ring.*
(iv) ([**1**, 1.41]) *Every right continuous regular ring is Baer.*

2. Utumi rings

Recall the notion that $Q^r(R)$ (resp., $Q^{\ell}(R)$) is the maximal right (resp., left) ring of quotients of a ring R. For simplicity, in this paper we will always denote the maximal right ring of quotients $Q^r(R)$ as $Q(R)$.

DEFINITION 2.1. A ring R is said to be *right Utumi* if it is both right nonsingular and right cononsingular. A *left Utumi ring* is defined similarly. An *Utumi ring* is a left and right Utumi ring.

EXAMPLE 2.2. (i) The product of fields is an Utumi ring.

(ii) Every finite matrix ring over a right and left self-injective regular ring is an Utumi ring.

(iii) Every right and left Ore domain is an Utumi ring.

(iv) Any semiprime PI-ring is an Utumi ring. (From [**2**, Theorem 3.3.4], a semiprime PI-ring is right and left nonsingular and $Q(R) = Q^{\ell}(R)$. So, from Lemma 2.5(i) every semiprime PI-ring is right and left cononsingular.) A ring R (not necessarily with identity) is said to satisfy a *polynomial identity* or simply, *PI*, if there is a polynomial $f(x_1,\dots,x_n)$ with integer coefficients and not necessarily commuting indeterminates $x_1,\dots,x_n$ such that the coefficient of one of the monomials in $f(x_1,\dots,x_n)$ of maximal degree is 1 and $f(a_1,\dots,a_n) = 0$ for all $a_1,\dots,a_n$ in R.

Note that every left Ore domain is left Utumi. However, a left Ore domain need not be right Utumi in general as shown in the next example:

EXAMPLE 2.3. Let σ be an endomorphism of a division ring R with $\sigma(R) \neq R$, and $S = R[x;\sigma]$ with $rx = x\sigma^{-1}(r)$. Then S is a left noetherian domain, so S is a left Ore domain, but as we will see shortly, not right Ore: Let $t \in R \setminus \sigma(R)$, then $\{1,t\} \subseteq R$ is right linearly independent over $\sigma(R)$. Thus, $\{x,tx\}$ is right linearly independent over S. Hence S is not right Ore. On the other hand, S is a left Utumi ring but not right Utumi because S is not right cononsingular. In fact, $\mathbf{l}_S(xS) = 0$ and there is no $0 \neq s \in S$ such that $txs \in xS$.

PROPOSITION 2.4. *Every right Utumi ring is left nonsingular.*

PROOF. Let R be a right Utumi ring. Consider $x \in R$ such that $\mathbf{l}_R(x) \leq^{\text{ess}} {}_RR$. We claim that $R \cap I \neq 0$ for all nonzero left ideals I of $Q(R)$: Suppose not, let I be a nonzero left ideal of $Q(R)$ such that $R \cap I = 0$. Since $Q(R)$ is a regular ring, there exists $0 \neq e^2 = e \in I$. Thus, $Q(R)e \cap R = 0$. Since $(1-e)Q(R) \cap R \leq^{\text{ess}} (1-e)Q(R)_R$ and $Q(R)$ is nonsingular as a right R-module, from [**16**, Remark 3.16(i)]

$$\mathbf{l}_R((1-e)Q(R)\cap R) = \mathbf{l}_R((1-e)Q(R)) = \mathbf{l}_{Q(R)}((1-e)Q(R))\cap R = Q(R)e\cap R = 0.$$

Since $[(1-e)Q(R)\cap R]\cap[eQ(R)\cap R] = 0$ and $eQ(R)\cap R \neq 0$, it contradicts the fact that R is right cononsingular, proving the claim. Thus, $R \cap I \neq 0$. Hence, $\mathbf{l}_R(x) \cap R \cap I \neq 0$ because $\mathbf{l}_R(x) \leq^{\text{ess}} {}_RR$. Since $0 \neq \mathbf{l}_R(x) \cap R \cap I \subseteq Q(R)\mathbf{l}_R(x) \cap I$ for all nonzero left ideals I of $Q(R)$, $Q(R)\mathbf{l}_R(x) \leq^{\text{ess}} {}_{Q(R)}Q(R)$. As $Q(R)\mathbf{l}_R(x) \subseteq \mathbf{l}_{Q(R)}(x)$, $x = 0$ because $Q(R)$ is left nonsingular. Therefore R is left nonsingular. □

LEMMA 2.5. (i)([**26**, Theorem 3.3] or [**4**, Theorem 2.38]) *Let R be a nonsingular ring. Then $Q(R) = Q^{\ell}(R)$ if and only if R is a cononsingular ring.*
(ii)([**26**, Theorem 3.3]) *For an Utumi ring R, the maximal right and left rings of quotients of R coincide, i.e., $Q(R) = Q^{\ell}(R)$.*

Using Proposition 2.4, we extend Lemma 2.5(i)

THEOREM 2.6. *Let R be any one-sided nonsingular ring. Then $Q(R) = Q^{\ell}(R)$ if and only if R is a cononsingular ring.*

PROOF. Let R be right nonsingular. Then $Q(R)$ is a regular ring, hence it is left nonsingular as a $Q(R)$-module. Since $Q(R) = Q^{\ell}(R)$ and ${}_RR \leq Q^{\ell}(R)$, R is also left nonsingular. From Lemma 2.5(i), R is an Utumi ring.

Conversely, let R be cononsingular and right nonsingular. From Proposition 2.4, R is left nonsingular. Thus R is an Utumi ring. Therefore from Lemma 2.5(i), $Q(R) = Q^\ell(R)$. □

In contrast to Proposition 2.4, a ring being right Utumi does not imply being also left cononsingular in general, as shown in the following example; note that this happens even under the additional condition of being left nonsingular.

EXAMPLE 2.7. Consider $R = \mathrm{CFM}_{\mathbb{N}}(F)$, the column finite matrix ring over a field F. Then R is a right self-injective regular ring, hence R is a right Utumi ring which implies that R is left nonsingular from Proposition 2.4. However, from Theorem 1.1(ii)(iii), it is easy to see that R is not left cononsingular because it is not unit-regular.

COROLLARY 2.8. *Let R be a right nonsingular ring, but not left nonsingular. Then the maximal right and left rings of quotients of R are different, i.e., $Q(R) \neq Q^\ell(R)$.*

The next examples illustrate Corollary 2.8.

EXAMPLE 2.9 ([**12**, Example 7.6(5)]). Let the ring $R = \left[\begin{smallmatrix} \mathbb{Z}_2 & \mathbb{Z}_2 \\ 0 & \mathbb{Z} \end{smallmatrix}\right]$. Then R is left nonsingular but not right nonsingular with a right singular ideal $\left[\begin{smallmatrix} \bar{0} & \mathbb{Z}_2 \\ 0 & 0 \end{smallmatrix}\right]$. Then from [**12**, Example 12.19(2)],

$$Q^r_{c\ell}(R) = Q^\ell_{c\ell}(R) = \left[\begin{smallmatrix} \mathbb{Z}_2 & \mathbb{Z}_2 \\ 0 & \mathbb{Z}_{(2)} \end{smallmatrix}\right]$$

where $Q^r_{c\ell}(R)$ (resp., $Q^\ell_{c\ell}(R)$) is the classical right (resp., left) ring of quotients and $\mathbb{Z}_{(2)} = \{\frac{m}{n} \in \mathbb{Q} \,|\, m, n \in \mathbb{Z} \text{ and } (n, 2) = 1\}$. By the calculation, we can see that $Q(R) \cong Q^r_{c\ell}(R) = Q^\ell_{c\ell}(R)$ (see [**24**, Theorem]).

However, $Q(R) \neq Q^\ell(R)$ because $Q(R)$ is not left injective, while $Q^\ell(R)$ is a left self-injective ring: For, consider a left ideal $I = \left[\begin{smallmatrix} \bar{0} & \mathbb{Z}_2 \\ 0 & 0 \end{smallmatrix}\right]$ and define $\varphi : I \to R$ by $\varphi \left[\begin{smallmatrix} \bar{0} & \bar{a} \\ 0 & 0 \end{smallmatrix}\right] = \left[\begin{smallmatrix} \bar{a} & \bar{0} \\ 0 & 0 \end{smallmatrix}\right]$. This is easily checked to be a left R-homomorphism. If φ can be extended to R, there would exist a matrix $\left[\begin{smallmatrix} \bar{x} & \bar{y} \\ 0 & z \end{smallmatrix}\right] \in R$ such that

$$\varphi \left[\begin{smallmatrix} \bar{0} & \bar{a} \\ 0 & 0 \end{smallmatrix}\right] = \left[\begin{smallmatrix} \bar{0} & \bar{a} \\ 0 & 0 \end{smallmatrix}\right] \left[\begin{smallmatrix} \bar{x} & \bar{y} \\ 0 & z \end{smallmatrix}\right] = \left[\begin{smallmatrix} \bar{0} & \bar{a}z \\ 0 & 0 \end{smallmatrix}\right],$$

a contradiction. Thus, the maximal right and left rings of quotients of R are different. Note: It is also an example of a PI-ring which is not an Utumi ring because R is a PI-ring but not semiprime.

The next example from [**12**, Example 7.14c] and [**12**, Example 13.26(4)] shows that the converse of Corollary 2.8 is not true. That is, it shows that for even a right and left nonsingular ring R, the maximal right and left rings of quotients of R might be different.

EXAMPLE 2.10 ([**12**, Example 7.14c] and [**12**, Example 13.26(4)]). Let $R = \left[\begin{smallmatrix} F & F & F \\ 0 & F & 0 \\ 0 & 0 & F \end{smallmatrix}\right]$ be a ring with a field F. Then R is a right and left nonsingular ring. Also, $Q(R) \cong \mathrm{Mat}_2(F) \times \mathrm{Mat}_2(F) \not\cong \mathrm{Mat}_3(F) \cong Q^\ell(R)$.

We conclude this section with results for "directly finite property" of Utumi rings. Recall that a ring R is said to be *directly finite* if $xy = 1$ implies $yx = 1$ for all $x, y \in R$.

EXAMPLE 2.11. Let $R = \mathrm{CFM}_{\mathbb{N}}(F)$ as in Example 2.7. Let e_{ij} be a matrix unit, that is, (i,j)-entry is 1, otherwise 0. Let $A = \sum_{i=1} e_{i,i+1} = \begin{bmatrix} 0 & 1 & 0 & \cdots \\ 0 & 0 & 1 & \cdots \\ \vdots & \vdots & \vdots & \vdots \\ 0 & 0 & 0 & \cdots \end{bmatrix}$ and $B = \sum_{i=1} e_{i+1,i} = \begin{bmatrix} 0 & 0 & 0 & \cdots \\ 1 & 0 & 0 & \cdots \\ 0 & 1 & 0 & \cdots \\ \vdots & \vdots & \vdots & \vdots \\ 0 & 0 & 0 & \cdots \end{bmatrix}$. Then $AB = I$ but $BA = \begin{bmatrix} 0 & 0 & 0 & \cdots \\ 0 & 1 & 0 & \cdots \\ 0 & 0 & 1 & \cdots \\ \vdots & \vdots & \vdots & \vdots \\ 0 & 0 & 0 & \cdots \end{bmatrix} \neq I$. Thus, R is not directly finite. Note from Example 2.7 that R is a right self-injective regular ring but not left cononsingular.

PROPOSITION 2.12 ([**21**, Lemma 2.13]). *Every right extending ring is right cononsingular.*

As seen in Example 2.11, a right self-injective regular ring does not have the directly finite property, in general. However, we show that under the additional condition of left cononsingularity, such a ring must be directly finite.

THEOREM 2.13. *Every left cononsingular, right extending ring is directly finite.*

PROOF. Let R be a left cononsingular, right extending ring. Assume that $ab = 1 \neq ba$ for some $a, b \in R$. Then by the standard observation of Shepherdson [**22**] the elements

$$e_{ij} = b^i(1 - ba)a^j \ (i, j \geq 1)$$

are nonzero elements of R satisfying the matrix units' equations: $e_{ij}e_{k\ell} = \delta_{jk}e_{i\ell}$. Since R is right extending, there exists $g^2 = g \in R$ such that $\sum_{i=1} e_{ii}R \leq^{\text{ess}} gR_R$. Take $e_{00} = 1 - g$. Note that $e_{00}e_{ii} = 0$ and $ge_{ii} = e_{ii}$ for all $i \geq 1$. Then $\sum_{i=0} e_{ii}R \leq^{\text{ess}} R_R$.

Claim 1. $\mathbf{r}_R\left(Re_{00} + \sum_{i=1} Re_{ii}g\right) = 0$:
Let $x \in \sum_{i=0} e_{ii}R \cap \mathbf{r}_R\left(Re_{00} + \sum_{i=1} Re_{ii}g\right)$ be arbitrary. Then $x = \sum_{i=0}^{n} e_{ii}x_i$ for $x_i \in R$, $n \in \mathbb{N}$, and $\left(Re_{00} + \sum_{i=1} Re_{ii}g\right)x = 0$. Thus, $0 = e_{00}x = e_{00}x_0$ and $0 = e_{jj}gx = e_{jj}g\sum_{i=0}^{n} e_{ii}x_i = e_{jj}x_j$ for all $1 \leq j \leq n$, hence $x = 0$. Therefore

$$\sum_{i=0} e_{ii}R \cap \mathbf{r}_R\left(Re_{00} + \sum_{i-1} Re_{ii}g\right) = 0.$$

Since $\sum_{i=0} e_{ii}R \leq^{\text{ess}} R_R$, $\mathbf{r}_R\left(Re_{00} + \sum_{i=1} Re_{ii}g\right) = 0$, which proves Claim 1.

From the left cononsingularity of R, $Re_{00} + \sum_{i=1} Re_{ii}g \leq^{\text{ess}} {}_RR$. In a similar way to Claim 1, it is easy to see that $\mathbf{l}_R\left(\sum_{j=0} e_{jj}R\right) = 0$. Also, since R is right extending, $e_{00}R + \sum_{j=2}(e_{1j} + e_{jj})R \leq^{\text{ess}} fR_R$ for some $f^2 = f \in R$.

Claim 2. $f = 1$:
Assume that $1 - f \neq 0$. Then since $Re_{00} + \sum_{i=1} Re_{ii}g \leq^{\text{ess}} {}_RR$, there exists $y \in R$ such that $0 \neq y(1 - f) \in Re_{00} + \sum_{i=1} Re_{ii}g$, that is, $0 \neq y(1 - f) = y_0e_{00} + \sum_{i=1}^{m} y_ie_{ii}g$ for $y_0, y_i \in R$ and some $m \in \mathbb{N}$. Thus, $y(1 - f)e_{m+1,m+1} = 0$. Also, $y(1 - f)(e_{00}R + \sum_{j=2}(e_{1j} + e_{jj})R) \leq y(1 - f)fR = 0 \Rightarrow y(1 - f)e_{00} = 0$ and $y(1 - f)(e_{1j} + e_{jj}) = 0$ for all $j \geq 2$. As $y(1 - f)e_{m+1,m+1} = 0$, $y(1 - f)e_{1,m+1} = 0$. Hence

$$0 = y(1 - f)e_{1,m+1}e_{m+1,j} = y(1 - f)e_{1j} \text{ for all } j \geq 1.$$

Therefore, $y(1 - f)e_{jj} = 0$ for all $j \geq 0$. Since $\mathbf{l}_R\left(\sum_{j=0} e_{jj}R\right) = 0$, $y(1 - f) = 0$, a contradiction. Thus $f = 1$, proving Claim 2.

Since $e_{00}R + \sum_{j=2}(e_{1j} + e_{jj})R \leq^{\text{ess}} R$ by Claim 2 and $e_{11} \neq 0$, there exists $z \in R$ such that $0 \neq e_{11}z = e_{00}z_0 + \sum_{j=2}^{\ell}(e_{1j} + e_{jj})z_j$ for some $z_0, z_j \in R$, $\ell \in \mathbb{N}$. Thus, $0 = e_{ii}e_{11}z = e_{ii}(e_{00}z_0 + \sum_{j=2}^{\ell}(e_{1j} + e_{jj})z_j) = e_{ii}z_i$ and $e_{1i}z_i = e_{1i}e_{ii}z_i = 0$ for all $2 \leq i \leq \ell$. Hence $0 \neq e_{11}z = e_{00}z_0$. However, $e_{11}z = ge_{11}z = ge_{00}z_0 = 0$, a contradiction. Therefore R is directly finite. □

Example 2.7 shows that the "left cononsingular" condition in Theorem 2.13 is not superfluous. In addition, the next example shows that the converse of Theorem 2.13 does not hold true, in general.

EXAMPLE 2.14 ([**6**, Proposition 15]). A ring $R = \left[\begin{smallmatrix} \mathbb{R} & \mathbb{C} \\ 0 & \mathbb{C} \end{smallmatrix}\right]$ is right extending and is directly finite. However, it is not left cononsingular because $\mathbf{r}_R\left(\left[\begin{smallmatrix} \mathbb{R} & \mathbb{R} \\ 0 & 0 \end{smallmatrix}\right]\right) = 0$ and there is no $0 \neq r \in R$ such that $r\left[\begin{smallmatrix} 0 & i \\ 0 & 0 \end{smallmatrix}\right] \in \left[\begin{smallmatrix} \mathbb{R} & \mathbb{R} \\ 0 & 0 \end{smallmatrix}\right]$. Also, it is a right Utumi ring but not left Utumi.

COROLLARY 2.15 ([**12**, Theorem 6.48]). *Every right and left extending ring is directly finite.*

PROPOSITION 2.16. *Every Utumi ring is directly finite.*

PROOF. Let R be an Utumi ring. Then by Lemma 2.5(ii), $Q(R) = Q^{\ell}(R)$ is a right and left self-injective regular ring. Thus, $Q(R)$ is directly finite from Theorem 2.13. So, R is directly finite from [**17**, Lemma 2.28(2)]. □

From Corollary 2.15 and Proposition 2.16, we may raise the following question:

QUESTION 1. Is every right and left cononsingular ring directly finite?

3. Continuous regular Baer rings

It is well known that every right and left continuous regular ring is unit-regular. Recall that a ring R is said to be *unit-regular* if for any $x \in R$, there exists an invertible element $u \in R$ such that $x = xux$.

In this section, we focus on relations between right and left continuous regular rings and unit-regular Baer rings. First of all, we characterize right and left continuous regular rings in terms of regular Baer rings (Theorem 3.6). We show that every strongly regular ring is an Utumi ring (Proposition 3.15). We then use these results to prove that every strongly regular Baer ring is right and left continuous regular (Corollary 3.17).

THEOREM 3.1. *Every left cononsingular, right nonsingular, right continuous ring is a unit-regular Baer ring.*

PROOF. Let R be a left cononsingular, right nonsingular, right continuous ring. In view of [**3**, Theorem 2.1] (see Theorem 1.1(iii)), R is a Baer ring. According to [**14**, Corollary 3.18], R is (von Neumann) regular. Moreover, R is directly finite by Theorem 2.13. Hence $E(R)$ is a directly finite R-module by [**2**, Lemma 2.2.10(ii)].

We claim that R_R has the internal cancellation property (see also [**17**, Theorem 2.33]): Suppose that $R_R = I_1 \oplus J_1 = I_2 \oplus J_2$ with $I_1 \cong I_2$. Take an injective hull of both sides: $E(R) = E(I_1) \oplus E(J_1) = E(I_2) \oplus E(J_2)$ with $E(I_1) \cong E(I_2)$. $E(R)$ has the internal cancellation property because $E(R)$ is directly finite (see [**17**, Proposition 1.28]). So, we have $E(J_1) \cong E(J_2)$. As R is right quasi-continuous, $J_1 \cong J_2$ from [**17**, Theorem 2.31], proving the claim.

Therefore R is a unit-regular ring because R is a regular ring with the internal cancellation property (see [**18**, Corollary 48]). □

Note that every left cononsingular, right nonsingular, right continuous ring is an Utumi ring.

COROLLARY 3.2 ([**5**, Corollary 13.23]). *Every right and left continuous regular ring is unit-regular.*

In addition, the converse of Corollary 3.2 does not hold true, in general as shown in Example 3.9(ii).

COROLLARY 3.3. *The maximal right ring of quotients of an Utumi ring is a right and left self-injective unit-regular ring (hence, is also Utumi).*

PROOF. Let R be an Utumi ring. Then $Q(R)$ (resp., $Q^\ell(R)$) is a right (resp., left) self-injective regular ring. Since $Q(R) = Q^\ell(R)$ from Theorem 2.6, $Q(R)$ is also left self-injective. So, from Theorem 3.1 $Q(R)$ is unit-regular. Therefore $Q(R)$ is a right and left self-injective unit-regular ring. □

The next example shows that the converse of Corollary 3.3 does not hold true, in general.

EXAMPLE 3.4 ([**4**, Exercises 2.C.3]). Let F be a field, K a proper subfield of F, and $R = \left[\begin{smallmatrix} K & F \\ 0 & F \end{smallmatrix}\right]$ a ring. While $Q(R) = \left[\begin{smallmatrix} F & F \\ F & F \end{smallmatrix}\right]$ is a right and left self-injective unit-regular ring, R is not an Utumi ring because $Q(R) \neq Q^\ell(R)$. In fact, for $\left[\begin{smallmatrix} 1 & 0 \\ 0 & 0 \end{smallmatrix}\right]$, $\left[\begin{smallmatrix} \alpha & 0 \\ 0 & 0 \end{smallmatrix}\right] \in Q(R)$ where $\alpha \in F \setminus K$, there is no $\begin{bmatrix} x & \beta \\ 0 & \gamma \end{bmatrix} \in R$ such that $0 \neq \begin{bmatrix} x & \beta \\ 0 & \gamma \end{bmatrix} \left[\begin{smallmatrix} 1 & 0 \\ 0 & 0 \end{smallmatrix}\right]$ and $\begin{bmatrix} x & \beta \\ 0 & \gamma \end{bmatrix} \left[\begin{smallmatrix} \alpha & 0 \\ 0 & 0 \end{smallmatrix}\right] \in R$, i.e., R is not dense in $Q(R)$ as a left R-submodule. Note that R is a right Utumi ring but not left Utumi.

REMARK 3.5. As a special case of Example 3.4, the ring $R = \left[\begin{smallmatrix} \mathbb{R} & \mathbb{C} \\ 0 & \mathbb{C} \end{smallmatrix}\right]$ is a right Utumi ring but not left Utumi. According to [**24**, Theorem], one can calculate $Q(R) = \left[\begin{smallmatrix} \mathbb{C} & \mathbb{C} \\ \mathbb{C} & \mathbb{C} \end{smallmatrix}\right]$ and $Q^\ell(R) \cong \mathrm{Mat}_3(\mathbb{R})$.

We now characterize the right and left continuous regular rings in terms of Utumi rings.

THEOREM 3.6. *Let R be a ring. Then the following conditions are equivalent:*

(a) *R is a right (left) continuous and Utumi ring;*
(b) *R is a right and left continuous regular ring;*
(c) *R is a right (left) nonsingular and right and left continuous ring;*
(d) *R is a regular Baer ring satisfying $Q(R) = Q^\ell(R)$.*

PROOF. Note that every regular ring and every Baer ring are nonsingular and every right continuous ring is right cononsingular. (a)⇔(b)⇔(c) follow from the previous note and Theorem 1.1(iii). (a)⇒(d) From Lemma 2.5(ii), $Q(R) = Q^\ell(R)$. Also, every right nonsingular right continuous ring is a regular Baer ring by Theorem 1.1(iii). Thus, R is a regular Baer ring satisfying $Q(R) = Q^\ell(R)$. (d)⇒(a) Since R is nonsingular and $Q(R) = Q^\ell(R)$, R is a cononsingular ring from Theorem 2.6. Also, from Theorem 1.1(iii), every cononsingular Baer ring is extending. Therefore, R is a right and left continuous Utumi ring. □

REMARK 3.7. If R is a PI-ring, then R is a regular Baer ring iff R is a right and left continuous regular ring because of Theorem 3.6 and $Q(R) = Q^{\ell}(R)$ when R is a semiprime PI-ring (see Example 2.2(vi) for detail).

We already showed that every directly finite right continuous regular ring is unit-regular in [**27**, Theorem 22]. The following result provides another situation for a regular ring to be unit-regular.

COROLLARY 3.8. *Every regular Baer ring satisfying $Q(R) = Q^{\ell}(R)$ is a unit-regular ring.*

PROOF. It directly follows from Theorems 3.1 and 3.6. □

The next examples show that the "$Q(R) = Q^{\ell}(R)$" condition in Corollary 3.8 is not superfluous and the converse of Corollary 3.8 does not hold true, in general, that is, there is a unit-regular ring, which is not Baer.

EXAMPLE 3.9.
(i) Let V_F be a countably infinite-dimensional vector space over a field F. Then $\mathrm{End}_F(V)$ is a right self-injective regular ring (hence, a regular Baer ring), which is not a unit-regular ring (see [**19**, Corollary]).
(ii) Let $R = \{(a_n) \in \prod_{n=1}^{\infty} \mathbb{Z}_2 \mid a_n \text{ is eventually constant}\}$. Then R is a unit-regular ring but not a Baer ring because there is no idempotent $e \in R$ such that $\mathbf{r}_R(I) = eR$ for $I = \{e_{2i-1} \mid i \in \mathbb{N}\}$ where e_i is the ith unit vector "$(0, \ldots, 1, 0, \ldots)$".

REMARK 3.10.
(i) ([**2**, Corollary 3.3.3]) When R is a right nonsingular ring satisfying $Q(R) = Q^{\ell}(R)$, R is a Baer ring iff R is a left (right) extending ring.
(ii) ([**5**, Theorem 13.13] and Theorem 3.6) When R is a regular ring satisfying $Q(R) = Q^{\ell}(R)$, R is a Baer ring iff R contains all the idempotents of $Q(R)$.

Using Theorem 3.6, we show that there is an Utumi unit-regular ring, which is neither (right) continuous nor Baer.

EXAMPLE 3.11 ([**5**, Example 13.8]). Let F be a field, which has a proper subfield K. Set $A = \{(a_n) \in \prod_{n=1}^{\infty} F \mid a_n \in K \text{ eventually}\}$, a subring of $\prod_{n=1}^{\infty} F$. Then A is a commutative continuous regular ring which is not a self-injective ring, while $Q(A) = \prod_{n=1}^{\infty} F$ is a commutative, regular, self-injective (hence, Baer) ring.

Consider the ring $R = \mathsf{Mat}_2(A)$, which is not a continuous ring. Since A is a unit-regular ring, so is R. Also, $Q(R) = \mathsf{Mat}_2(Q(A)) = Q^{\ell}(R)$. Since R is a nonsingular ring, from Theorem 2.6 R is an Utumi ring. However, by Theorem 3.6, R is not a Baer ring.

We conclude this paper with results for strongly regular rings. A von Neumann regular ring is said to be *strongly regular* if all its idempotents are central. Since a ring is said to be *abelian* if all its idempotents are central, it is easy to see that a ring R is abelian regular if and only if R is a strongly regular ring.

Next, we show that the maximal right ring of quotients of an abelian semisimple I-ring is strongly regular. Recall that a ring R is said to be a *semisimple I-ring* if every nonzero right ideal of R contains a nonzero idempotent [**13**]. Originally, the notion of I-rings is defined by Kaplansky (also called *Zorn rings*, see [**7**]). In fact, a ring R is called an *I-ring* if every nonnil right ideal of R contains a nonzero idempotent, which is defined by Jacobson [**10**]. We see that a ring R is an I-ring if and only if every nonnil left ideal contains a nonzero idempotent. By Kaplansky [**8**], it is shown that if R is an I-ring, then R contains an infinite number of orthogonal idempotents or else R is semilocal (see also [**2**, Theorem 1.2.16]).

PROPOSITION 3.12. *If R is an abelian semisimple I-ring then $Q(R)$ is strongly regular.*

PROOF. Note that every semisimple I-ring is right nonsingular from [**25**, 4.10]. Let R be an abelian semisimple I-ring. Since R is right nonsingular, $Q(R)$ is a right self-injective regular ring. So, it suffices to show that $Q(R)$ is abelian.

We claim that every nonzero right ideal of $Q(R)$ contains a nonzero central idempotent: Let $e \in R$ be an idempotent and $x \in Q(R)$ be any element. Assume that $ex \neq xe$. Then there exists $r \in R$ such that $(ex - xe)r \neq 0$ and $xr \in R$. However, $exr = xre = xer$ because R is abelian, a contradiction. So, $ex = xe$. Thus, all idempotents in R are central in $Q(R)$. Consider J a nonzero right ideal of $Q(R)$. Then $J \cap R_R \neq 0$. Since R is a semisimple I-ring, $J \cap R$ contains a nonzero idempotent in R which is central in $Q(R)$ from the previous argument, which proves the claim.

Let $f \in Q(R)$ be an idempotent. Let K be the right ideal of $Q(R)$ generated by all central idempotents of $Q(R)$ which lie in $fQ(R)$. Note that K is a 2-sided ideal and $K \leq^{\text{ess}} fQ(R)$ by the claim. Given any $x \in Q(R)$, we have $xK \leq K \leq fQ(R)$ and so $(1-f)xK = 0$. Define $\varphi : fQ(R)/K \to Q(R)$ via $fy + K \mapsto (1-f)xfy$ for all $y \in Q(R)$. Since $fQ(R)/K$ is singular, the image $(1-f)xfQ(R)$ is also singular. However, since $Q(R)$ is nonsingular, $(1-f)xf = 0$ for all $x \in Q(R)$. Thus, $(1-f)Q(R)f = 0$. Therefore, f is central in $Q(R)$ as $Q(R)$ is semiprime. □

The next example exhibits that the "semisimple I-ring" condition is not superfluous.

EXAMPLE 3.13 ([**12**, Example 13.28]). Let $R = \mathbb{Q}\langle a, b\rangle$ (the free $\mathbb{Q}$-algebra on a, b), which is a domain. Then R is abelian and right nonsingular. Hence $Q(R)$ is a right self-injective regular ring. However, $Q(R)$ is not abelian.

COROLLARY 3.14 ([**26**, Corollary 1.2]). *The maximal right ring of quotients of every strongly regular ring is strongly regular.*

PROOF. It follows from the fact that a strongly regular ring is an abelian semisimple I-ring and Proposition 3.12. □

PROPOSITION 3.15. *Every strongly regular ring is an Utumi ring.*

PROOF. Although the result directly follows from [**5**, Theorem 3.8] and Theorem 2.6, we include the proof for the convenience of the reader. Since every strongly regular ring is right and left nonsingular, it suffices to show that every strongly regular ring is right and left crononsingular. Note that $E(R) = Q(R)$ where $E(R)$ is an injective hull of R_R. Let R be strongly regular and I be a right ideal of R with $\mathbf{l}_R(I) = 0$. Consider K as a closure of I in R_R.

We claim that $K = \mathbf{r}_R(\mathbf{l}_R(K))$: Assume that $K \neq \mathbf{r}_R(\mathbf{l}_R(K))$. Then there exists a nonzero right ideal $J \leq \mathbf{r}_R(\mathbf{l}_R(K))$ such that $K \cap J = 0$ because K is closed. Let L_R be a complement of J_R such that $K \subseteq L \leq R$. Since $L \leq^{\text{ess}} (E(L) \cap R)_R$ and L_R is closed, $L = E(L) \cap R$. Since $E(L) \leq^{\oplus} E(R)_R$ and $E(R)$ is strongly regular from Corollary 3.14, there exists a central idempotent $e \in \text{End}_R(E(R))$ such that

$E(L) = eE(R)$. So, $L = eE(R) \cap R$. Since $R \leq^{\text{ess}} E(R)_R$, $E(R)(1-e) \cap R = (1-e)E(R) \cap R \neq 0$. Let $0 \neq x \in E(R)(1-e) \cap R$. Since $xL = 0$, $xK = 0$. Thus, $x \in \mathbf{l}_R(K) = \mathbf{l}_R(\mathbf{r}_R(\mathbf{l}_R(K)))$, so $x\mathbf{r}_R(\mathbf{l}_R(K)) = 0$. Hence $xJ = 0$. Thus $x(L \oplus J) = 0$. Since $L \oplus J \leq^{\text{ess}} R_R$ and R is right nonsingular, $x = 0$, a contradiction, proving the claim.

Since $0 = \mathbf{l}_R(I) \supseteq \mathbf{l}_R(K)$ and $K = \mathbf{r}_R(\mathbf{l}_R(K))$ by the claim, $K = R$. Hence $I \leq^{\text{ess}} R_R$. Thus R is right cononsingular. Similarly, R is left cononsingular. □

However, the next example shows that an abelian endoregular module may not be $\mathcal{K}$-cononsingular, even though a strongly regular ring is a cononsingular ring as shown in Proposition 3.15.

EXAMPLE 3.16 ([**15**, Example 4.10]). Consider the ring $R = \left[\begin{smallmatrix} \mathbb{R} & \mathbb{C} \\ 0 & \mathbb{R} \end{smallmatrix}\right]$ and the idempotent $e = \left[\begin{smallmatrix} 1 & 0 \\ 0 & 0 \end{smallmatrix}\right]$. Then it is easy to see that $M = eR = \left[\begin{smallmatrix} \mathbb{R} & \mathbb{C} \\ 0 & 0 \end{smallmatrix}\right]$ is an abelian endoregular Baer module. Take $N = \left[\begin{smallmatrix} 0 & \mathbb{R} \\ 0 & 0 \end{smallmatrix}\right]$ as a submodule of M. Then $\mathbf{l}_S(N) = 0$ and there is no element $0 \neq r \in R$ such that $\left[\begin{smallmatrix} 0 & i \\ 0 & 0 \end{smallmatrix}\right] r \in N = \left[\begin{smallmatrix} 0 & \mathbb{R} \\ 0 & 0 \end{smallmatrix}\right]$. Thus, N is not $\mathcal{K}$-cononsingular.

COROLLARY 3.17. *Every strongly regular Baer ring R is a right and left continuous regular ring. In addition, R contains all the idempotents of $Q(R)$.*

PROOF. It directly follows from Lemma 2.5(ii), Theorem 3.6, and Proposition 3.15. Also, see Remark 3.10(ii). □

The next example shows that Corollary 3.17, in rings-theoretic terms, does not extend to module-theoretic setting (with the usual definitions for the module properties therein).

EXAMPLE 3.18. Let R and M be as in Example 3.16. Then M is an abelian endoregular Baer module. However, M is not a continuous module.

Acknowledgments

The authors express their sincere gratitude to the referee for a prompt and thorough report. The authors are very thankful to Research Institute of Mathematical Sciences, Chungnam National University (CNU-RIMS), Republic of Korea, for the support of this research work. The first author gratefully acknowledges the support of this research work by the National Research Foundation of Korea (NRF) grant funded by the Korea government(MSIT)(2019R1F1A1059883). The last author is supported by the National Natural Science Foundation of China (Grant No. 11871145) and the Qinglan Project of Jiangsu Province of China.

References

[1] Sterling K. Berberian, *Baer *-rings*, Die Grundlehren der mathematischen Wissenschaften, Band 195, Springer-Verlag, New York-Berlin, 1972. MR0429975

[2] Gary F. Birkenmeier, Jae Keol Park, and S. Tariq Rizvi, *Extensions of rings and modules*, Birkhäuser/Springer, New York, 2013, DOI 10.1007/978-0-387-92716-9. MR3099829

[3] A. W. Chatters and S. M. Khuri, *Endomorphism rings of modules over nonsingular CS rings*, J. London Math. Soc. (2) **21** (1980), no. 3, 434–444, DOI 10.1112/jlms/s2-21.3.434. MR577719

[4] K. R. Goodearl, *Ring theory: Nonsingular rings and modules*, Pure and Applied Mathematics, No. 33, Marcel Dekker, Inc., New York-Basel, 1976. MR0429962

[5] K. R. Goodearl, *von Neumann regular rings*, Monographs and Studies in Mathematics, vol. 4, Pitman (Advanced Publishing Program), Boston, Mass.-London, 1979. MR533669

[6] Dinh Van Huynh and S. Tariq Rizvi, *On some classes of Artinian rings*, J. Algebra **223** (2000), no. 1, 133–153, DOI 10.1006/jabr.1999.8080. MR1738256

[7] Irving Kaplansky, *Rings of operators*, W. A. Benjamin, Inc., New York-Amsterdam, 1968. MR0244778

[8] Irving Kaplansky, *Topological representation of algebras. II*, Trans. Amer. Math. Soc. **68** (1950), 62–75, DOI 10.2307/1990539. MR32612

[9] Nathan Jacobson, *The Theory of Rings*, American Mathematical Society Mathematical Surveys, Vol. II, American Mathematical Society, New York, 1943. MR0008601

[10] Nathan Jacobson, *Structure of rings*, American Mathematical Society Colloquium Publications, Vol. 37, American Mathematical Society, Providence, RI, 1956. MR0081264

[11] R. E. Johnson, *The extended centralizer of a ring over a module*, Proc. Amer. Math. Soc. **2** (1951), 891–895, DOI 10.2307/2031703. MR45695

[12] T.Y. Lam, *Lectures on Modules and Rings*, GTM 189, Springer Verlag, Berlin-Heidelberg-New York (1999).

[13] J. Levitzki, *A theorem on polynomial identities*, Proc. Amer. Math. Soc. **1** (1950), 334–341, DOI 10.2307/2032378. MR35758

[14] Gangyong Lee, S. Tariq Rizvi, and Cosmin S. Roman, *Rickart modules*, Comm. Algebra **38** (2010), no. 11, 4005–4027, DOI 10.1080/00927872.2010.507232. MR2764847

[15] Gangyong Lee, S. Tariq Rizvi, and Cosmin Roman, *Modules whose endomorphism rings are von Neumann regular*, Comm. Algebra **41** (2013), no. 11, 4066–4088, DOI 10.1080/00927872.2012.700979. MR3169507

[16] Gangyong Lee, S. Tariq Rizvi, and Cosmin S. Roman, *$\mathfrak{L}$-Rickart modules*, Comm. Algebra **43** (2015), no. 5, 2124–2151, DOI 10.1080/00927872.2014.888563. MR3316842

[17] Saad H. Mohamed and Bruno J. Müller, *Continuous and discrete modules*, London Mathematical Society Lecture Note Series, vol. 147, Cambridge University Press, Cambridge, 1990, DOI 10.1017/CBO9780511600692. MR1084376

[18] W. K. Nicholson and E. Sánchez Campos, *Morphic modules*, Comm. Algebra **33** (2005), no. 8, 2629–2647, DOI 10.1081/AGB-200064348. MR2159493

[19] W. K. Nicholson and K. Varadarajan, *Countable linear transformations are clean*, Proc. Amer. Math. Soc. **126** (1998), no. 1, 61–64, DOI 10.1090/S0002-9939-98-04397-4. MR1452816

[20] D. G. Northcott, *Ideal theory*, Cambridge Tracts in Mathematics and Mathematical Physics, No. 42, Cambridge, at the University Press, 1953. MR0058575

[21] S. Tariq Rizvi and Cosmin S. Roman, *Baer and quasi-Baer modules*, Comm. Algebra **32** (2004), no. 1, 103–123, DOI 10.1081/AGB-120027854. MR2036224

[22] J. C. Shepherdson, *Inverses and zero divisors in matrix rings*, Proc. London Math. Soc. (3) **1** (1951), 71–85, DOI 10.1112/plms/s3-1.1.71. MR41831

[23] Bo Stenström, *Rings of quotients: An introduction to methods of ring theory*, Die Grundlehren der mathematischen Wissenschaften, Band 217, Springer-Verlag, New York-Heidelberg, 1975. MR0389953

[24] Bo Stenström, *The maximal ring of quotients of a triangular matrix ring*, Math. Scand. **34** (1974), 162–166, DOI 10.7146/math.scand.a-11517. MR357459

[25] Yuzo Utumi, *On quotient rings*, Osaka Math. J. **8** (1956), 1–18. MR78966

[26] Y. Utumi, *On rings of which any one-sided quotient rings are two-sided*, Proc. Amer. Math. Soc. **14** (1963), 141–147, DOI 10.2307/2033975. MR142568

[27] Xiaoxiang Zhang and Gangyong Lee, *Modules whose endomorphism rings are unit-regular*, Comm. Algebra **44** (2016), no. 2, 697–709, DOI 10.1080/00927872.2014.984839. MR3449948

Department of Mathematics Education, Chungnam National University, Daejeon 34134, Republic of Korea

Email address: lgy999@cnu.ac.kr

Department of Mathematics, The Ohio State University, Lima, OH 45804, USA

Email address: cosmin@math.ohio-state.edu

Faculty of Mathematics and Computer Science, University of Science, Vietnam National University, Ho Chi Minh City, Vietnam

Email address: nktung@hcmus.edu.vn

School of Mathematics, Southeast University, Nanjing 211189, P. R. China

Email address: z990303@seu.edu.cn

Contemporary Mathematics
Volume **785**, 2023
https://doi.org/10.1090/conm/785/15783

Matrices representable as product of conjugates of a singular matrix

S. K. Jain and André Leroy

Dedicated to Professor Syed Tariq Rizvi for his contribution to Ring and Module Theory

ABSTRACT. Using semigroup techniques, J. Araújo and F.C. Silva [Commun. Algebra 28, No. 8 (2000), 3679–3689] proved that a singular matrix $B \in M_n(F)$ with coefficients in a field F is a product of conjugates of any matrix A with $\operatorname{rank}(B) \leq \operatorname{rank}(A)$. We prove this result over an algebraically closed field, in an elementary short way.

We start with the following short and elementary proof of a classical fact related to the decomposition of a matrix into products of idempotent matrices. Let us stress that this result is true for matrices with coefficients in a general division ring. The decomposition of a singular matrix into the product of idempotent matrices is an interesting topic in connection with many other branches of mathematics (Cf. [**2**]).

The symbol $\mathbb{N}$ stands for all positive integer numbers.

THEOREM 1. *Let K be a division ring and $B \in M_n(K)$ be an $n \times n$ singular matrix. Then $B = \prod_{i=1}^{k} E_i$ for some idempotent matrices E_i and some $k \in \mathbb{N}$ such that* $\operatorname{rank}(B) = \operatorname{rank}(E_i)$ *for all i.*

PROOF. The fact that B is singular implies that there exists an invertible matrix U such that UBU^{-1} is of the form:

$$\begin{pmatrix} D & C \\ 0 & 0 \end{pmatrix}, \tag{1}$$

where D is an $(n-1) \times (n-1)$ matrix and C is a column matrix. We argue by induction on $n \geq 1$. If $n = 1$, then $B = 0$ and the result is obvious.

Now, let us assume that $n \geq 2$ and that the result is true for singular matrices of size less than n.

CASE 1. D is singular. Then by the induction hypothesis D is a product of idempotent matrices, say $D = E_1 \cdots E_r$, for some $r \in \mathbb{N}$, where $\operatorname{rank}(E_i) = \operatorname{rank}(D)$ for all i. This gives

$$B = \begin{pmatrix} D & C \\ 0 & 0 \end{pmatrix} = \begin{pmatrix} E_1 & C \\ 0 & 0 \end{pmatrix} \begin{pmatrix} E_2 & 0 \\ 0 & 1 \end{pmatrix} \cdots \begin{pmatrix} E_r & 0 \\ 0 & 1 \end{pmatrix}. \tag{2}$$

2020 *Mathematics Subject Classification.* Primary 15A23; Secondary 15B33.
Key words and phrases. Idempotent, Singular Matrix.

Let us consider the following two subcases.

(a) $\operatorname{rank}(D) < \operatorname{rank}(B)$. Then we must have $\operatorname{rank}(B) = \operatorname{rank}(D)+1$. Note that the matrices of the right-hand side in (2) are all idempotent matrices and clearly each of them has rank greater than or equal to $\operatorname{rank}(B)$. The first matrix has rank smaller or equal to $\operatorname{rank}(E_1) + 1 = \operatorname{rank}(D) + 1 = \operatorname{rank}(B)$. Similarly, all other matrices of the right-hand side of (2) have the same rank as $\operatorname{rank}(B)$.

(b) $\operatorname{rank}(D) = \operatorname{rank}(B)$. Then the column C is a linear combination of the columns of D and we must have $C = DQ$ for some column Q. This gives

$$B = \begin{pmatrix} D & DQ \\ 0 & 0 \end{pmatrix} = \begin{pmatrix} E_1 & 0 \\ 0 & 0 \end{pmatrix} \begin{pmatrix} E_2 & 0 \\ 0 & 0 \end{pmatrix} \cdots \begin{pmatrix} E_r & E_rQ \\ 0 & 0 \end{pmatrix}, \tag{3}$$

where the idempotent matrices on the right-hand side all have ranks equal to the rank of B.

CASE 2. D is invertible. Then straightforward block multiplications give that

$$B = \begin{pmatrix} D & C \\ 0 & 0 \end{pmatrix} = \begin{pmatrix} 0 & D \\ 0 & 0 \end{pmatrix} \begin{pmatrix} 0 & 0 \\ I_{n-1,n-1} & 0 \end{pmatrix} \begin{pmatrix} I_{n-1,n-1} & D^{-1}C \\ 0 & 0 \end{pmatrix}. \tag{4}$$

The first matrix on the right-hand side of this equality is of the form treated in the previous case because the 1-1 top left corner block is a singular matrix. For the second matrix, we introduce

$$P = \begin{pmatrix} 0 & I_{n-1,n-1} \\ 1 & 0 \end{pmatrix}.$$

We then have

$$P \begin{pmatrix} 0 & 0 \\ I_{n-1,n-1} & 0 \end{pmatrix} P^{-1} = \begin{pmatrix} 0 & I_{n-1 \times n-1} \\ 0 & 0 \end{pmatrix} = \begin{pmatrix} D' & C' \\ 0 & 0 \end{pmatrix},$$

where D' is singular. Again this matrix was considered in the first case above. The last matrix on the right-hand side of (4) is itself an idempotent matrix with rank is equal to the rank of B. This shows that the singular matrix B is a product of idempotent matrices of rank equal to that of B.

□

We are now ready to prove the main result of this paper in the following theorem.

THEOREM 2. *Let F be an algebraically closed field and A, B be $n \times n$ singular matrices over F such that* $\operatorname{rank}(B) \leq \operatorname{rank}(A)$. *Then B is a product of conjugates of A.*

PROOF.

STEP 1. We first consider the case when the two matrices A and B are idempotent matrices. Since idempotent matrices are similar to diagonal matrices having only zeros and ones on the diagonal, we may assume that

$$A = \begin{pmatrix} I_r & 0 \\ 0 & 0 \end{pmatrix} \quad \text{and} \quad B = \begin{pmatrix} I_s & 0 \\ 0 & 0 \end{pmatrix},$$

with $s \leq r$. If $r = s$, we are done. Consider the case when $s < r$. Let P be the matrix

$$P_1 = \begin{pmatrix} I_{r-1} & 0 & 0 & 0 \\ 0 & 0 & 1 & 0 \\ 0 & 1 & 0 & 0 \\ 0 & 0 & 0 & I_{n-r-1} \end{pmatrix} = P_1^{-1}.$$

The matrix $A(P_1AP_1^{-1}) = \begin{pmatrix} I_{r-1} & 0 \\ 0 & 0 \end{pmatrix}$ is a product of conjugates of A. Repeating this process with $\begin{pmatrix} I_{r-1} & 0 \\ 0 & 0 \end{pmatrix}$ we get $\begin{pmatrix} I_{r-2} & 0 \\ 0 & 0 \end{pmatrix}$ is a product of conjugates of $\begin{pmatrix} I_{r-1} & 0 \\ 0 & 0 \end{pmatrix}$, and hence a product of conjugates of A. We finally reach that $B = \begin{pmatrix} I_s & 0 \\ 0 & 0 \end{pmatrix}$ is a product of conjugates of A. This shows that the theorem is true in the case A and B are idempotent matrices. Remark that so far we did not use the hypothesis that F is algebraically closed.

STEP 2. We will now use the Jordan canonical form and so we are now assuming that F is algebraically closed. By Theorem 1, we can write $B = E_1 \cdots E_n$ where all E_i are idempotent matrices such that $\operatorname{rank}(E_i) = \operatorname{rank}(B) \leq \operatorname{rank}(A)$ and we set $\operatorname{rank}(A) = r$. We have proved above, in step 1, that each E_i is a product of conjugates of $\operatorname{diag}(I_r, 0)$. Therefore, it is enough to prove that the idempotent matrix $\operatorname{diag}(I_r, 0)$ is a product of matrices similar (conjugates) to A. The Jordan canonical form of A gives that A is similar to a matrix of the form $\begin{pmatrix} A_1 & 0 \\ 0 & A_0 \end{pmatrix}$, where A_1 consists of Jordan blocks associated to nonzero eigenvalues, and $A_0 = \operatorname{diag}(J_1, \ldots, J_s)$ corresponds to all the Jordan blocks $J_1, \ldots, J_s$ associated to the eigenvalue zero. For each $1 \leq i \leq s$ we fix J_i to be an $n_i \times n_i$ matrix of rank $n_i - 1$. Then for any $1 \leq i \leq s$ we have

$$J_iP_iJ_iP_i^{-1} = \operatorname{diag}(1, \ldots, 1, 0), \quad \text{where} \quad P_i = \begin{pmatrix} 0 & \cdots & 0 & 1 \\ 0 & \cdots & 1 & 0 \\ 0 & \cdot^{\cdot^{\cdot}} & 0 & 0 \\ 1 & \cdots & 0 & 0 \end{pmatrix}. \tag{5}$$

Here, we point out that $\operatorname{rank}(J_iP_iJ_iP_i^{-1}) = \operatorname{rank}(J_i) = n_i - 1$. We put $Q := \operatorname{diag}(P_1, \ldots, P_s)$. From (5) we see that the matrix

$$A_0QA_0Q^{-1} = \begin{pmatrix} J_1P_1J_1P_1^{-1} & 0 & \cdots & 0 \\ 0 & J_2P_2J_2P_2^{-1} & \cdots & 0 \\ \cdots & \cdots & \cdots & \cdots \\ 0 & 0 & \cdots & J_sP_sJ_sP_s^{-1} \end{pmatrix},$$

is a diagonal matrix with only one's and zero's on the diagonal. Also, we have the $\operatorname{rank}(A_0QA_0Q^{-1}) = \operatorname{rank}(A_0)$. For a suitable permutation matrix X we will have $XA_0QA_0Q^{-1}X^{-1}$ is of the form $\operatorname{diag}(1, \ldots, 1, 0, \ldots, 0)$, where the number of one's is equal to the $\operatorname{rank}(A_0)$. Recall that $A = \operatorname{diag}(A_1, A_0)$, where A_1 is an upper triangular with nonzero elements on the diagonal. Hence, a product of conjugates of A will be of the form

$$\operatorname{diag}(A_1^2, XA_0QA_0Q^{-1}X^{-1}) = \operatorname{diag}(A_1^2, 1, 1, \ldots, 1, 0, \ldots, 0) = (A', 0),$$

where $A' = \operatorname{diag}(A_1^2, 1, \ldots, 1)$ is an upper triangular with (nonzero) elements $\alpha_1, \ldots, \alpha_r$ on the diagonal. Let us remark that $r = \operatorname{rank}(A)$.

We consider the following two invertible matrices:

$$T_V = \begin{pmatrix} I_r & V \\ 0 & I_{n-r} \end{pmatrix} \quad \text{and} \quad T_W = \begin{pmatrix} I_r & 0 \\ W & I_{n-r} \end{pmatrix},$$

where V and W are aribitrary $r \times (n-r)$ and $(n-r) \times r$ matrices respectively. Then

$$(T_V^{-1} A T_V)(T_W^{-1} A T_W) = \begin{pmatrix} A'^2 - A'VWA' & 0 \\ 0 & 0 \end{pmatrix}.$$

For each $1 \leq i \leq r$, and for any element $\lambda_i \in F$, we put $V_i = \lambda_i e_{i1} \in M_{r,n-r}(F)$ and $W_i = e_{1i} \in M_{n-r,r}(F)$. Then $V_i W_i = \lambda_i e_{i,i} = \operatorname{diag}(0, \ldots, 0, \lambda_i, 0 \ldots, 0) \in M_r(F)$. Moreover, a direct computation shows that, for any $1 \leq i \leq r$, the matrix $A_i' := A'^2 - A'V_iW_iA'$ is an upper triangular with its diagonal entries is equal to

$$(\alpha_1^2, \ldots, \alpha_{i-1}^2, (1-\lambda_i)\alpha_i^2, \alpha_{i+1}^2, \ldots, \alpha_r^2).$$

Set $A'' = A_1' \cdots A_r'$. This is an upper triangular matrix with diagonal entries equal $(1-\lambda_i)\alpha_i^{2r}$, for $i = 1, \ldots, r$. Let $\beta_1, \ldots, \beta_r$ be any distinct elements of the base field F. Then, choosing $\lambda_i = \frac{\alpha_i^{2r} - \beta_i}{\alpha_i^{2r}}$, we get the diagonal entries of A'' are $\beta_1, \ldots, \beta_r$. Hence the $r \times r$ matrix A'' is diagonalizable, and so there is an invertible matrix $P \in M_r(F)$ such that $PA''P^{-1} = \operatorname{diag}(\beta_1, \ldots, \beta_r)$. This gives that the $n \times n$ matrix $\operatorname{diag}(\beta_1, \ldots, \beta_r, 0, \ldots, 0)$ is a product of conjugates of A. The same is true for the matrices $\operatorname{diag}(\beta_1^{-1}, \ldots, \beta_r^{-1}, 0)$, and so for their product $\operatorname{diag}(I_r, 0)$, as desired.

□

Using some semigroup techniques, we may show that the above theorem is true for matrices with coefficients in a non-algebraically closed field (cf. [**1**]).

References

[1] João Araújo and Fernando C. Silva, *Semigroups of linear endomorphisms closed under conjugation*, Comm. Algebra **28** (2000), no. 8, 3679–3689, DOI 10.1080/00927870008827049. MR1767581

[2] S. K. Jain and A. Leroy, *Decomposition of singular elements of an algebra into product of idempotents, a survey*, Contributions in algebra and algebraic geometry, Contemp. Math., vol. 738, Amer. Math. Soc., [Providence], RI, [2019] ©2019, pp. 57–74, DOI 10.1090/conm/738/14878. MR4014964

Department of Mathematics, Ohio University, and King Andulaziz University Jeddah, SA

Email address: jain@ohio.edu

Univ. Artois, UR 2462, Laboratoire de Mathématiques de Lens (LML), F-62300 Lens, France

Email address: andre.leroy@univ-artois.fr

Contemporary Mathematics
Volume **785**, 2023
https://doi.org/10.1090/conm/785/15784

Almost Projective Modules Over Non-Hereditary Algebras

Fatma Kaynarca, Gabriella D'Este, and Derya Keskin Tütüncü

This paper is dedicated to Professor Syed Tariq Rizvi for his contribution to Ring and Module Theory

Abstract. We investigate almost projective modules over non-hereditary algebras. The proper submodules of these modules have a projective type behavior. We prove that there exists a non-hereditary algebra such that the proper submodules of their almost projective modules are actually projective. We also show that an algebra of finite representation type and its opposite algebra do not always have the same number of indecomposable almost projective modules.

1. Introduction

Recall that Problem 6 in [**1**] asks to determine the artin algebras with finitely many almost projective modules up to isomorphism. In the hereditary case a solution to this problem was obtained by [**2**] and [**3**]. In this note we investigate non-hereditary algebras admitting either infinitely many or very few indecomposable almost projective modules.

Looking at [**1**] and [**2**] we see that there are two different definitions for the almost projective modules. Let M be a non-projective module.

(i) According to [**1**]: M is called *almost projective* if the induced epimorphism $\tilde{g} = g \mid_{g^{-1}(X)}: g^{-1}(X) \to X$ is split for all proper submodules X of M, where $g : P \to M$ is a projective cover of M. Note that in this definition it is supposed that M has the projective cover.

(ii) According to [**2**]: If every proper submodule of M is projective, then M is called *almost projective*.

Clearly, $(ii) \Rightarrow (i)$. Indeed, assume that every proper submodule of M is projective. Let X be a proper submodule of M and let $g : P \to M$ be the projective cover of M. Take the induced epimorphism $g \mid_{g^{-1}(X)}: g^{-1}(X) \to X$. Say $\tilde{g} = g \mid_{g^{-1}(X)}$. By assumption, X is projective. Therefore there exists a homomorphism $\alpha : X \to g^{-1}(X)$ such that $\tilde{g}\alpha = 1_X$, where $1_X : X \to X$ is the identity homomorphism. This means that $\tilde{g}$ is split, as desired. It is not hard to see that $(i) \Rightarrow (ii)$ if R is a left hereditary ring. Indeed, let $g : P \to M$ be the projective cover of M and let X be a proper submodule of M. We want to show that X is projective. By

2020 *Mathematics Subject Classification.* Primary 16G20; Secondary 16D10.
Key words and phrases. Projective module, almost projective module, projective cover, representation type.

assumption, $\tilde{g} : g^{-1}(X) \to X$ is a split epimorphism. So $\operatorname{Ker}\tilde{g}$, which is $\operatorname{Ker} g$ as well, is a direct summand of $g^{-1}(X)$. Hence $g^{-1}(X) = \operatorname{Ker}\tilde{g} \oplus L$, for some submoulde L of $g^{-1}(X)$. Since R is left hereditary, $g^{-1}(X)$ (also, L) is projective. Therefore X is projective since $g^{-1}(X)/\operatorname{Ker}\tilde{g} \cong X \cong L$. In this paper we will use the definition in (i). We should note that there exist several modules satisfying (i) but not satisfying (ii) (see, for instance the modules constructed in Example 2.2 and Example 2.3 or the injective module constructed in Example 2.4). However there are non-hereditary algebras with the property that all the proper submodules of their indecomposable almost projective non projective modules are actually projective modules (Corollary 2.7). We also show that an algebra of finite representation type and its opposite algebra do not always admit the same number of indecomposable almost projective modules (Example 2.6). We finally prove that a non-projective sum of two indecomposable projective modules is not necessarily almost projective (Example 2.10).

We refer to [**3**] and [**4**] for other definitions of almost projective modules.

All modules will be unitary left modules and K will be a field. For any module M we will denote the radical of M by $\operatorname{Rad} M$ which is the intersection of the maximal submodules of M. Note that in our examples A is always an artin algebra. Hence every A-module has a projective cover. By dealing with a K-algebra A given by a quiver with vertices $1, \dots, n$, the symbols $1, \dots, n$ will denote also the simple left A-modules corresponding to the vertices $1, \dots, n$. Finally pictures of the form

$$\begin{matrix}1\\1\end{matrix},\ \begin{matrix}1\\2\end{matrix},\ \begin{matrix}1\\2\,2\end{matrix},\ \begin{matrix}1\,2\\2\end{matrix},\ \begin{matrix}1\,2\,3\\4\end{matrix},\ \begin{matrix}1\,2\,3\\4\,4\end{matrix}$$

will describe the structure of the indecomposable modules in the usual way [**5**].

We sum up the properties of some almost projective modules M constructed in the sequel as follows:

- M is uniserial (Example 2.2, Example 2.3 and Example 2.5).
- M admits a uniserial radical and $M/\operatorname{Rad} M$ is decomposable (Example 2.4).
- Both $\operatorname{Rad}M$ and $M/\operatorname{Rad} M$ are decomposable (Example 2.8 with $M = \begin{matrix}1\,2\,3\\4\,4\end{matrix}$).
- $M/\operatorname{Rad} M$ is simple (Example 2.9).

This paper is organized as follows. In Section 1 we recall the useful definitions and conventions. In Section 2 we collect all the proofs and examples.

2. Results and Examples

We begin with a preliminary lemma that may be known in the literature.

LEMMA 2.1. *Let M be a non-projective module such that if X is a proper submodule of M, then one of the following conditions holds:*

(i) *X is projective;*

(ii) *If $g : P \to M$ is the projective cover, then $g^{-1}(X)$ is semisimple.*

Then M is almost projective.

PROOF. Let $g : P \to M$ be the projective cover of M, let X be a proper submodule of M and let $\tilde{g} : g^{-1}(X) \to X$ be the induced epimorphism. Since either

X is projective or $g^{-1}(X)$ is semisimple, it follows that $\tilde{g}$ is a split epimorphism. Hence M is almost projective. □

EXAMPLE 2.2. Let K be a field. Then there exist a K-algebra A and $|K|$ pairwise non-isomorphic almost projective uniserial modules of dimension two and of infinite projective dimension.

CONSTRUCTION. Let A be the algebra $K\langle x,y\rangle/I$ where I is the ideal generated by the monomials x^2, xy, yx, y^2. For any $\lambda \in K$, let M_λ denote the uniserial module of dimension two with a basis v_1, v_2 such that $xv_1 = v_2$, $yv_1 = \lambda v_2$. Then we clearly have $M_\lambda \not\cong M_\mu$ if $\lambda \neq \mu$. Moreover $\mathrm{Soc}M_\lambda = \langle v_2\rangle$ is the unique proper submodule of M_λ. If $g : P \to M_\lambda$ is the projective cover, then we have $P \cong A$ and $g^{-1}(\mathrm{Soc}\, M_\lambda) = \mathrm{Soc}\, P$. Hence condition (ii) of Lemma 2.1 guarantees that M_λ is almost projective. Since A is a local algebra of dimension three it follows that the projective dimension of M_λ is equal to infinity.

EXAMPLE 2.3. Let K be a field. Then there exist a K-algebra A and $|K|$ pairwise non-isomorphic almost projective uniserial modules of dimension two and of projective dimension two.

CONSTRUCTION. Let A be the finite dimensional algebra given by the quiver

$$1 \underset{b}{\overset{a}{\rightrightarrows}} 2 \xrightarrow{c} 3$$

with relations $ca = 0$ and $cb = 0$. For any $\lambda \in K$, let M_λ denote the uniserial module $P_1/(a - \lambda b)$ where $P_1 = \begin{smallmatrix} 1 \\ 2\ 2 \end{smallmatrix}$. Then we clearly have $M_\lambda \not\cong M_\mu$ if $\lambda \neq \mu$. Let $g : P \to M_\lambda$ be the projective cover. Then we have $P \cong P_1$ and $g^{-1}(\mathrm{Soc}\, M_\lambda) = \mathrm{Soc}\, P$. This remark and condition (ii) of Lemma 2.1 imply that M_λ is a uniserial and almost projective module of dimension two. Since there is an exact sequence of the form

$$0 \longrightarrow 3 \longrightarrow \begin{smallmatrix} 2 \\ 3 \end{smallmatrix} \longrightarrow \begin{smallmatrix} 1 \\ 2\ 2 \end{smallmatrix} \longrightarrow M_\lambda \longrightarrow 0$$

it follows that the projective dimension of M_λ is equal to two.

It is easy to construct an indecomposable module of dimension three such that exactly two (resp. one) of its proper submodules satisfy condition (i) (resp. (ii)) of Lemma 2.1.

EXAMPLE 2.4. There exist an algebra A and an indecomposable injective module I with the following properties:

(a) Every maximal submodule of I is projective.
(b) If $g : P \to I$ is the projective cover of I and L is a proper non-maximal submodule of I, then $L = \mathrm{Soc}\, I$ and $g^{-1}(L)$ is semisimple.
(c) The projective dimension of I is equal to infinity.

CONSTRUCTION. Let A be the algebra given by the quiver $1 \xrightarrow{a} 2 \circlearrowright b$

with relations $ba = 0$ and $b^2 = 0$. Then the dimension of A is equal to four and the Auslander-Reiten quiver of A is as follows:

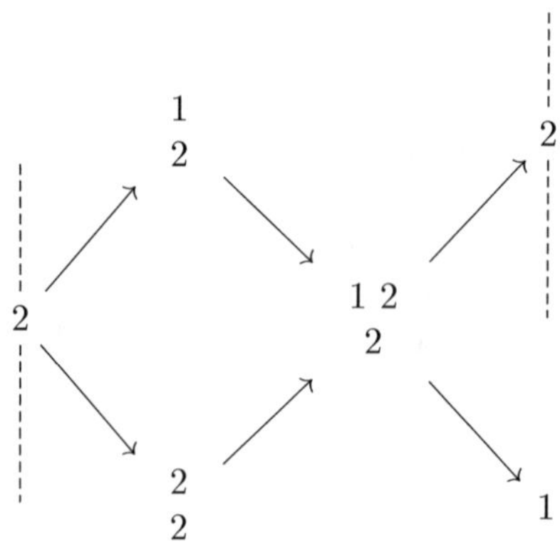

Next let $I = \begin{smallmatrix}1\,2\\2\end{smallmatrix}$. We first note that the maximal submodules of I are $\begin{smallmatrix}1\\2\end{smallmatrix}$ and $\begin{smallmatrix}2\\2\end{smallmatrix}$. Hence (a) holds. Moreover $\mathrm{Soc}I = 2$ is the unique proper non-maximal submodule of I and if $g : P \to M$ is the projective cover, then we have $P = \begin{smallmatrix}1\\2\end{smallmatrix} \oplus \begin{smallmatrix}2\\2\end{smallmatrix}$. Since $g^{-1}(2) = 2 \oplus 2 = \mathrm{Soc}\, P$, it follows that also (b) holds. Finally the projective resolution of I is of the form

$$\cdots \longrightarrow \begin{smallmatrix}2\\2\end{smallmatrix} \longrightarrow \begin{smallmatrix}2\\2\end{smallmatrix} \longrightarrow \begin{smallmatrix}2\\2\end{smallmatrix} \longrightarrow \begin{smallmatrix}1\\2\end{smallmatrix} \oplus \begin{smallmatrix}2\\2\end{smallmatrix} \longrightarrow I \longrightarrow 0\ .$$

Hence also (c) holds.

The algebra A considered in the previous example has the property that the unique indecomposable module which is neither projective nor simple is injective and almost projective. The next example shows that its opposite algebra has the same property.

EXAMPLE 2.5. There is an algebra A such that every indecomposable module which is neither projective nor simple is injective and almost projective and its projective dimension is either one or infinity.

CONSTRUCTION. Let A be the algebra given by the quiver $a \circlearrowleft 1 \xrightarrow{\ b\ } 2$ with relations $a^2 = 0$ and $ba = 0$. Then the dimension of A is equal to four and the

Auslander-Reiten quiver of A is as follows:

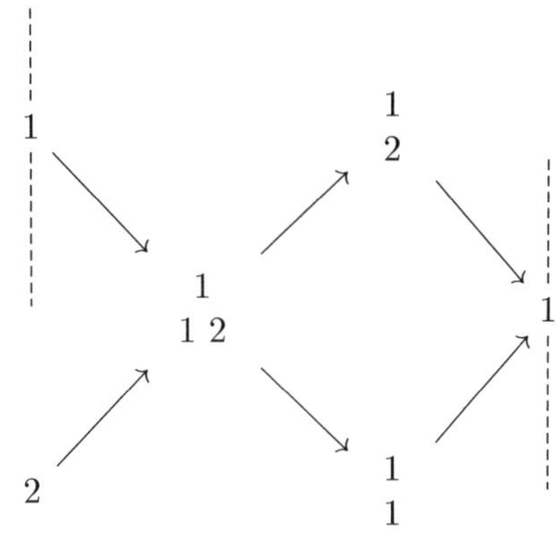

Hence there exist obvious projective covers of the form $g : \begin{smallmatrix} 1 \\ 1\,2 \end{smallmatrix} \longrightarrow \begin{smallmatrix} 1 \\ 1 \end{smallmatrix}$ and $h : \begin{smallmatrix} 1 \\ 1\,2 \end{smallmatrix} \longrightarrow \begin{smallmatrix} 1 \\ 2 \end{smallmatrix}$. Since $g^{-1}(1) = 1 \oplus 2 = h^{-1}(2)$, we deduce from condition (ii) of Lemma 2.1 that the injective modules $\begin{smallmatrix} 1 \\ 1 \end{smallmatrix}$ and $\begin{smallmatrix} 1 \\ 2 \end{smallmatrix}$ are almost projective. Moreover the Auslander-Reiten sequence $0 \longrightarrow 2 \longrightarrow \begin{smallmatrix} 1 \\ 1\,2 \end{smallmatrix} \longrightarrow \begin{smallmatrix} 1 \\ 1 \end{smallmatrix} \longrightarrow 0$ implies that the projective dimension of $\begin{smallmatrix} 1 \\ 1 \end{smallmatrix}$ is equal to one. On the other hand the projective resolution

$$\cdots \longrightarrow \begin{smallmatrix} 1 \\ 1\,2 \end{smallmatrix} \oplus 2 \longrightarrow \begin{smallmatrix} 1 \\ 1\,2 \end{smallmatrix} \oplus 2 \longrightarrow \begin{smallmatrix} 1 \\ 1\,2 \end{smallmatrix} \longrightarrow \begin{smallmatrix} 1 \\ 1\,2 \end{smallmatrix} \longrightarrow \begin{smallmatrix} 1 \\ 2 \end{smallmatrix} \longrightarrow 0$$

implies that the projective dimension of $\begin{smallmatrix} 1 \\ 2 \end{smallmatrix}$ is equal to infinity.

Example 2.4 and Example 2.5 suggest that we compare the number of indecomposable almost projective modules of two opposite algebras of finite representation type.

EXAMPLE 2.6. There is an algebra A of finite representation type such that A and A^{op} do not admit the same number of indecomposable almost projective modules.

CONSTRUCTION. Let A be the algebra given by the quiver

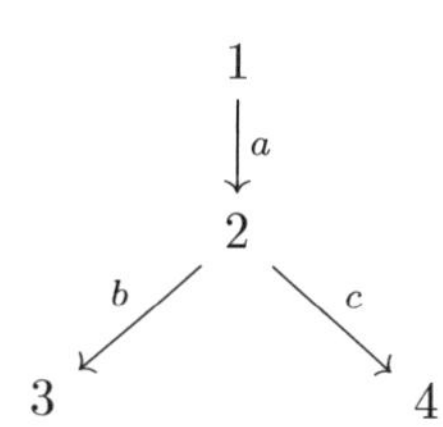

with relation $ca = 0$. Note that A is a finite dimensional algebra. Then the

Auslander-Reiten quiver of A is as follows:

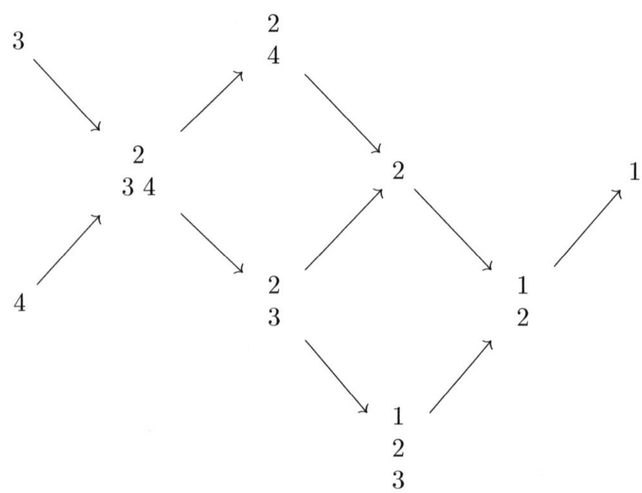

Since the canonical epimorphism $g : \begin{smallmatrix}1\\2\\3\end{smallmatrix} \to \begin{smallmatrix}1\\2\end{smallmatrix}$ is the projective cover of $\begin{smallmatrix}1\\2\end{smallmatrix}$, it follows that the epimorphism $g^{-1}(2) = \begin{smallmatrix}2\\3\end{smallmatrix} \to 2$ induced by g is not a split epimorphism. Hence $\begin{smallmatrix}1\\2\end{smallmatrix}$ is not almost projective. This implies that

(1) The indecomposable almost projective non projective A-modules are 1, 2, $\begin{smallmatrix}2\\3\end{smallmatrix}$, $\begin{smallmatrix}2\\4\end{smallmatrix}$.

On the other hand A^{op} is the algebra given by the quiver

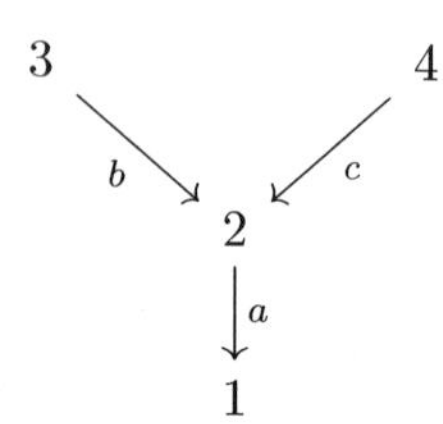

with

relation $ac = 0$. Then the Auslander-Reiten quiver of A^{op} is as follows:

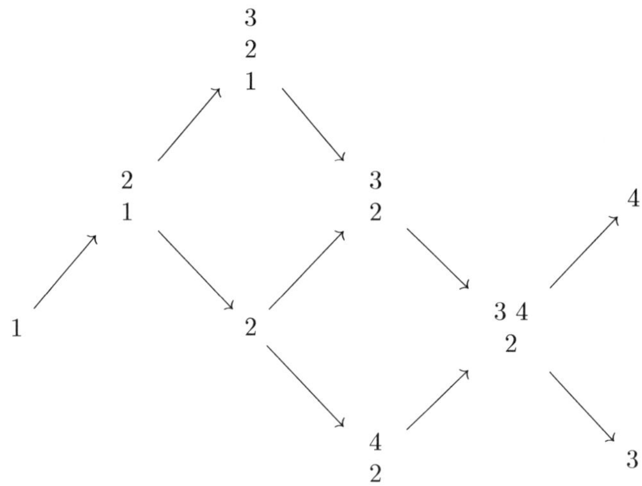

Hence the canonical epimorphisms $g : \begin{smallmatrix}3\\2\\1\end{smallmatrix} \to \begin{smallmatrix}3\\2\end{smallmatrix}$ and $h : \begin{smallmatrix}3\\2\\1\end{smallmatrix} \oplus \begin{smallmatrix}4\\2\end{smallmatrix} \to \begin{smallmatrix}3\ 4\\2\end{smallmatrix}$ are projective covers. Moreover the epimorphisms $g^{-1}(2) = \begin{smallmatrix}2\\1\end{smallmatrix} \to 2$ and $h^{-1}\begin{pmatrix}3\\2\end{pmatrix} = \begin{smallmatrix}3\\2\\1\end{smallmatrix} \oplus 2 \to \begin{smallmatrix}3\\2\end{smallmatrix}$ induced by g and h do not split. Consequently $\begin{smallmatrix}3\\2\end{smallmatrix}$ and $\begin{smallmatrix}3\ 4\\2\end{smallmatrix}$ are not almost projective modules. This implies that

(2) The indecomposable almost projective non projective A^{op}-modules are $2, 3$ and 4.

Putting (1) and (2) together, we conclude that A and A^{op} do not have the same number of indecomposable almost projective modules.

COROLLARY 2.7. *There exists an algebra A of finite representation type with the following properties:*

(i) *A and A^{op} admit exactly one indecomposable projective left module with a non-projective submodule.*
(ii) *The proper submodules of indecomposable almost projective non projective left modules over A and A^{op} are projective.*

PROOF. Let A be the algebra considered in Example 2.6. Then $\begin{smallmatrix}1\\2\\3\end{smallmatrix}$ (resp. $\begin{smallmatrix}4\\2\end{smallmatrix}$) is the unique indecomposable projective left A-module (resp. A^{op}-module) with a non-projective submodule, namely its radical. Hence (i) holds. On the other hand 1, 2, $\begin{smallmatrix}2\\3\end{smallmatrix}$ and $\begin{smallmatrix}2\\4\end{smallmatrix}$ (resp. 2, 3 and 4) are the indecomposable almost projective non projective left A-modules (resp. left A^{op}-modules). This implies that also (ii) holds. □

Up to now, many almost projective modules have the property that all their proper submodules are uniserial. The next example shows that uniserial submodules give few information also in the hereditary case.

EXAMPLE 2.8. There are an algebra A and an indecomposable injective module I with the following properties:

(i) Every uniserial submodule of I is projective;
(ii) I is the unique indecomposable A-module which is not almost projective.
(iii) I is the sum of three uniserial projective modules and also of two indecomposable almost projective modules.

CONSTRUCTION. Let A be the finite dimensional hereditary algebra given by the quiver

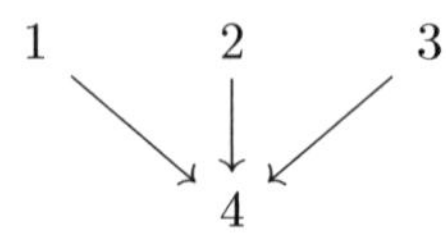

and let I be the indecomposable injective module $\begin{smallmatrix}1&2&3\\&4&\end{smallmatrix}$. Then the uniserial submodules of I are $\begin{smallmatrix}1\\4\end{smallmatrix}, \begin{smallmatrix}2\\4\end{smallmatrix}, \begin{smallmatrix}3\\4\end{smallmatrix}$, 4 and the Auslander-Reiten quiver of A is as follows:

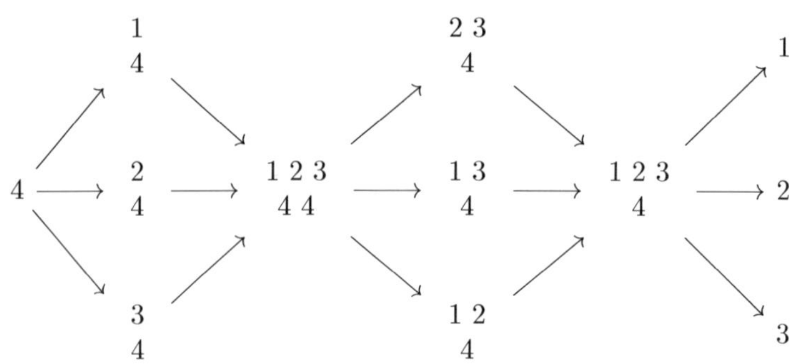

Hence (i) holds. We first note that the three maximal submodules $\begin{smallmatrix}1&2\\&4\end{smallmatrix}, \begin{smallmatrix}1&3\\&4\end{smallmatrix}$ and $\begin{smallmatrix}2&3\\&4\end{smallmatrix}$ of I are not projective. Since A is hereditary, it follows that I is not almost projective. We finally note that if M is uniserial, then M is either projective or simple. On the other hand if $\dim_K M = 3$, then any proper submodule of M is projective. We also note that any proper submodule of $\begin{smallmatrix}1&2&3\\4&4&\end{smallmatrix}$ is projective. Since $\begin{smallmatrix}1&2&3\\&4&\end{smallmatrix}$ and $\begin{smallmatrix}1&2&3\\4&4&\end{smallmatrix}$ are the unique indecomposable modules of dimension > 3, it follows that also (ii) holds. We clearly have $I = \begin{smallmatrix}1&2&3\\&4&\end{smallmatrix} = \begin{smallmatrix}1\\4\end{smallmatrix} + \begin{smallmatrix}2\\4\end{smallmatrix} + \begin{smallmatrix}3\\4\end{smallmatrix} = \begin{smallmatrix}1&2\\&4\end{smallmatrix} + \begin{smallmatrix}2&3\\&4\end{smallmatrix}$. Consequently (iii) holds.

The almost projective modules in Examples 2.2, 2.3, 2.4, 2.5, and 2.8 have Loewy length at most two. In the next example we construct almost projective modules of length and Loewy length n with $n \geq 3$.

EXAMPLE 2.9. For any $n \geq 3$ there are an algebra A and two uniserial almost projective modules L and M of length n with the following properties:

(a) L is injective and the projective dimension of L is equal to one.
(b) M is injective and the projective dimension of M is equal to infinity.

CONSTRUCTION. Let A be the finite dimensional algebra given by the quiver

$$\circlearrowleft 1 \longrightarrow 2 \cdots n-1 \longrightarrow n$$

with relations such that the indecomposable projective modules are of the form $\begin{smallmatrix} 1 & \\ 1 & 2 \\ \vdots & \vdots \\ 1 & n \end{smallmatrix}, \begin{smallmatrix} 2 \\ \vdots \\ n \end{smallmatrix}, \ldots, \begin{smallmatrix} n\text{-}1 \\ n \end{smallmatrix}, n$. Next let L and M denote the almost projective and injective modules $\begin{smallmatrix} 1 \\ 1 \\ \vdots \\ 1 \end{smallmatrix}$ and $\begin{smallmatrix} 1 \\ 2 \\ \vdots \\ n \end{smallmatrix}$ respectively. Then the projective resolution of L is of the form $0 \to \begin{smallmatrix} 2 \\ \vdots \\ n \end{smallmatrix} \to \begin{smallmatrix} 1 & \\ 1 & 2 \\ \vdots & \vdots \\ 1 & n \end{smallmatrix} \to \begin{smallmatrix} 1 \\ 1 \\ \vdots \\ 1 \end{smallmatrix} \to 0$. We finally note that the projective resolution of M is of the form

$$\cdots \to \begin{smallmatrix} 1 & \\ 1 & 2 \\ \vdots & \vdots \\ 1 & n \end{smallmatrix} \oplus \begin{smallmatrix} 2 \\ \vdots \\ n \end{smallmatrix} \to \begin{smallmatrix} 1 & \\ 1 & 2 \\ \vdots & \vdots \\ 1 & n \end{smallmatrix} \oplus \begin{smallmatrix} 2 \\ \vdots \\ n \end{smallmatrix} \to \begin{smallmatrix} 1 & \\ 1 & 2 \\ \vdots & \vdots \\ 1 & n \end{smallmatrix} \to \begin{smallmatrix} 1 & \\ 1 & 2 \\ \vdots & \vdots \\ 1 & n \end{smallmatrix} \to \begin{smallmatrix} 1 \\ 2 \\ \vdots \\ n \end{smallmatrix} \to 0.$$

Hence (a) and (b) hold.

The module $\begin{smallmatrix} 1 & 2 \\ & 2 \end{smallmatrix}$ in Example 2.4 and the modules $\begin{smallmatrix} 1 & 2 \\ 4 & \end{smallmatrix}, \begin{smallmatrix} 1 & 3 \\ 4 & \end{smallmatrix}, \begin{smallmatrix} 2 & 3 \\ 4 & \end{smallmatrix}$ in Example 2.8 are almost projective modules of the form $P + Q$ with P and Q indecomposable projective modules. However it is easy to see that the sum of two indecomposable projective modules is not necessarily almost projective.

EXAMPLE 2.10. There exists an indecomposable module M over a non-hereditary algebra with the following properties:

(a) $M = P + Q$ with P and Q indecomposable projective modules.
(b) M is not almost projective.
(c) M has exactly one proper non-projective submodule which does not satisfy condition (ii) of Lemma 2.1.
(d) The maximal submodules of M are either projective or almost projective.

CONSTRUCTION. Let A be the finite dimensional algebra given by the quiver

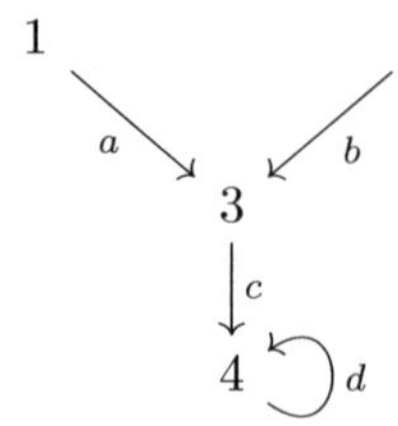

with relations $cb = 0, dc = 0, d^2 = 0$. Then the indecomposable projective modules are $\begin{smallmatrix}1\\3\\4\end{smallmatrix}$, $\begin{smallmatrix}2\\3\end{smallmatrix}$, $\begin{smallmatrix}3\\4\end{smallmatrix}$ and $\begin{smallmatrix}4\\4\end{smallmatrix}$. Hence 4 is a non-projective submodule of $\begin{smallmatrix}1\\3\\4\end{smallmatrix}$, $\begin{smallmatrix}3\\4\end{smallmatrix}$ and $\begin{smallmatrix}4\\4\end{smallmatrix}$, while 3 is a non-projective submodule of $\begin{smallmatrix}2\\3\end{smallmatrix}$. The Auslander-Reiten quiver of A is as follows:

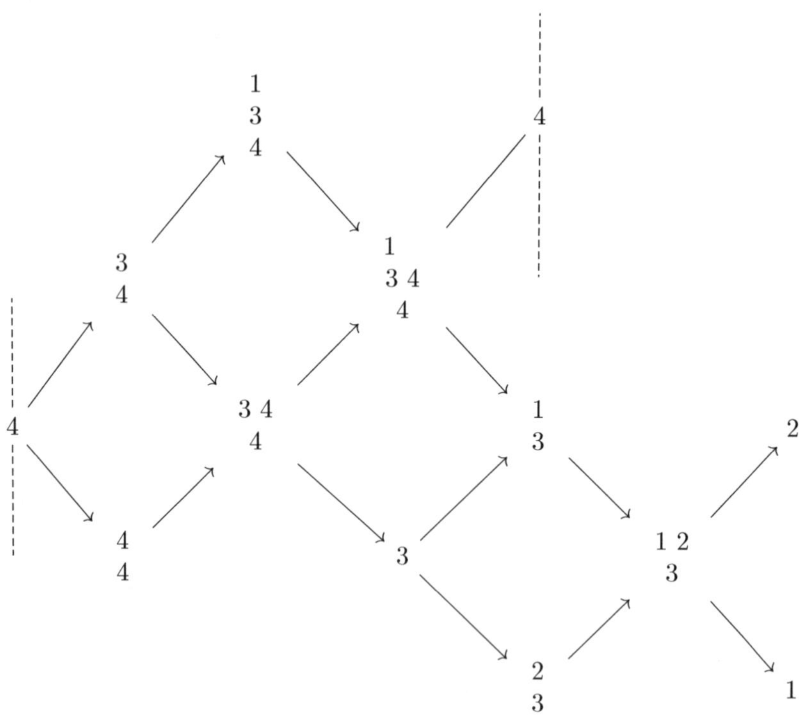

Let M be the indecomposable injective module $\begin{smallmatrix}1\\3\ 4\\4\end{smallmatrix}$. Then we have $M = P + Q$ with $P \cong \begin{smallmatrix}1\\3\\4\end{smallmatrix}$ and $Q \cong \begin{smallmatrix}4\\4\end{smallmatrix}$. Hence (a) holds. On the other hand the projective cover of M is the canonical epimorphism $g : \begin{array}{c}1\\3\\4\end{array} \oplus \begin{array}{c}4\\4\end{array} \to M$ and $X = \begin{array}{c}3\ 4\\4\end{array}$ is a proper submodule of M such that $g^{-1}(X) = \begin{array}{c}3\\4\end{array} \oplus \begin{array}{c}4\\4\end{array}$. This implies that the epimorphism $g^{-1}(X) \to X$ induced by g does not split. Consequently M is not almost projective and so (b) holds. We finally note that 4 and $\begin{array}{c}3\ 4\\4\end{array}$ are the unique proper submodules of M which are not projective. Moreover we clearly have $g^{-1}(4) = 4 \oplus 4$. Hence the module 4 satisfies condition (ii) of Lemma 2.1. This observation and the proof of (b) imply that also (c) holds. Since $\begin{array}{c}1\\3\\4\end{array}$ and $\begin{array}{c}3\ 4\\4\end{array}$ are the two maximal submodules of M, it follows that (d) holds.

Acknowledgments

We would like to thank Professor Roberto Martinez -Villa, who kindly sent us his papers on almost projective modules.

References

[1] Maurice Auslander, Idun Reiten, and Sverre O. Smalø, *Representation theory of Artin algebras*, Cambridge Studies in Advanced Mathematics, vol. 36, Cambridge University Press, Cambridge, 1997. Corrected reprint of the 1995 original. MR1476671

[2] Roberto Martínez-Villa, *Almost projective modules over hereditary algebras*, An. Inst. Mat. Univ. Nac. Autónoma México **20** (1980), no. 1, 1–89 (1982). MR684702

[3] Roberto Martínez-Villa, *Almost projective modules and almost split sequences with indecomposable middle term*, Comm. Algebra **8** (1980), no. 12, 1123–1150, DOI 10.1080/00927878008822512. MR579792

[4] Park J. S., *Almost projective modules over Artin algebras*, J. Chungcheong Math. Soc. 1 (1988), 43-53.

[5] Ralf Schiffler, *Quiver representations*, CMS Books in Mathematics/Ouvrages de Mathématiques de la SMC, Springer, Cham, 2014, DOI 10.1007/978-3-319-09204-1. MR3308668

Department of Mathematics, University of Afyon Kocatepe,03200 Afyonkarahisar, Turkey

Email address: fkaynarca@aku.edu.tr

Department of Mathematics, University of Milano, Milano, Italy

Email address: gabriella.deste@unimi.it

Department of Mathematics, University of Hacettepe, 06800 Beytepe, Ankara, Turkey

Email address: keskin@hacettepe.edu.tr

Contemporary Mathematics
Volume **785**, 2023
https://doi.org/10.1090/conm/785/15785

On prime ideals with generalized derivations in rings with involution

Mohammad Salahuddin Khan, Adnan Abbasi, Shakir Ali, and Mohammed Ayedh

Dedicated to Professor Syed Tariq Rizvi for his contribution to Ring and Module Theory

ABSTRACT. The main objective of this paper is to study about the action of generalized derivations on prime ideals of an arbitrary ring with involution. In particular, apart from proving some other results, we establish that if (F, d) is a generalized derivation of R satisfying the condition $[F(x), x^*] \in P$ for all $x \in R$, then one of the following holds:

(1) $char(R/P) = 2$
(2) R/P is a commutative integral domain
(3) $F(R) \subseteq P$.

INTRODUCTION AND NOTATIONS

Throughout this article, R will represent an associative ring with center $Z(R)$. Recall that an ideal P of R is said to be prime if $P \neq R$ and for $x, y \in R$, $xRy \subseteq P$ implies that $x \in P$ or $y \in P$. Therefore, R is called a prime ring if and only if (0) is the prime ideal of R. For any $x, y \in R$, the symbol $[x, y]$ will denote the commutator $xy - yx$; while the symbol $x \circ y$ will stand for the anticommutator $xy + yx$. An additive mapping $x \mapsto x^*$ satisfying $(xy)^* = y^*x^*$ and $(x^*)^* = x$ is called an involution. A ring equipped with an involution is known as ring with involution or $*$-ring. An element x in a ring with involution $*$ is said to be hermitian if $x^* = x$ and skew-hermitian if $x^* = -x$. The sets of all hermitian and skew-hermitian elements of R will be denoted by $H(R)$ and $S(R)$, respectively. If R is 2-torsion free then every $x \in R$ can be uniquely represented in the form $2x = h + k$, where $h \in H(R)$ and $k \in S(R)$. The involution is said to be of the first kind if $Z(R) \subseteq H(R)$, otherwise it is said to be of the second kind. In the second case $S(R) \cap Z(R) \neq \{0\}$. We refer the reader to [**9**] for justification and amplification for the above mentioned notations and key definitions.

A map $d : R \to R$ is a derivation of a ring R if d is additive and satisfies $d(xy) = d(x)y + xd(y)$ for all $x, y \in R$. A map $F : R \to R$ is a generalized

2020 *Mathematics Subject Classification.* 16N60, 16W10, 16W25.

Key words and phrases. Associative ring, derivation, generalized derivation, involution, prime ideal.

Corresponding author: Shakir Ali(shakir.ali.mm@amu.ac.in).

The third author was supported by SERB-DST MATRIC Project (Grant No.: MTR/2019/000603), India.

derivation of a ring R with a derivation d if F is additive and satisfies $F(xy) = F(x)y + xd(y)$ for all $x, y \in R$. We denote a generalized derivation F associated with the derivation d by (F, d). An additive mapping $T : R \to R$ is said to a left centralizer if $T(xy) = T(x)y$ for all $x, y \in R$ (see [**17, 18**] for details). Recently, many authors have obtained commutativity of prime and semiprime rings admitting suitably constrained additive mappings, such as automorphisms, derivations, skew derivations and generalized derivations acting on appropriate subsets of the rings (see [**2, 4–8**] and [**12**] for details). We first recall that for a nonempty subset S of R, a mapping $f : S \to R$ is called centralizing if $[f(x), x] \in Z(R)$ for all $x \in S$, in the special case where $[f(x), x] = 0$ for all $x \in S$, the mapping f is said to be commuting on S. In [**15**], Posner proved that if a prime ring R admits a nonzero derivation d such that $[d(x), x] \in Z(R)$ for all $x \in R$, then R is commutative. Over the last few decades, several authors have subsequently refined and extended this classical theorem in various directions (see [**1**], [**12**], [**13**] and [**14**] where further references can be found).

In the present paper, we investigate various functional identities on prime ideal of a $*$-ring and describe the structure of the quotient ring R/P. In particular, we prove that if a ring R with involution $*$ admits a generalized derivation (F, d) satisfying any one of the following differential identities:

(i) $[F(x), x^*] \in P$
(ii) $[F(x), F(x^*)] \pm F([x, x^*]) \in P$
(iii) $F([x, x^*]) \pm [F(x), x^*] \pm [F(x), F(x^*)] \in P$
(iv) $F([x, x^*]) \pm [F(x), x^*] \in P$
(v) $[F(x), F(x^*)] \pm [x, x^*] \in P$
(vi) $[F(x), x^*] \pm [x, x^*] \in P$,

for all $x \in R$, then one of the following holds:

(1) $char(R/P) = 2$
(2) R/P is a commutative integral domain
(3) $F(R) \subseteq P$ or $d(R) \subseteq P$.

1. The Results

We shall do a great deal of calculation with commutators and anti-commutators, routinely using the following basic identities: For all $s, t, w \in R$;

$$[st, w] = s[t, w] + [s, w]t \text{ and } [s, tw] = t[s, w] + [s, t]w$$

$$s \circ (tw) = (s \circ t)w - t[s, w] = t(s \circ w) + [s, t]w$$

$$(st) \circ w = s(t \circ w) - [s, w]t = (s \circ w)t + s[t, w].$$

We start our investigation with the following lemmas which will be used frequently throughout the discussions.

LEMMA 1.1. [**10**, Theorem 1] *Let R be a ring, P a prime ideal of R. If F is a generalized derivation of R satisfies the condition $[F(x), x] \in Z(R/P)$ for all $x \in R$, then $d(R) \subseteq P$ or R/P is a commutative integral domain.*

LEMMA 1.2. [**11**, Lemma 2.3] *Let R be a ring with involution $*$ of the second kind, P a prime ideal of R such that $S(R) \cap Z(R) \not\subseteq P$ and $char(R/P) \neq 2$. If $[x, x^*] \in P$ for all $x \in R$, then R/P is a commutative integral domain.*

LEMMA 1.3. *Let R be a ring with involution $*$ of the second kind and P a prime ideal of R such that $S(R) \cap Z(R) \nsubseteq P$. If $x \circ x^* \in P$ for all $x \in R$, then $char(R/P) = 2$.*

PROOF. On the contrary, we suppose that $char(R/P) \neq 2$. By the hypothesis, we have

$$x \circ x^* \in P \tag{1.1}$$

for all $x \in R$. Direct linearization of (1.1) gives

$$x \circ y^* + y \circ x^* \in P \tag{1.2}$$

for all $x, y \in R$. Replacing x by xk in (1.2), where $0 \neq k \in S(R) \cap Z(R)$, we get

$$k(x \circ y^* - y \circ x^*) \in P$$

for all $x, y \in R$. Since $S(R) \cap Z(R) \nsubseteq P$, it follows that

$$x \circ y^* - y \circ x^* \in P \tag{1.3}$$

for all $x, y \in R$. Combining (1.3) and (1.2), we obtain

$$2(x \circ y^*) \in P$$

for all $x, y \in R$. Taking $y = k$, we get

$$-4kx \in P$$

for all $x \in R$. Using the conditions $S(R) \cap Z(R) \nsubseteq P$ and $char(R/P) \neq 2$, we conclude that $x \in P$ for all $x \in R$, which is not possible. Thus, we have $char(R/P) = 2$. This proves the lemma. □

THEOREM 1.4. *Let R be a ring with involution $*$ of the second kind, P a prime ideal of R such that $S(R) \cap Z(R) \nsubseteq P$. If (F, d) is a generalized derivation of R satisfying the condition $[F(x), x^*] \in P$ for all $x \in R$, then one of the following holds:*

(1) *$char(R/P) = 2$*
(2) *R/P is a commutative integral domain*
(3) *$F(R) \subseteq P$.*

PROOF. Assume that $char(R/P) \neq 2$. By the given hypothesis, we have

$$[F(x), x^*] \in P \text{ for all } x \in R. \tag{1.4}$$

Linearization of equation (1.4) gives that

$$[F(x), y^*] + [F(y), x^*] \in P \text{ for all } x, y \in R. \tag{1.5}$$

Replacing y by yh in (1.5), where $h \in H(R) \cap Z(R)$, we obtain

$$[F(x), y^*]h + [F(y), x^*]h + [y, x^*]d(h) \in P \text{ for all } x, y \in R. \tag{1.6}$$

In view of (1.5), the expression (1.6) reduces to $[y, x^*]d(h) \in P$ for all $x, y \in R$ and $h \in H(R) \cap Z(R)$. Now using the primeness of P, we obtain either $[y, x^*] \in P$ or $d(h) \in P$. In view of Lemma 1.2, the first case gives R/P is a commutative integral domain. Now consider $d(h) \in P$ for all $h \in H(R) \cap Z(R)$. Replacing h by k^2, where $k \in S(R) \cap Z(R)$, we get $2d(k)k \in P$ for all $k \in S(R) \cap Z(R)$. Since $S(R) \cap Z(R) \nsubseteq P$ and $char(R/P) \neq 2$, we conclude that $d(k) \in P$ for all $k \in S(R) \cap Z(R)$. Now substituting yk for y in (1.5), where $k \in S(R) \cap Z(R)$, we get

$$-[F(x), y^*]k + [F(y), x^*]k + [y, x^*]d(k) \in P$$

$x, y \in R$ and $k \in S(R) \cap Z(R)$. Now using $d(k) \in P$, we arrive at

$$-[F(x), y^*]k + [F(y), x^*]k \in P \tag{1.7}$$

for all $x, y \in R$. Since $S(R) \cap Z(R) \not\subseteq P$, the above relation becomes

$$-[F(x), y^*] + [F(y), x^*] \in P \tag{1.8}$$

for all $x, y \in R$. Now combining (1.5) and (1.8), we get $2[F(y), x^*] \in P$ for all $x, y \in R$. This implies that

$$[F(y), x] \in P \text{ for all } x, y \in R. \tag{1.9}$$

Substituting yx for y in (1.9), we find that

$$y[d(x), x] + [y, x]d(x) \in P \tag{1.10}$$

for all $x, y \in P$. Replacing y by ry in (1.10), we get

$$ry[d(x), x] + r[y, x]d(x) + [r, x]yd(x) \in P$$

for all $x, y, r \in R$. Application of (1.10) yields

$$[r, x]yd(x) \in P \tag{1.11}$$

for all $x, y, r \in R$. Then for a fixed element x applying the primeness of P, we obtain either $[r, x] \in P$ or $d(x) \in P$. Let $A = \{x \in R : [r, x] \in P\}$ and $B = \{x \in R : d(x) \in P\}$. It can be easily verified that A and B are the additive subgroups of R whose union is R. But a group cannot be written as a union of its two proper subgroups. Thus, we have $A = R$ or $B = R$. If $A = R$, then R/P is a commutative integral domain in view of Lemma 1.2. If $B = R$, then $d(R) \subseteq P$. Now, replacing y by yz in (1.9) and using the fact $d(R) \subseteq P$, we obtain

$$F(y)[z, x] \in P$$

for all $x, y, z \in R$, which further implies

$$F(y)R[z, x] \subseteq P$$

for all $x, y, z \in R$. The primeness of P gives the required results. Thereby the proof of theorem is complete now. □

COROLLARY 1.5. *Let R be a ring with involution $*$ of the second kind, P a prime ideal of R such that $S(R) \cap Z(R) \not\subseteq P$. If d is a derivation of R satisfying the condition $[d(x), x^*] \in P$ for all $x \in R$, then then one of the following holds:*

1. *$char(R/P) = 2$*
2. *R/P is a commutative integral domain*
3. *$d(R) \subseteq P$.*

COROLLARY 1.6. *Let R be a prime ring of $char(R) \neq 2$ with involution $*$ of the second kind. If (F, d) is a generalized derivation of R satisfying the condition $[F(x), x^*] = 0$ for all $x \in R$, then R is a commutative integral domain or $F = 0$.*

COROLLARY 1.7. *Let R be a prime ring of $char(R) \neq 2$ with involution $*$ of the second kind. If d is a derivation of R satisfying the condition $[d(x), x^*] = 0$ for all $x \in R$, then R is a commutative integral domain or $d = 0$.*

THEOREM 1.8. *Let R be a ring with involution $*$ of the second kind, P a prime ideal of R such that $S(R) \cap Z(R) \not\subseteq P$. If (F, d) is a generalized derivation of R satisfying the condition $[F(x), F(x^*)] \pm F([x, x^*]) \in P$ for all $x \in R$, then one of* *the following holds:*

(1) $char(R/P) = 2$
(2) R/P *is a commutative integral domain*
(3) $d(R) \subseteq P$.

PROOF. Suppose $char(R/P) \neq 2$. By the assumption, we have

$$[F(x), F(x^*)] \pm F([x, x^*]) \in P \tag{1.12}$$

for all $x \in R$. Linearization of (1.12) yields that

$$[F(x), F(y^*)] + [F(y), F(x^*)] \pm F([x, y^*]) \pm F([y, x^*]) \in P \tag{1.13}$$

for all $x, y \in R$. Substituting xh for x in (1.13), where $h \in H(R) \cap Z(R)$, we get

$$([x, F(y^*)] + [F(y), x^*] \pm [x, y^*] \pm [y, x^*])d(h) \in P$$

for all $x, y \in R$. Using the primeness of P, we conclude that $[x, F(y^*)] + [F(y), x^*] \pm [x, y^*] \pm [y, x^*] \in P$ or $d(h) \in P$. In the first case, we have

$$[x, F(y^*)] + [F(y), x^*] \pm [x, y^*] \pm [y, x^*] \in P \tag{1.14}$$

for all $x, y \in R$. Replacing x by xk in (1.14) and using the condition $S(R) \cap Z(R) \not\subseteq P$, we find that

$$[x, F(y^*)] - [F(y), x^*] \pm [x, y^*] \mp [y, x^*] \in P$$

for all $x, y \in R$. Combining the last relation with (1.14), we obtain

$$2([x, F(y^*)] \pm [x, y^*]) \in P$$

for all $x, y \in R$. This implies that

$$[x, F(y)] \pm [x, y] \in P \text{ for all } x, y \in R. \tag{1.15}$$

Taking $y = x$ in (1.15), we get $[F(x), x] \in P$ for all $x \in R$. Therefore by Lemma 1.1, R/P is a commutative integral domain or $d(R) \subseteq P$. Now, consider the case $d(h) \in P$ for all $h \in H(R) \cap Z(R)$. Taking $h = k^2$, we have $2kd(k) \in P$, which gives that $d(k) \in P$ for all $k \in S(R) \cap Z(R)$. Replacing x by xk in (1.13) and using the fact that $d(k) \in P$, we obtain

$$k([F(x), F(y^*)] - [F(y), F(x^*)] \pm F([x, y^*]) \mp F([y, x^*])) \in P$$

for all $x, y \in R$. Since $S(R) \cap Z(R) \not\subseteq P$, it follows that

$$[F(x), F(y^*)] - [F(y), F(x^*)] \pm F([x, y^*]) \mp F([y, x^*]) \in P \tag{1.16}$$

for all $x, y \in R$. From (1.13) and (1.16), we conclude that

$$2([F(x), F(y^*)] \pm F([x, y^*])) \in P$$

for all $x, y \in R$. This gives that

$$[F(x), F(y)] \pm F([x, y]) \in P \tag{1.17}$$

for all $x, y \in R$. Replacing y by yt in (1.17) and using it, we get

$$F(y)[F(x), t] + y[F(x), d(t)] + [F(x), y]d(t) \pm [x, y]d(t) \pm F(y)[x, t] \pm yd([x, t]) \in P$$

for all $x, y, t \in R$. This can be written as

$$F(y)([F(x), t] \pm [x, t]) + y([F(x), d(t)] \pm d([x, t])) + ([F(x), y] \pm [x, y])d(t) \in P$$

for all $x, y, t \in R$. This implies that

$$F(y)[F(x) \pm x, t] + y([F(x), d(t)] \pm d([x, t])) + [F(x) \pm x, y]d(t) \in P \tag{1.18}$$

for all $x, y, t \in R$. Replacing y by ry in (1.18), we get

$$\begin{aligned}F(ry)[F(x) \pm x, t] + ry([F(x), d(t)] \pm d([x, t]))\\ + r[F(x) \pm x, y]d(t) + [F(x) \pm x, r]yd(t) \in P\end{aligned} \tag{1.19}$$

for all $x, y, t, r \in R$. The left multiplication of (1.18) by r gives that

$$rF(y)[F(x) \pm x, t] + ry([F(x), d(t)] \pm d([x, t])) + r[F(x) \pm x, y]d(t) \in P \tag{1.20}$$

for all $x, y, t, r \in R$. Subtracting (1.20) from (1.19), we obtain

$$(F(ry) - rF(y))[F(x) \pm x, t] + [F(x) \pm x, r]yd(t) \in P \tag{1.21}$$

for all $x, y, t, r \in R$. Replacing t by ts in (1.21) and using it, we get

$$(F(ry) - rF(y))t[F(x) \pm x, s] + [F(x) \pm x, r]ytd(s) \in P \tag{1.22}$$

for all $x, y, t, r, s \in R$. That is,

$$(\overline{F(ry) - rF(y)})\overline{t}\overline{[F(x) \pm x, s]} + \overline{[F(x) \pm x, r]}\overline{y}\overline{t}\overline{d(s)} = \overline{0}.$$

In view of [**9**, Lemma 1.3.2], we conclude that

(i) $(\overline{F(ry) - rF(y)}]) = \overline{[F(x) \pm x, r]y} = \overline{0}$ or
(ii) $\overline{[F(x) \pm x, s]} = \overline{d(s)} = \overline{0}$
(iii) $\lambda_1\overline{[F(x) \pm x, s]} + \lambda_2\overline{d(s)} = \overline{0}$

for some $\lambda_1, \lambda_2 \in \overline{C}$, where $\overline{C}$ is the extended centroid of $\overline{R} = R/P$. From (i), we have $\overline{[F(x) \pm x, r]} = \overline{0}$, which means $[F(x) \pm x, r] \in P$ for all $x, r \in R$. Taking $r = x$, we have $[F(x), x] \in P$ and hence result follows from Lemma 1.1.
In (ii), $\overline{d(s)} = \overline{0}$ gives that $d(R) \subseteq P$.
In (iii), suppose $\lambda_2 \neq \overline{0}$, then

$$\overline{d(s)} + \lambda_2^{-1}\lambda_1\overline{[F(x) \pm x, s]} = \overline{0}.$$

In particular, for $x = 0$ the above expression reduces to $\overline{d(s)} = \overline{0}$. This gives $d(R) \subseteq P$. Thereby the proof of theorem is complete now. □

COROLLARY 1.9. *Let R be a ring with involution $*$ of the second kind, P a prime ideal of R such that $S(R) \cap Z(R) \not\subseteq P$. If d is a derivation of R satisfying the condition $[d(x), d(x^*)] \pm d([x, x^*]) \in P$ for all $x \in R$, then one of the following holds:*

(1) *$char(R/P) = 2$*
(2) *R/P is a commutative integral domain*
(3) *$d(R) \subseteq P$.*

COROLLARY 1.10. *Let R be a prime ring of $char(R) \neq 2$ with involution $*$ of the second kind. If (F, d) is a generalized derivation of R satisfying the condition $[F(x), F(x^*)] \pm F([x, x^*]) = 0$ for all $x \in R$, then one of the following holds:*

(1) *R is a commutative integral domain*
(2) *F is a left centralizer of R.*

COROLLARY 1.11. *Let R be a prime ring of $char(R) \neq 2$ with involution $*$ of the second kind. If d is a derivation of R satisfying the condition $[d(x), d(x^*)] \pm d([x, x^*]) = 0$ for all $x \in R$, then R is a commutative integral domain or $d = 0$.*

THEOREM 1.12. *Let R be a ring with involution $*$ of the second kind, P a prime ideal of R such that $S(R) \cap Z(R) \not\subseteq P$. If (F, d) is a generalized derivation of R satisfying the condition $F([x, x^*]) \pm [F(x), x^*] \pm [F(x), F(x^*)] \in P$ for all $x \in R$, then one of the following holds:*

(1) *$char(R/P) = 2$*
(2) *R/P is a commutative integral domain*
(3) *$d(R) \subseteq P$.*

PROOF. Suppose $char(R/P) \neq 2$. First we assume that

$$F([x, x^*]) + [F(x), x^*] + [F(x), F(x^*)] \in P \tag{1.23}$$

for all $x \in R$. Interchanging the role of x and x^* in (1.23), we get

$$-F([x, x^*]) + [F(x^*), x] - [F(x), F(x^*)] \in P \tag{1.24}$$

for all $x \in R$. Combination of (1.23) and (1.24) gives that

$$[F(x), x^*] + [F(x^*), x] \in P \text{ for all } x \in R.$$

Linearizing the last relation, we obtain

$$[F(x), y^*] + [F(y), x^*] + [F(x^*), y] + [F(y^*), x] \in P \text{ for all } x, y \in R. \tag{1.25}$$

Substituting xh for x in (1.25), where $h \in H(R) \cap Z(R)$, we get

$$([x, y^*] + [x^*, y])d(h) \in P \text{ for all } x, y \in R.$$

The primeness of P yields $[x, y^*] + [x^*, y] \in P$ or $d(h) \in P$. First, we consider the case

$$[x, y^*] + [x^*, y] \in P \text{ for all } x, y \in R. \tag{1.26}$$

Replacing x by xk in (1.26) where $k \in S(R) \cap Z(R)$, we find that

$$[x, y^*] - [x^*, y] \in P \text{ for all } x, y \in R. \tag{1.27}$$

Adding (1.26) and (1.27), we get

$$2[x, y^*] \in P \text{ for all } x, y \in R.$$

This implies that $[x, x^*] \in P$ for all $x \in R$. Therefore, in view of Lemma 1.2, R/P is a commutative integral domain. On the other hand, if $d(h) \in P$ for all $h \in H(R) \cap Z(R)$. Replacing h by k^2 in the last relation, we get $2kd(k) \in P$, which further implies $d(k) \in P$ for all $k \in S(R) \cap Z(R)$. Substituting xk for x in (1.25) and using the fact $d(k) \in P$, we obtain

$$k([F(x), y^*] - [F(y), x^*] - [F(x^*), y] + [F(y^*), x]) \in P \text{ for all } x, y \in R.$$

This gives that

$$[F(x), y^*] - [F(y), x^*] - [F(x^*), y] + [F(y^*), x] \in P \text{ for all } x, y \in R. \tag{1.28}$$

Combining (1.25) and (1.28), we get

$$2([F(x), y^*] + [F(y^*), x]) \in P \text{ for all } x, y \in R.$$

This further implies that

$$[F(x), y] + [F(y), x] \in P \text{ for all } x, y \in R.$$

In particular, for $y = x$ we have $[F(x), x] \in P$ for all $x \in R$. Thus by Lemma 1.1, R/P is a commutative integral domain or $d(R) \subseteq P$. Similarly, we can prove the result for the case $F([x,x^*]) - [F(x),x^*] - [F(x),F(x^*)] \in P$ for all $x \in R$. This proves the theorem. □

COROLLARY 1.13. *Let R be a ring with involution $*$ of the second kind, P a prime ideal of R such that $S(R) \cap Z(R) \not\subseteq P$. If d is a derivation of R satisfying the condition $d([x,x^*]) \pm [d(x),x^*] \pm [d(x),d(x^*)] \in P$ for all $x \in R$, then one of the following holds:*

(1) *$char(R/P) = 2$*
(2) *R/P is a commutative integral domain*
(3) *$d(R) \subseteq P$.*

COROLLARY 1.14. *Let R be a prime ring of $char(R) \neq 2$ with involution $*$ of the second kind. If (F,d) is a generalized derivation of R satisfying the condition $F([x,x^*]) \pm [F(x),x^*] \pm [F(x),F(x^*)] = 0$ for all $x \in R$, then one of the following holds:*

(1) *R is a commutative integral domain*
(2) *F is a left centralizer of R.*

COROLLARY 1.15. *Let R be a prime ring of $char(R) \neq 2$ with involution $*$ of the second kind. If d is a derivation of R satisfying the condition $d([x,x^*]) \pm [d(x),x^*] \pm [d(x),d(x^*)] = 0$ for all $x \in R$, then R is a commutative integral domain or $d = 0$.*

THEOREM 1.16. *Let R be a ring with involution $*$ of the second kind, P a prime ideal of R such that $S(R) \cap Z(R) \not\subseteq P$. If (F,d) is a generalized derivation of R satisfying the condition $F([x,x^*]) \pm [F(x),x^*] \in P$ for all $x \in R$, then one of the following holds:*

(1) *$char(R/P) = 2$*
(2) *R/P is a commutative integral domain*
(3) *$d(R) \subseteq P$.*

PROOF. Suppose $char(R/P) \neq 2$. We have

$$F([x,x^*]) \pm [F(x),x^*] \in P \tag{1.29}$$

for all $x \in R$. Linearization of (1.29) yields that

$$F([x,y^*]) + F([y,x^*]) \pm [F(x),y^*] \pm [F(y),x^*] \in P \tag{1.30}$$

for all $x,y \in R$. Substituting xh for x in (1.30), where $0 \neq h \in H(R) \cap Z(R)$, we get

$$([x,y^*] + [y,x^*] \pm [x,y^*])d(h) \in P \text{ for all } x,y \in R.$$

Since P is a prime ideal of R, so we have $[x,y^*] + [y,x^*] \pm [x,y^*] \in P$ or $d(h) \in P$. Consider the first case

$$[x,y^*] + [y,x^*] \pm [x,y^*] \in P \tag{1.31}$$

for all $x,y \in R$. Now, we have two cases with positive or negative sign of the third term in (1.31).

CASE 1. If we take positive sign in (1.31), we have

$$2[x,y^*] + [y,x^*] \in P \text{ for all } x,y \in R. \tag{1.32}$$

Replacing y by yk in (1.32), where $k \in S(R) \cap Z(R)$ and using the condition $S(R) \cap Z(R) \not\subseteq P$, we find that

$$-2[x,y^*] + [y,x^*] \in P \text{ for all } x,y \in R. \tag{1.33}$$

Adding (1.32) and (1.33), we get

$$2[y, x^*] \in P \text{ for all } x, y \in R,$$

which yields $[x, x^*] \in P$ for all $x \in R$. Thus by Lemma 1.2, R/P is a commutative integral domain.

CASE 2. If we take negative sign in (1.31), then result follows from Lemma 1.2.

On the other hand, if $d(h) \in P$ for all $h \in H(R) \cap Z(R)$. This further gives that $d(k) \in P$ for all $k \in S(R) \cap Z(R)$. Now, substituting xk in place of x in (1.30) and using the fact $d(k) \in P$, we obtain

$$k(F([x, y^*]) - F([y, x^*]) \pm [F(x), y^*] \mp [F(y), x^*]) \in P$$

for all $x, y \in R$. Since $S(R) \cap Z(R) \not\subseteq P$, we have

$$F([x, y^*]) - F([y, x^*]) \pm [F(x), y^*] \mp [F(y), x^*] \in P \tag{1.34}$$

for all $x, y \in R$. Combining (1.30) and (1.34), we obtain

$$2(F([x, y^*] \pm [F(x), y^*]) \in P$$

This implies that

$$F([x, y]) \pm [F(x), y] \in P \tag{1.35}$$

for all $x, y \in R$. For $y = x$, we have $[F(x), x] \in P$ for all $x \in R$. By Lemma 1.1, R/P is a commutative integral domain or $d(R) \subseteq P$. □

COROLLARY 1.17. *Let R be a ring with involution $*$ of the second kind, P a prime ideal of R such that $S(R) \cap Z(R) \not\subseteq P$. If d is a derivation of R satisfying the condition $d([x, x^*]) \pm [d(x), x^*] \in P$ for all $x \in R$, then one of the following holds:*

(1) *$char(R/P) = 2$*
(2) *R/P is a commutative integral domain*
(3) *$d(R) \subseteq P$.*

COROLLARY 1.18. *Let R be a prime ring of $char(R) \neq 2$ with involution $*$ of the second kind. If (F, d) is a generalized derivation of R satisfying the condition $F([x, x^*]) \pm [F(x), x^*] = 0$ for all $x \in R$, then one of the following holds:*

(1) *R is a commutative integral domain*
(2) *F is a left centralizer of R.*

COROLLARY 1.19. *Let R be a prime ring of $char(R) \neq 2$ with involution $*$ of the second kind. If d is a derivation of R satisfying the condition $d([x, x^*]) \pm [d(x), x^*] = 0$ for all $x \in R$, then R is a commutative integral domain or $d = 0$.*

THEOREM 1.20. *Let R be a ring with involution $*$ of the second kind, P a prime ideal of R such that $S(R) \cap Z(R) \not\subseteq P$. If (F, d) is a generalized derivation of R satisfying the condition $[F(x), F(x^*)] \pm [x, x^*] \in P$ for all $x \in R$, then one of the following holds:*

(1) *$char(R/P) = 2$*
(2) *R/P is a commutative integral domain*
(3) *$d(R) \subseteq P$.*

PROOF. Suppose $char(R/P) \neq 2$. By the hypothesis, we have

$$[F(x), F(x^*)] \pm [x, x^*] \in P \tag{1.36}$$

for all $x \in R$. If $F(R) \subseteq P$, then result follows from Lemma 1.2. Thus, we may now assume that $F(R) \nsubseteq P$. Linearization of (1.36) yields that

$$[F(x), F(y^*)] + [F(y), F(x^*)] \pm [x, y^*] \pm [y, x^*] \in P \tag{1.37}$$

for all $x, y \in R$. Substituting xh for x in (1.37) and using it, where $h \in H(R) \cap Z(R)$, we get

$$([x, F(y^*)] + [F(y), x^*])d(h) \in P \text{ for all } x, y \in R.$$

Primeness of P gives that $[x, F(y^*)] + [F(y), x^*] \in P$ or $d(h) \in P$. Consider the first case

$$[x, F(y^*)] + [F(y), x^*] \in P \text{ for all } x, y \in R. \tag{1.38}$$

Replacing x by xk in (1.38) and using $S(R) \cap Z(R) \nsubseteq P$, we obtain

$$[x, F(y^*)] - [F(y), x^*] \in P \text{ for all } x, y \in R. \tag{1.39}$$

Addition of last two relations yields that $2[x, F(y^*)] \in P$. This further gives $[x, F(x)] \in P$ for all $x \in R$. Therefore, in view of Lemma 1.1 we get R/P is commutative or $d(R) \subseteq P$. In the other case, if $d(h) \in P$, then $d(k) \in P$ for all $k \in S(R) \cap Z(R)$. Replacing x by xk in (1.37) and using last relation, we get

$$k([F(x), F(y^*)] - [F(y), F(x^*)] \pm [x, y^*] \mp [y, x^*]) \in P \text{ for all } x, y \in R.$$

Since $S(R) \cap Z(R) \nsubseteq P$, so we have

$$[F(x), F(y^*)] - [F(y), F(x^*)] \pm [x, y^*] \mp [y, x^*] \in P \text{ for all } x, y \in R. \tag{1.40}$$

Combining (1.37), and (1.40), we obtain

$$2([F(x), F(y^*)] \pm [x, y^*]) \in P \text{ for all } x, y \in R.$$

This implies that

$$[F(x), F(y)] \pm [x, y] \in P \text{ for all } x, y \in R.$$

Thus, by [**16**, Theorem 1.4], we conclude that either R/P is a commutative integral domain or $d(R) \subseteq P$. □

COROLLARY 1.21. *Let R be a ring with involution $*$ of the second kind, P a prime ideal of R such that $S(R) \cap Z(R) \nsubseteq P$. If d is a derivation of R satisfying the condition $[d(x), d(x^*)] \pm [x, x^*] \in P$ for all $x \in R$, then one of the following holds:*

(1) *$char(R/P) = 2$*
(2) *R/P is a commutative integral domain.*

COROLLARY 1.22. *Let R be a prime ring of $char(R) \neq 2$ with involution $*$ of the second kind. If (F, d) is a generalized derivation of R satisfying the condition $[F(x), F(x^*)] \pm [x, x^*] = 0$ for all $x \in R$, then one of the following holds:*

(1) *R is a commutative integral domain*
(2) *F is a left centralizer of R.*

COROLLARY 1.23. [**3**, Theorem 1] *Let R be a prime ring of $char(R) \neq 2$ with involution $*$ of the second kind. If d is a derivation of R satisfying the condition $[d(x), d(x^*)] - [x, x^*] = 0$ for all $x \in R$, then R is a commutative integral domain.*

THEOREM 1.24. *Let R be a ring with involution $*$ of the second kind, P a prime ideal of R such that $S(R) \cap Z(R) \nsubseteq P$. If (F, d) is a generalized derivation of R satisfying the condition $F([x, x^*]) \in P$ for all $x \in R$, then one of the following* *holds:*

(1) $char(R/P) = 2$
(2) R/P *is a commutative integral domain*
(3) $F(R) \subseteq P$.

PROOF. Assume that $char(R/P) \neq 2$. By the given condition, we have

$$F([x, x^*]) \in P \text{ for all } x \in R. \tag{1.41}$$

Linearization of (1.41) yields that

$$F([x, y^*]) + F([y, x^*]) \in P \text{ for all } x, y \in R. \tag{1.42}$$

Substituting xh for x in (1.42) and using it, where $0 \neq h \in H(R) \cap Z(R)$, we get

$$([x, y^*] + [y, x^*])d(h) \in P \text{ for all } x, y \in R. \tag{1.43}$$

Primeness of P forces that $([x, y^*] + [y, x^*]) \in P$ or $d(h) \in P$. If $[x, y^*] + [y, x^*] \in P$, then for $y = x$ and using our assumption $char(R/P) \neq 2$, we have $[x, x^*] \in P$ for all $x \in R$. Therefore, R/P is a commutative integral domain in view Lemma 1.2. If $d(h) \in P$ for all $h \in H(R) \cap Z(R)$, then we can find $d(k) \in P$ for all $k \in S(R) \cap Z(R)$. Now, replacing x by xk in (1.42) and using $d(k) \in P$, we get

$$k(F([x, y^*]) - F([y, x^*])) \in P \text{ for all } x, y \in R.$$

This gives

$$F([x, y^*]) - F([y, x^*]) \in P \text{ for all } x, y \in R.$$

Combining the above relation with (1.42), we obtain

$$2F([x, y^*]) \in P \text{ for all } x, y \in R.$$

This implies that

$$F([x, y]) \in P \text{ for all } x, y \in R. \tag{1.44}$$

Substituting yt in place of y in (1.44) and using it, we find that

$$[x, y]d(t) + F(y)[x, t] + yd([x, t]) \in P \tag{1.45}$$

for all $x, y, t \in R$. Putting $t = x$, we have $[x, y]d(x) \in P$. Replacing y by yr in the last relation, we find that

$$[x, r]yd(x) \in P \text{ for all } x, y, r \in R$$

and so $[x, r] \in P$ or $d(x) \in P$. Set $A = \{x \in R \mid [x, r] \in P \text{ for all } r \in R\}$ and $B = \{x \in R \mid d(x) \in P\}$. Clearly, A and B are additive subgroups of R such that $A \cup B = R$. But, a group cannot be written as a union of its two proper subgroups, consequently $A = R$ or $B = R$. The first case with Lemma 1.2 gives that R/P is a commutative integral domain. If $B = R$, then $d(x) \in P$ for all $x \in R$. Hence (1.45) reduces to

$$F(y)[x, t] \in P \text{ for all } x, y, t \in R.$$

Replacing t by rt in the above, we obtain

$$F(y)r[x, t] \in P \text{ for all } x, y, t, r \in R.$$

By the primeness of P, we conclude that R/P is a commutative integral domain or $F(R) \subseteq P$. This completes the proof of theorem. □

COROLLARY 1.25. *Let R be a ring with involution $*$ of the second kind, P a prime ideal of R such that $S(R) \cap Z(R) \not\subseteq P$. If (F, d) is generalized derivation of R satisfying the condition $F([x, x^*]) \pm [x, x^*] \in P$ for all $x \in R$, then one of the following holds:*

(1) $char(R/P) = 2$
(2) R/P *is a commutative integral domain*
(3) $\overline{F(x)} = \mp \overline{x}$.

PROOF. We have

$$F([x, x^*]) \pm [x, x^*] \in P \text{ for all } x, y \in R.$$

This can be written as

$$(F \pm I_R)([x, x^*]) \in P \text{ for all } x, y \in R,$$

Set $G = F \pm I_R$, where I_R is the identity mapping on R. Then G is a generalized derivation of R associated with the derivation d and $G([x, x^*]) \in P$ for all $x \in R$. Application of Theorem 1.24 gives the required results. □

COROLLARY 1.26. *Let R be a ring with involution $*$ of the second kind, P a prime ideal of R such that $S(R) \cap Z(R) \not\subseteq P$. If d is a derivation of R satisfying the condition $d([x, x^*]) \in P$ for all $x \in R$, then one of the following holds:*

(1) $char(R/P) = 2$
(2) R/P *is a commutative integral domain*
(3) $d(R) \subseteq P$.

COROLLARY 1.27. *Let R be a prime ring of $char(R) \neq 2$ with involution $*$ of the second kind. If (F, d) is a generalized derivation of R satisfying the condition $F([x, x^*]) = 0$ for all $x \in R$, then R is a commutative integral domain or $F = 0$.*

COROLLARY 1.28. *Let R be a prime ring of $char(R) \neq 2$ with involution $*$ of the second kind. If d is a derivation of R satisfying the condition $d([x, x^*]) = 0$ for all $x \in R$, then R is a commutative integral domain or $d = 0$.*

THEOREM 1.29. *Let R be a ring with involution $*$ of the second kind, P a prime ideal of R such that $S(R) \cap Z(R) \not\subseteq P$. If (F, d) a generalized derivation of R satisfying the condition $F(x \circ x^*) \in P$ for all $x \in R$, then one of the following holds:*

(1) $char(R/P) = 2$
(2) $F(R) \subseteq P$.

PROOF. Assume that $char(R/P) \neq 2$. By the given hypothesis, we have

$$F(x \circ x^*) \in P \text{ for all } x \in R. \tag{1.46}$$

Linearization of (1.46) yields that

$$F(x \circ y^*) + F(y \circ x^*) \in P \text{ for all } x, y \in R. \tag{1.47}$$

Substituting yh for y in (1.47), where $h \in H(R) \cap Z(R)$, we arrive at

$$(x \circ y^* + y \circ x^*)d(h) \in P \tag{1.48}$$

for all $x, y \in R$. In particular taking $x = y = k$, where $k \in S(R) \cap Z(R)$ and using the hypothesis, we find that

$$d(h) \in P \text{ for all } h \in H(R) \cap Z(R). \tag{1.49}$$

Replacing h by k^2 in (1.49), we get $2kd(k) \in P$, this implies that

$$d(k) \in P \text{ for all } k \in S(R) \cap Z(R). \tag{1.50}$$

Now replacing x by xk in (1.47) and using (1.50), we obtain

$$F(x \circ y^*) - F(y \circ x^*) \in P \tag{1.51}$$

for all $x, y \in R$. Combining (1.47) and (1.51), we get $2(F(x \circ y^*)) \in P$ for all $x, y \in R$. Since $char(R/P) \neq 2$, it follows that

$$F(x \circ y^*) \in \text{ for all } x, y \in R.$$

In particular, for $y = k$, where $k \in S(R) \cap Z(R)$, we have $-2F(x)k - 2xd(k) \in P$ for all $x, y \in R$. Using (1.50) and the hypotheses of the theorem, we conclude that $F(R) \subseteq P$, which is the required result. This proves the theorem. □

COROLLARY 1.30. *Let R be a ring with involution $*$ of the second kind, P a prime ideal of R such that $S(R) \cap Z(R) \nsubseteq P$. If (F, d) a generalized derivation of R satisfying the condition $F(x \circ x^*) \pm x \circ x^* \in P$ for all $x \in R$, then one of the following holds:*

(1) $char(R/P) = 2$
(2) $\overline{F(x)} = \mp\overline{x}$.

PROOF. We have

$$F(x \circ x^*) \pm x \circ x^* \in P \tag{1.52}$$

for all $x \in R$. If $F(R) \subseteq P$, then (1.52) becomes $x \circ x^* \in P$ for all $x \in R$ and hence by Lemma 1.3, we get $char(R/P) = 2$. Now, we suppose that $F(R) \nsubseteq P$. From (1.52), we have

$$(F \pm I_R)(x \circ x^*) \in P$$

for all $x \in R$. Set $G = F \pm I_R$, where I_R is the identity mapping on R. Then G is a generalized derivation of R such that $G(x \circ x^*) \in P$ for all $x \in R$. Therefore, in view of Theorem 1.29, we get the required results. □

COROLLARY 1.31. *Let R be a ring with involution $*$ of the second kind, P a prime ideal of R such that $S(R) \cap Z(R) \nsubseteq P$. If d a derivation of R satisfying the condition $d(x \circ x^*) \in P$ for all $x \in R$, then one of the following holds.*

(1) $char(R/P) = 2$
(2) $d(R) \subseteq P$.

COROLLARY 1.32. *Let R be a prime ring of $char(R) \neq 2$ with involution $*$ of the second kind. If (F, d) is a generalized derivation of R satisfying the condition $F(x \circ x^*) = 0$ for all $x \in R$, then $F = 0$.*

COROLLARY 1.33. *Let R be a prime ring of $char(R) \neq 2$ with involution $*$ of the second kind. If d is a derivation of R satisfying the condition $d(x \circ x^*) = 0$ for all $x \in R$, then $d = 0$.*

THEOREM 1.34. *Let R be a ring with involution $*$ of the second kind, P a prime ideal of R such that $S(R) \cap Z(R) \nsubseteq P$. If (F, d) is a generalized derivation of R satisfying the condition $[F(x), x^*] \pm [x, x^*] \in P$ for all $x \in R$, then one of the following holds:*

(1) $char(R/P) = 2$
(2) R/P *is a commutative integral domain*

(3) $d(R) \subseteq P$.

PROOF. Suppose $char(R/P) \neq 2$. We have

$$[F(x), x^*] \pm [x, x^*] \in P \tag{1.53}$$

for all $x \in R$. If $F(R) \subseteq P$, then result follows from Lemma 1.2. We may now assume that $F(R) \nsubseteq P$. Writing $x + y$ instead of x in (1.53), we obtain

$$[F(x), y^*] + [F(y), x^*] \pm [x, y^*] \pm [y, x^*] \in P \tag{1.54}$$

for all $x, y \in R$. Now replacing y by yh in (1.54), where $h \in H(R) \cap Z(R)$. We get

$$([F(x), y^*] + [F(y), x^*] \mp [x, y^*] \pm [y, x^*])h + [y, x^*]d(h) \in P \tag{1.55}$$

for all $x, y \in R$. Using (1.54) in (1.55), we get

$$[y, x^*]d(h) \in P \text{ for all } x, y \in R. \tag{1.56}$$

This implies that $[y, x^*] \in P$ or $d(h) \in P$ for all $h \in H(R) \cap Z(R)$. If we consider $[y, x^*] \in P$ for all $x, y \in R$, then in view of Lemma 1.2, R/P is a commutative integral domain. Now consider $d(h) \in P$, this implies that $d(k) \in P$ for all $k \in S(R) \cap Z(R)$. Now replacing y by yk in (1.55) and using $d(k) \in P$, we may obtain

$$-[F(x), y^*] + [F(y), x^*] \mp [x, y^*] \pm [y, x^*] \in P \tag{1.57}$$

for all $x, y \in R$. Combining (1.54) and (1.57), we get

$$2([F(y), x^*] \pm [y, x^*]) \in P$$

for all $x, y \in R$. This implies that

$$[F(y), x] \pm [y, x] \in P \tag{1.58}$$

for all $x, y \in R$. Putting $y = x$, we get

$$[F(x), x] \in R \text{ for all } x \in R. \tag{1.59}$$

In view of Lemma 1.1, we get R/P is a commutative integral domain or $d(R) \subseteq P$. Hence the proof is completed now. □

THEOREM 1.35. *Let R be a ring with involution $*$ of the second kind, P a prime ideal of R such that $S(R) \cap Z(R) \nsubseteq P$. If (F, d) is a generalized derivation of R satisfying the condition $F(x \circ x^*) \pm [x, x^*] \in P$ for all $x \in R$, then one of the following holds:*

(1) *$char(R/P) = 2$*
(2) *R/P is a commutative integral domain.*

PROOF. Assume that $char(R/P) \neq 2$. By the given condition, we have

$$F(x \circ x^*) \pm [x, x^*] \in P \tag{1.60}$$

for all $x \in R$. If $F(R) \subseteq P$ then result follows from Lemma 1.2. Interchanging the role of x and x^* in (1.60), we obtain

$$F(x \circ x^*) \mp [x, x^*] \in P \tag{1.61}$$

for all $x \in R$. Combining (1.60) and (1.61), one can easily get $[x, x^*] \in P$ for all $x \in R$. Thus by Lemma 1.2, R/P is a commutative integral domain. □

THEOREM 1.36. *Let R be a ring with involution $*$ of the second kind, P a prime ideal of R such that $S(R) \cap Z(R) \not\subseteq P$. If (F, d) is a generalized derivation of R satisfying the condition $F([x, x^*]) \pm xx^* \in P$ for all $x \in R$, then one of the following holds:*

(1) *$char(R/P) = 2$*
(2) *R/P is a commutative integral domain.*

PROOF. Suppose $char(R/P) \neq 2$. We have

$$F([x, x^*]) \pm xx^* \in P \tag{1.62}$$

for all $x \in R$. If $F(R) \subseteq P$, then $xx^* \in P$ for all $x \in R$. Also, $x^*x \in P$ for all $x \in R$. This gives $[x, x^*] \in P$ for all $x \in R$ and hence by Lemma 1.2, R/P is a commutative integral domain. Now, assume that $F(R) \not\subseteq P$ Writing $x + y$ instead of x in (1.62), we obtain

$$F([x, y^*]) + F([y, x^*]) \pm xy^* \pm yx^* \in P \tag{1.63}$$

for all $x, y \in R$. Replacing x by xh in (1.63) and using it, we get

$$([x, y^*] + [y, x^*])d(h) \in P \tag{1.64}$$

for all $x, y \in R$, which is same as (1.43). Using the similar arguments, we find that R/P is a commutative integral domain or $d(k) \in P$ for all $k \in S(R) \cap Z(R)$. Consider the case $d(k) \in P$. Replacing x by xk in (1.63) and using $d(k) \in P$, we obtain

$$k(F([x, y^*]) - F([y, x^*]) \pm xy^* \mp yx^*) \in P.$$

Since $S(R) \cap Z(R) \not\subseteq P$, it follows that

$$F([x, y^*]) - F([y, x^*]) \pm xy^* \mp yx^* \in P \tag{1.65}$$

for all $x, y \in R$. Combining (1.63) and (1.65), we get

$$2(F([x, y^*]) \pm xy^*) \in P.$$

This implies

$$F[x, y] \pm xy \in P$$

for all $x, y \in R$. Taking $y = k$ and using the condition $S(R) \cap Z(R) \not\subseteq P$, we find that $x \in P$ for all $x \in R$, which is a contradiction. □

COROLLARY 1.37. *Let R be a ring with involution $*$ of the second kind, P a prime ideal of R such that $S(R) \cap Z(R) \not\subseteq P$. If d is a derivation of R satisfying the condition $d([x, x^*]) \pm xx^* \in P$ for all $x \in R$, then one of the following holds:*

(1) *$char(R/P) = 2$*
(2) *R/P is a commutative integral domain.*

COROLLARY 1.38. *Let R be a prime ring of $char(R) \neq 2$ with involution $*$ of the second kind. If (F, d) is a generalized derivation of R satisfying the condition $F([x, x^*]) \pm xx^* = 0$ for all $x \in R$, then R is a commutative integral domain.*

COROLLARY 1.39. *Let R be a prime ring of $char(R) \neq 2$ with involution $*$ of the second kind. If d is a derivation of R satisfying the condition $d([x, x^*]) \pm xx^* = 0$ for all $x \in R$, then R is a commutative integral domain.*

Similarly, we can prove the following result.

THEOREM 1.40. *Let R be a ring with involution $*$ of the second kind, P a prime ideal of R such that $S(R) \cap Z(R) \nsubseteq P$. If (F, d) is a generalized derivation of R satisfying the condition $F([x, x^*]) \pm x^*x \in P$ for all $x \in R$, then one of the following holds:*

(1) *$char(R/P) = 2$*
(2) *R/P is a commutative integral domain.*

COROLLARY 1.41. *Let R be a ring with involution $*$ of the second kind, P a prime ideal of R such that $S(R) \cap Z(R) \nsubseteq P$. If d is a derivation of R satisfying the condition $d([x, x^*]) \pm x^*x \in P$ for all $x \in R$, then one of the following holds:*

(1) *$char(R/P) = 2$*
(2) *R/P is a commutative integral domain.*

COROLLARY 1.42. *Let R be a prime ring of $char(R) \neq 2$ with involution $*$ of the second kind. If (F, d) is a generalized derivation of R satisfying the condition $F([x, x^*]) \pm x^*x = 0$ for all $x \in R$, then R is a commutative integral domain.*

COROLLARY 1.43. *Let R be a prime ring of $char(R) \neq 2$ with involution $*$ of the second kind. If d is a derivation of R satisfying the condition $d([x, x^*]) \pm x^*x = 0$ for all $x \in R$, then R is a commutative integral domain.*

Acknowledgment

The authors are deeply indebted to the learned referees for their careful reading of the manuscript and constructive comments.

References

[1] Shakir Ali and Nadeem Ahmed Dar, *On $*$-centralizing mappings in rings with involution*, Georgian Math. J. **21** (2014), no. 1, 25–28, DOI 10.1515/gmj-2014-0006. MR3245455

[2] Shakir Ali and Huang Shuliang, *On derivations in semiprime rings*, Algebr. Represent. Theory **15** (2012), no. 6, 1023–1033, DOI 10.1007/s10468-011-9271-9. MR2994014

[3] Shakir Ali, Nadeem Ahmad Dar, and Abdul Nadim Khan, *On strong commutativity preserving like maps in rings with involution*, Miskolc Math. Notes **16** (2015), no. 1, 17–24, DOI 10.18514/mmn.2015.1297. MR3384583

[4] Mohammad Ashraf and Nadeem-ur Rehman, *On commutativity of rings with derivations*, Results Math. **42** (2002), no. 1-2, 3–8, DOI 10.1007/BF03323547. MR1934218

[5] Howard E. Bell and Mohamad Nagy Daif, *On commutativity and strong commutativity-preserving maps*, Canad. Math. Bull. **37** (1994), no. 4, 443–447, DOI 10.4153/CMB-1994-064-x. MR1303669

[6] Mohamad Nagy Daif, *Commutativity results for semiprime rings with derivations*, Internat. J. Math. Math. Sci. **21** (1998), no. 3, 471–474, DOI 10.1155/S0161171298000660. MR1620326

[7] Qing Deng and Mohammad Ashraf, *On strong commutativity preserving mappings*, Results Math. **30** (1996), no. 3-4, 259–263, DOI 10.1007/BF03322194. MR1419942

[8] I. N. Herstein, *A note on derivations*, Canad. Math. Bull. **21** (1978), no. 3, 369–370, DOI 10.4153/CMB-1978-065-x. MR506447

[9] I. N. Herstein, *Rings with involution*, Chicago Lectures in Mathematics, University of Chicago Press, Chicago, Ill.-London, 1976. MR0442017

[10] Moulay Abdallah Idrissi and Lahcen Oukhtite, *Structure of a quotient ring R/P with generalized derivations acting on the prime ideal P and some applications*, Indian J. Pure Appl. Math. **53** (2022), no. 3, 792–800, DOI 10.1007/s13226-021-00173-x. MR4474990

[11] Mohammad Salahuddin Khan, Shakir Ali, and Mohammed Ayedh, *Herstein's theorem for prime ideals in rings with involution involving pair of derivations*, Comm. Algebra **50** (2022), no. 6, 2592–2603, DOI 10.1080/00927872.2021.2014513. MR4413854

[12] Charles Lanski, *Differential identities, Lie ideals, and Posner's theorems*, Pacific J. Math. **134** (1988), no. 2, 275–297. MR961236

[13] A. Mamouni, B. Nejjar, and L. Oukhtite, *Differential identities on prime rings with involution*, J. Algebra Appl. **17** (2018), no. 9, 1850163, 11 pp., DOI 10.1142/S0219498818501633. MR3846411

[14] B. Nejjar, A. Kacha, A. Mamouni, and L. Oukhtite, *Commutativity theorems in rings with involution*, Comm. Algebra **45** (2017), no. 2, 698–708, DOI 10.1080/00927872.2016.1172629. MR3562533

[15] Edward C. Posner, *Derivations in prime rings*, Proc. Amer. Math. Soc. **8** (1957), 1093–1100, DOI 10.2307/2032686. MR95863

[16] Nadeem ur Rehman, Motoshi Hongan, and Hafedh M. Alnoghashi, *On generalized derivations involving prime ideals*, Rend. Circ. Mat. Palermo (2) **71** (2022), no. 2, 601–609, DOI 10.1007/s12215-021-00639-1. MR4453340

[17] Joso Vukman and Irena Kosi-Ulbl, *On centralizers of semiprime rings with involution*, Studia Sci. Math. Hungar. **43** (2006), no. 1, 61–67, DOI 10.1556/SScMath.43.2006.1.4. MR2218678

[18] Borut Zalar, *On centralizers of semiprime rings*, Comment. Math. Univ. Carolin. **32** (1991), no. 4, 609–614. MR1159807

Department of Applied Mathematics, Z. H. College of Engineering & Technology, Aligarh Muslim University, Aligarh-202002, India
Email address: salahuddinkhan50@gmail.com

Department of Mathematics, Madanapalle Institute of Technology & Science, Madanapalle-517325, A. P., India
Email address: adnan.abbasi001@gmail.com

Department of Mathematics, Aligarh Muslim University, Aligarh-202002, India
Email address: shakir.ali.mm@amu.ac.in,drshakir1971@gmail.com

Department of Mathematics, Aligarh Muslim University, Aligarh-202002, India
Email address: abo.cother@gmail.com

Contemporary Mathematics
Volume **785**, 2023
https://doi.org/10.1090/conm/785/15786

On nil algebras and a problem of Passman concerning nilpotent free algebras

Jan Krempa

Dedicated to Professor Syed Tariq Rizvi for his contribution to Ring and Module Theory
This paper is also dedicated to the memory of Professor Edmund R. Puczyłowski, who passed away in October 2021

Abstract. We provide here the full solution of a problem of Donald Passman concerning nilpotent free algebras. It will be done with the help of examples constructed by Agata Smoktunowicz. Our paper on this problem from 1989 year will also be used. In Section 3 we add some comments, questions and results inspired by this problem.

From the editors

Professor Jan Krempa passed away in February 2023. His talent, kindness, and humility leave a profound wound in the heart of everyone who has had the chance to know him.

1. Some auxiliary notions and results

Let K be a commutative, associative, unital ring. Associative K-algebras will be named here algebras, and tensor products will usually be considered over K.

Radicals will be understood in the sense of Kurosh and Amitsur, (see [**GW04**] for details). Some classical radicals used here are:

$\mathcal{B}$ - the Bear radical,
$\mathcal{L}$ - the Levitzki radical,
$\mathcal{K}$ - the Koethe radical,
$\mathcal{J}$ - the Jacobson radical.
$\mathcal{U}$ - the Brown-McCoy radical.

In the sequel inclusions of classes will often be denoted as inequalities. Thus we have:

$$\mathcal{B} < \mathcal{L} < \mathcal{K} < \mathcal{J} < \mathcal{U}.$$

2020 *Mathematics Subject Classification.* Primary 16N20, 16N40, 16N80, 16S20, 16S36, 16S50, 16S85, 16S99.

Key words and phrases. Nilpotent free algebra, eventually semisimple algebra, tensor product of algebras, scalar extensions.

Many thanks to my wife Irena for her continuous help and care and to dr Małgorzata Jastrzębska for her substantial assistance in the preparation of this paper.

Several new interesting radicals can be defined with the help of tensor product, in the following way:

DEFINITION 1.1. Let $\mathcal{R}$ be a radical class, B be an algebra and $\mathcal{R}_B$ be the class of all algebras A such that $A \otimes B$ is in the class $\mathcal{R}$.

It was checked in [**Pu77**], that $\mathcal{R}_B$ is a radical class. In this note we apply the above definition only to radicals hereditary on ideals and to unital algebras B.

A radical $\mathcal{R}$ will be named here:

- *B-normal* if for every algebra A we have $\mathcal{R}(A \otimes B) = I \otimes B$, for an ideal I of A,
- *B-invariant* if $\mathcal{R}$ is B-normal and $\mathcal{R}_B = \mathcal{R}$.

Under our restrictions, for B-normal radical $\mathcal{R}$ and for any algebra A we have:

$$\mathcal{R}_B(A) = A \cap \mathcal{R}(A \otimes B) \text{ and } \mathcal{R}(A \otimes B) \geq \mathcal{R}_B(A) \otimes B. \tag{1.1}$$

EXAMPLE 1.2. For any $n \geq 1$ let $P_n = K[x_1, \ldots, x_n]$ and $P_\infty = K[x_1, x_2, \ldots]$ be the algebras of polynomials in commuting indeterminates.

Further $\mathcal{R}_{P_n}$ will be abbreviated to $\mathcal{R}_n$. The case of $n = 1$ is of special interest, because we have *an iterating formula:*

$$P_{n+1} = P_n \otimes P_1.$$

The following important result is due to Amitsur:

THEOREM 1.3 ([**Am56a**]). *Classical radicals are P_n-normal, while $\mathcal{B}$ and $\mathcal{L}$ are even P_n-invariant. Moreover, $\mathcal{K} \geq \mathcal{J}_1$.*

Now we will concentrate on the interplay of $\mathcal{J}$ and $\mathcal{K}$. An obvious inequality $\mathcal{J} > \mathcal{K}$ and Theorem 1.3 with the help of the iterating formula induce the following infinite sequence of inequalities (see [**Am73, Kr74b**]):

$$\begin{aligned} &\mathcal{J} > \mathcal{K} \geq \mathcal{J}_1 \geq \mathcal{K}_1 \geq \mathcal{J}_2 \geq \mathcal{K}_2 \geq \ldots \\ &\qquad \ldots \geq \wedge_{n=1}^{\infty} \mathcal{J}_n = \wedge_{n=1}^{\infty} \mathcal{K}_n \geq \mathcal{J}_\infty = \mathcal{K}_\infty > \mathcal{L}. \end{aligned} \tag{1.2}$$

Some further relations are possible due to results below.

LEMMA 1.4 ([**Kr72**]). *Let A be an algebra. Then A is $\mathcal{J}_1$-radical iff the matrix algebra $M_r(A)$ is a nil algebra for every $r \geq 1$.*

THEOREM 1.5 ([**Kr72**]). *The following conditions are equivalent:*

(1) *For every algebra A and any its onesided nil ideal I we have $I \subseteq \mathcal{K}(A)$, (Koethe conjecture);*
(2) *For every nil algebra A the matrix algebra $M_2(A)$ is also nil;*
(3) *For every nil algebra A and for every $r \geq 1$, the matrix algebra $M_r(A)$ is also nil;*
(4) $\mathcal{K} = \mathcal{J}_1$.

REMARK 1.6. The above lemma, theorem, and some other their consequences related to the Koethe conjecture were published as a preprint of Institute of Mathematics of the Polish Academy of Sciences in 1971 year. This preprint was later published as [**Kr72**].

2. A problem of Passman

In the sequel K will be a field, $\overline{K}$ its algebraic closure and $\tilde{K}$, the purely inseparable closure of K in $\overline{K}$. For further results and notions from field theory see [**La02**]. Under this agreements we go to connections of radicals with scalar extensions. Our goal will be to show the solution of a problem of Passman concerning nilpotent free algebras. For this purpose we invoke some needed definitions, and results (see [**Pa71, Pa77**]).

DEFINITION 2.1 (see [**Pa71**]). An algebra A is *nilpotent free* if $A \otimes L$ is semiprime for every field extension L of K.

If K is a perfect field, then every semiprime K-algebra is nilpotent free. However, if K is not perfect, then even a field need not to be nilpotent free.

EXAMPLE 2.2. Let K be a nonperfect field, Then $K \neq \tilde{K}$. Let L be a subfield of $\tilde{K}$ properly containing K. Then it is easy to see that $L \otimes L$ is a local algebra with the maximal ideal M such that M is nil and $L \otimes L/M \simeq L$.

If $\mathcal{R}$ is any classical radical, then $\mathcal{R}(L \otimes L) = \mathcal{B}(L \otimes L) = M \neq 0$.

Thus $\mathcal{R}$ is neither L-normal nor L-invariant. Also the field L is not nilpotent free as K-algebra.

The following result is well known:

THEOREM 2.3. [**Pa71, Pa77, Be89, Kr79a**] *Let L be a separable extension of $K, \mathcal{R}$ be a classical radical and A be an algebra. If A is $\mathcal{R}$-semisimple then $\mathcal{R}(A \otimes L) = \mathcal{B}(A \otimes L)$.*
If A is $\mathcal{R}$-semisimple and nilpotent free, then $A \otimes L$ is $\mathcal{R}$-semisimple.
If $A \otimes L$ is $\mathcal{R}$-semisimple, and $\mathcal{R} \neq \mathcal{K}$, then A is $\mathcal{R}$-semisimple.
Moreover, $\mathcal{R}(A \otimes L) = \mathcal{R}_L(A) \otimes L$ if and only if $A/\mathcal{R}_L(A)$ is nilpotent free.

Now let us consider some nonalgebraic field extensions.

EXAMPLE 2.4. Let $1 \leq n \leq \infty$, $Q_n = K(x_1, \ldots, x_n)$ be the field of rational functions of n commuting indeterminates. Then Q_n is a nilpotent free algebra. If $\mathcal{R}$ is any classical radical, then the induced radical $\mathcal{R}_{Q_n}$ we will abbreviate to $\mathcal{R}_{(n)}$.

The following results on radicals in the case of nonalgebraic scalar extensions will be used:

THEOREM 2.5 ([**Am56b**]). *Let L be a purely transcendental field extension of K, and $\mathcal{R}$ be any classical radical. Then $\mathcal{R}$ is L-normal. If $\mathcal{R} \neq \mathcal{U}$, then $\mathcal{R}_L \leq \mathcal{K}$.*

LEMMA 2.6 ([**Kr79b**]). *For every $1 \leq n \leq \infty$ the radicals $\mathcal{J}_n$ and $\mathcal{J}_{(n)}$ coincide.*

Passman introduced the following interesting notion:

DEFINITION 2.7 ([**Pa71**]). A K-algebra A is *eventually semisimple* if A is nilpotent free and there exists a field extension L of K such that $A \otimes L$ is $\mathcal{J}$-semisimple.

From the results quoted above, or [**Pa71**], we have

LEMMA 2.8. *Let A be a nilpotent free algebra. Then the following conditions are equivalent:*

(1) *A is eventually semisimple;*
(2) *A is $\mathcal{J}_\infty$-semisimple;*
(3) *$A \otimes Q_\infty$ is $\mathcal{J}$-semisimple.*

If A is $\mathcal{J}_\infty$-radical algebra then A is not eventually semisimple.

PROOF. (1) $\Rightarrow$ (2). Let A be eventually semisimple algebra with suitable field L. By Theorems 2.3 and 2.5 we know that the algebra A has no nontrivial nil ideals. Thus, by Theorem 1.3, $A \otimes P_\infty$ is $\mathcal{J}$-semisimple.

$2 \Rightarrow 3$. This implication follows from Lemma 2.6, while the implication (3) $\Rightarrow$ (1) is obvious. The final statement is also evident. $\square$

Let mtd(A), (the $\mathcal{J}$-degree of A), be equal to the minimum of transcendence degrees of L over K, for $A \otimes L$ being $\mathcal{J}$-semisimple.

Thus mtd(A)=0 if A is $\mathcal{J}$-semisimple and, for $n > 0$, mtd(A)=n if and only if A is $\mathcal{J}_n$-semisimple but A is not $\mathcal{J}_{(n-1)}$-semisimple.

PROBLEM 2.9 (Pa71, Problem 25). Let A be an eventually semisimple algebra over K. What are possible values for the $\mathcal{J}$-degree of A?

Examples of algebras with transcendence degrees 0 and 1 are indicated in [**Pa71**]. Passman also indicated, after Amitsur [**Am56b**], that his problem is not interesting in the case of uncountable field K, because in this case we have $\mathcal{J}_1 = \mathcal{J}_\infty$. To solve this problem in the case, when K is countable we use the Lemma 2.6 and the following nontrivial result mainly due to Agata Smoktunowicz from [**SP01, Sm00, Sm02b**]:

THEOREM 2.10. *Let K be any countable field and $1 \leq n \leq \infty$. Then there exists such a K-algebra R_n that $R_n[x_1, \ldots, x_n]$ is $\mathcal{J}$-radical, but not nil. Hence, if $R_\infty = \oplus_{n=1}^\infty R_n$, then R_∞ is $\mathcal{J}_\infty$-semisimple, but not $\mathcal{J}_n$-semisimple for any $n \leq \infty$.*

We need the following consequence of the above result:

COROLLARY 2.11. *For every countable field K and for every $1 \leq n < \infty$ there exists a nontrivial algebra A_n such that $\mathcal{J}_n(A_n) = A_n$, but A_n is $\mathcal{J}_{n+1}$-semisimple.*

PROOF. Let K, n and R_n be as in Theorem 2.10. Put $I_n = \mathcal{K}(R_n[x_1, \ldots, x_n])$ and $B_n = R_n[x_1, \ldots, x_n]/I_n$. By Theorem 2.10 B_n is nontrivial and has no nonzero nil ideals.

By the Theorem 1.3 it follows that $I_n = I'_n[x_1, \ldots, x_n]$ for an ideal I'_n of R_n. Thus

$$B_n = R_n[x_1, \ldots, x_n]/I'_n[x_1, \ldots, x_n] \simeq (R_n/I'_n)[x_1, \ldots, x_n].$$

Now let $A_n = R_n/I'_n$. Then A_n is nonzero and $A_n[x_1, \ldots, x_n]$ is $\mathcal{J}$-radical as a homomorphic image of $\mathcal{J}$-radical algebra $R_n[x_1, \ldots, x_n]$. Moreover, since B_n has no nonzero nil ideals, then, by Theorem 1.3, $B_n[x] = A_n[x_1, \ldots, x_{n+1}]$ is $\mathcal{J}$-semisimple. Hence mtd$(A_n) = n + 1$. $\square$

As an immediate consequence of the above corollary and Lemmas 2.8 and 2.6 we have:

THEOREM 2.12. *Let K be any countable field. Every natural number $1 \leq n \leq \infty$ can be a degree of eventually semisimple algebra. For $n = 1$ it is enough to take any $\mathcal{J}$-radical, but $\mathcal{K}$-semisimple algebra. If $n > 1$ then, for example, $n + 1 = mtd(A_n)$. Infinity can also occur as a degree because for $A_\infty = \oplus_{n=1}^{\infty} A_n$ we have $mtd(A_\infty) = \infty$.*

3. On some related questions

One can adopt some results and ideas from previous section to other classical radicals. Nilpotent freeness seems to fit well, without any modification.

It is known that an algebra A is nilpotent free iff $A \otimes \tilde{K}$ is semiprime ($\mathcal{B}$-semisimple). Thus nilpotent free algebras form a semisimple class in the sense of radical theory (see [**Pu77**, **GW04**]). Several other properties of nilpotent free algebras one can find for example in [**Pa71**, **Pa77**, **Be89**, **Kr79a**].

In the case of eventual semisimplicity and a problem of Passman one can replace the radical $\mathcal{J}$ by any (classical) radical $\mathcal{R}$. In this way one obtains eventual $\mathcal{R}$-semisimplicity and a transcendence $\mathcal{R}$-degree. The cases of $\mathcal{R} = \mathcal{B}$ and $\mathcal{R} = \mathcal{L}$ are trivial, because these radicals are Q_n-invariant. Now we do this replacement for $\mathcal{R} = \mathcal{K}$. The $\mathcal{K}$-version of Lemma 2.6 is true, because $A \otimes Q_n$ is a central localization of $A \otimes P_n$, for every algebra A and any $n \leq \infty$. Hence $\mathcal{K}_n = \mathcal{K}_{(n)}$ for every $n \leq \infty$. However, we are not able to prove a full $\mathcal{K}$-version of Lemma 2.8 because in Theorem 2.3 the radical $\mathcal{K}$ is an exception. Fortunately, we can adopt Theorem 2.10 to obtain all possible $\mathcal{K}$-degrees:

EXAMPLE 3.1. Let $n \geq 1$ and A_n be as in Corollary 2.11. Then $A_n[x_1, \dots, x_n]$ is $\mathcal{K}$-semisimple, but $\mathcal{J}$-radical. However, we have $A_n[x_1, \dots, x_n] = (A_n[x_1 \dots, x_{n-1}])[x_n]$. Thus, by Theorem 1.3 we obtain, that $A_n[x_1, \dots, x_{n-1}]$ is a nil algebra and $\mathcal{K}$-degree of A_n is equal to $n - 1$. Hence, every natural number $n \geq 0$ can be the transcendence $\mathcal{K}$-degree. Infinity is also transcendence $\mathcal{K}$-degree, for $A_\infty = \oplus_{n=1}^{\infty} A_n$.

For making transcendent $\mathcal{K}$-degree more important one should consider the following problem

PROBLEM 3.2. Let A be a nil algebra and let L be a finite field extension of K. Is $A \otimes L$ nil?

This problem was mentioned in [**Pu91**] as Question 9 and in [**Sm01**] as Question 16, with an interesting comments on its connection to Koethe conjecture. The case when L is a Galois extension of K was discussed, for example, in [**FK83**]. The case of $L \subseteq \tilde{K}$ seems to be of special interest.

With the help of algebras R_n from Theorem 2.10 we have $\mathcal{J}_n \neq \mathcal{K}_n$ and $\wedge_{n=1}^{\infty} \mathcal{J}_n = \wedge_{n=1}^{\infty} \mathcal{K}_n \neq \mathcal{J}_\infty = \mathcal{K}_\infty$. and the sequence (1.2) can be modified to the following:

$$\begin{aligned} &\mathcal{J} > \mathcal{K} \geq \mathcal{J}_1 > \mathcal{K}_1 \geq \mathcal{J}_2 > \mathcal{K}_2 \geq \dots \\ &\qquad \dots \geq \wedge_{n=1}^{\infty} \mathcal{J}_n = \wedge_{n=1}^{\infty} \mathcal{K}_n > \mathcal{J}_\infty = \mathcal{K}_\infty > \mathcal{L}. \end{aligned} \tag{3.1}$$

We already know, from Theorem 1.5, that the condition $\mathcal{K} = \mathcal{J}_1$ is equivalent to the Koethe conjecture. For some newer results related to the Koethe conjecture see for example [**Pu91**, **Sm01**, **Fe01**, **CKLP18**, **RS21**], and references there.

Lemma 1.4 has several constructive proofs (see [**Am73**, **Kr74a**]). In connection with this lemma it is reasonable to consider the following example:

EXAMPLE 3.3. For every $r \geq 1$ let $M_r = M_r(K)$, be the algebra of matrices of degree r over K. Then classical radicals are M_r-normal. It seems to be still not known, whether $\mathcal{K}$ is M_r-invariant (see Theorem 1.5), while the other classical radicals are.

Thus the decreasing sequence of radicals $\mathcal{K}_{M_r} = \mathcal{K}_{[r]}$ is interesting:

$$\mathcal{K} = \mathcal{K}_{[1]} \geq \mathcal{K}_{[2]} \geq \ldots \geq \mathcal{K}_{[\infty]} = \wedge_{r=1}^{\infty} \mathcal{K}_{[r]} = \mathcal{J}_1. \tag{3.2}$$

Now, if in the above sequence we replace $\mathcal{K}$ by $\mathcal{K}_n$, then we obtain an infinite sequence of radicals

$$(c_n) = (\mathcal{K}_n)_{[r]}$$

between $\mathcal{K}_n$ and $\mathcal{J}_{n+1}$.

We can include the sequences (c_n) between $\mathcal{K}_n$ and $\mathcal{J}_{n+1}$. In this way we obtain:

$$\begin{aligned} \mathcal{J} > \mathcal{K} \geq (c_1) \geq \mathcal{J}_1 > \mathcal{K}_1 \geq (c_2) \geq \mathcal{J}_2 > \mathcal{K}_2 \geq \ldots \\ \ldots \geq \mathcal{J}_\infty = \mathcal{K}_\infty > \mathcal{L}. \end{aligned} \tag{3.3}$$

This sequence should help to prove whether all inequalities in the sequence (3.1) are strict.

A. Smoktunowicz and E.R. Puczyłowski initiated studying unital homomorphic images of polynomials with coefficients in nil algebras (see §4 of [**Sm01**] and [**FW03**]). Results and methods used in these papers could help to solve the following problem:

PROBLEM 3.4. Consider eventually $\mathcal{U}$-semisimple algebras and a problem of Passman for these algebras.

E. S. Golod and I. R. Shafarevich in [**GS64**] gave a criterion for infinite dimension of some finitely generated algebras and applied it to 2-generated algebras over countable fields. Using some additional arguments Golod proved in fact the following result:

THEOREM 3.5 ([**Go64**]). *For every field K and for every $d > 1$ there exists a d-generated K-algebra which is infinite dimensional, but belongs to $\mathcal{J}_\infty$.*

Some further applications of the methods from [**GS64**, **Go64**] one can find for example in [**Go68**].

It is known, and easy, that every simple, locally nilpotent algebra has trivial multiplication. Due to A. Smoktunowicz (see [**Sm02a**]) we know, that over every countable field K there exists a nontrivial, simple nil K-algebra, say S_K. We ask:

PROBLEM 3.6. Is every algebra S_K eventually $\mathcal{J}$-semisimple? If yes, then what is the transcendence $\mathcal{J}$-degree of S_K? The same question we ask also for eventual $\mathcal{K}$-semisimplicity and $\mathcal{K}$-degree of S_K.

In **[GKW07]**, Questions 3 and 4, we asked about the existence of simple $\mathcal{J}_\infty$-algebra over any field. Due to Theorem 3.5 we know that for every field K there exists, even finitely generated, $\mathcal{J}_\infty$-algebra over K, which is prime, but not locally nilpotent. If K is uncountable, then every nil algebra over K is $\mathcal{J}_\infty$-algebra. Thus the above mentioned questions seem to be well motivated.

References

[Am56a] S. A. Amitsur, *Radicals of polynomial rings*, Canadian J. Math. **8** (1956), 355–361, DOI 10.4153/CJM-1956-040-9. MR78345

[Am56b] A. S. Amitsur, *Algebras over infinite fields*, Proc. Amer. Math. Soc. **7** (1956), 35–48, DOI 10.2307/2033240. MR75933

[Am73] — , *Nil radicals. Historical notes and some new results*, pp. 47-65 in *Rings, Modules and Radicals* (Proc. Colloq. Kesthely, 1971), North-Holland, 1973. MR0347873 (50 #374)

[Be89] George M. Bergman, *Radicals, tensor products, and algebraicity*, Ring theory 1989 (Ramat Gan and Jerusalem, 1988/1989), Israel Math. Conf. Proc., vol. 1, Weizmann, Jerusalem, 1989, pp. 150–192. MR1029309

[BPR13] Gary F. Birkenmeier, Jae Keol Park, and S. Tariq Rizvi, *Extensions of rings and modules*, Birkhäuser/Springer, New York, 2013, DOI 10.1007/978-0-387-92716-9. MR3099829

[CKLP18] M. Chebotar, W.-F. Ke, P.-H. Lee, and E. R. Puczyłowski, *On polynomial rings over nil rings in several variables and the central closure of prime nil rings*, Israel J. Math. **223** (2018), no. 1, 309–322, DOI 10.1007/s11856-017-1616-6. MR3773063

[Fe01] Miguel Ferrero, *An introduction to Köthe's conjecture and polynomial rings*, Resenhas **5** (2001), no. 2, 139–148. IX Algebra Meeting USP/UNICAMP/UNESP (Portuguese) (São Pedro, 2001). MR1945298

[FW03] Miguel Ferrero and Robert Wisbauer, *Unitary strongly prime rings and related radicals*, J. Pure Appl. Algebra **181** (2003), no. 2-3, 209–226, DOI 10.1016/S0022-4049(02)00311-0. MR1975299

[FK83] Joe W. Fisher and Jan Krempa, *"R^G is nil implies R is nil" is equivalent to the "Koethe conjecture"*, Houston J. Math. **9** (1983), no. 2, 177–180. MR703266

[GKW07] B. J. Gardner, Jan Krempa, and R. Wiegandt, *Open problems in radical theory (ICOR-2006)*, Algebra Discrete Math. **3** (2007), 15–17. MR2421773

[GW04] B. J. Gardner and R. Wiegandt, *Radical theory of rings*, Monographs and Textbooks in Pure and Applied Mathematics, vol. 261, Marcel Dekker, Inc., New York, 2004. MR2015465

[Go64] E. S. Golod, *On nil-algebras and finitely approximable p-groups* (Russian), Izv. Akad. Nauk SSSR Ser. Mat. **28** (1964), 273–276. MR0161878

[Go68] —, *Some problems of Burnside type*, pp. 284–289 in *Proc. I.C.M., Moscow, 1966*, Moscow, 1968. MR0238880 (39 #240)

[GS64] E. S. Golod and I. R. Šafarevič, *On the class field tower* (Russian), Izv. Akad. Nauk SSSR Ser. Mat. **28** (1964), 261–272. MR0161852

[Kr72] Jan Krempa, *Logical connections between some open problems concerning nil rings*, Fund. Math. **76** (1972), no. 2, 121–130, DOI 10.4064/fm-76-2-121-130. MR306251

[Kr74a] J. Krempa, *On the Jacobson radical of polynomial rings* (English, with Russian summary), Bull. Acad. Polon. Sci. Sér. Sci. Math. Astronom. Phys. **22** (1974), 887–890. MR393105

[Kr74b] Jan Krempa, *Radicals of semi-group rings*, Fund. Math. **85** (1974), no. 1, 57–71, DOI 10.4064/fm-85-1-57-71. MR360898

[Kr79a] — , *On semisimplicity of tensor products*, in *Ring theory* (Proc. Antwerp Conf., Univ. Antwerp, Antwerp, 1978), pp. 105–122, Dekker, New York, 1979. MR563289 (82b:16008)

[Kr79b] Jan Krempa, *On Passman's problem concerning nilpotent free algebras* (English, with Russian summary), Bull. Acad. Polon. Sci. Sér. Sci. Math. **27** (1979), no. 9, 645–648. MR600714

[La02] Serge Lang, *Algebra*, 3rd ed., Graduate Texts in Mathematics, vol. 211, Springer-Verlag, New York, 2002, DOI 10.1007/978-1-4613-0041-0. MR1878556

[Pa71] Donald S. Passman, *Infinite group rings*, Pure and Applied Mathematics, vol. 6, Marcel Dekker, Inc., New York, 1971. MR0314951

[Pa77] — , *The algebraic structure of group rings*, Wiley-Interscience [John Wiley & Sons], New York, 1977. MR0470211 (81d:16001)

[Pu77] Edmund R. Puczyłowski, *Action of algebras on radical properties* (English, with Russian summary), Bull. Acad. Polon. Sci. Sér. Sci. Math. Astronom. Phys. **25** (1977), no. 7, 617–622. MR485961

[Pu91] — , *Some questions concerning radicals of associative rings,* Theory of radicals, Szekszard (Hungary) 1991, Colloq. Math. Soc. János Bolyai 61, 216–227. MR1243913 (94j:16033)

[SP01] Agata Smoktunowicz and E. R. Puczyłowski, *A polynomial ring that is Jacobson radical and not nil*, Israel J. Math. **124** (2001), 317–325, DOI 10.1007/BF02772627. MR1856524

[RS21] Armando Reyes and Héctor Suárez, *Radicals and Köthe's conjecture for skew PBW extensions*, Commun. Math. Stat. **9** (2021), no. 2, 119–138, DOI 10.1007/s40304-019-00189-0. MR4258034

[Sm00] Agata Smoktunowicz, *Polynomial rings over nil rings need not be nil*, J. Algebra **233** (2000), no. 2, 427–436, DOI 10.1006/jabr.2000.8451. MR1793911

[Sm01] Agata Smoktunowicz, *On some results related to Köthe's conjecture*, Serdica Math. J. **27** (2001), no. 2, 159–170. MR1879880

[Sm02a] Agata Smoktunowicz, *A simple nil ring exists*, Comm. Algebra **30** (2002), no. 1, 27–59, DOI 10.1081/AGB-120006478. MR1880660

[Sm02b] Agata Smoktunowicz, *Amitsur's conjecture on polynomial rings in n commuting indeterminates*, Math. Proc. R. Ir. Acad. **102A** (2002), no. 2, 205–213, DOI 10.3318/PRIA.2002.102.2.205. MR1961638

Institute of Mathematics, University of Warsaw, Banacha 2, 02-097 Warszawa, Poland

Email address: jkrempa@mimuw.edu.pl

Contemporary Mathematics
Volume **785**, 2023
https://doi.org/10.1090/conm/785/15787

Rings with transitive chaining of idempotents

Xavier Mary

Dedicated to Professor Syed Tariq Rizvi for his contribution to Ring and Module Theory

ABSTRACT. We prove that transitivity of strong or weak 2-chaining of idempotents is equivalent to centrality of idempotents modulo the Jacobson radical.

1. Introduction

It has recently been proved that apparently very different concepts such as semicentrality (of idempotents), perspectivity (of modules) or special cleanness (of regular elements) can be handled by means of chains of associated idempotents ([**2**], [**6**], [**7**], [**8**], [**9**], [**10**], [**14**], [**13**], [**15**]). While many of the global ring-theoretical results focus on the question whether isomorphic idempotents are related by a chain of association of a certain length, others study transitivity of perspectivity, or equivalently of right 3-chaining. The purpose of this article is to consider rings where 1 or 2-chaining is transitive.

We quickly review a few of the basics concepts here.

Perspectivity is historically a lattice concept, and can be defined in any complemented lattice [**1**]. Perspectivity in modules traces back J. Von Neumann [**17**] and his studies on continuous geometries. It has then been reconsidered in the 60's and 70's by L. Fuchs [**3**] and D. Handelman [**5**], in link with cancellation and substitution properties. The study of perspective modules and rings in full generality is more recent [**4**].

Let R be a ring, M be a (right) module and $A, A' \subseteq^{\oplus} M$ be direct summands.

(1) the two direct summands A and A' are *perspective* (and we note $A \sim_{\oplus} A'$) if they share a common complementary summand ($A \oplus X = M = A' \oplus X$ for some $X \subseteq^{\oplus} M$);
(2) The module M is perspective if any two isomorphic direct summands are perspective (for any two $A, A' \subseteq^{\oplus} M$, $A \simeq A' \Rightarrow A \sim_{\oplus} A'$);
(3) The ring R is perspective if the right R-module R_R is perspective;
(4) The module M has *perspectivity transitive* if $A \sim_{\oplus} B \sim_{\oplus} C \Rightarrow A \sim_{\oplus} C$.

An element x of a ring R is said to be *regular* if there exists some $y \in R$ such that $xyx = x$. The element y is called an *inner inverse* of x. When the pair of equalities $xyx = x$ and $yxy = y$ hold, we say that x and y are *reflexive inverses* of each other. We denote by $V(x)$ the set of reflexive inverses of x. It is well known that every

---

2020 *Mathematics Subject Classification.* Primary 16D70, Secondary 16E50, 16U99.
Key words and phrases. chains of associated idempotents, (transitivity of) perspectivity.

regular element has a reflexive inverse. An element $x \in R$ is *strongly regular* if it admits a commuting inner inverse, and *unit-regular* if it admits an inner inverse that is also a unit. A ring such that every element is regular (resp. strongly regular, unit-regular) is called a *regular ring* (resp. *strongly regular ring*, *unit-regular ring*). An element $x \in R$ is *perspective* if xR and yR are perspective summands of R_R, for any $y \in V(x)$ [**13**]. It is *special clean* if it admits a decomposition $x = e+u = xu^{-1}x$ for some $e \in \text{idem}(R)$ (set of idempotents of R) and $u \in \text{U}(R)$ (set of units of R). Special cleanness relates to strongly regular elements and perspectivity as follows: x is special clean iff it admits a strongly regular reflexive inverse iff xR and yR are perspective summands of R_R, for some $y \in V(x)$ ([**7**, Theorem 2.13], [**12**, Theorem 4.1], [**14**, Proposition 4.20], [**16**, Theorem 6.1]).

We now turn our attention to idempotents. An idempotent $e \in \text{idem}(R)$ is *left semicentral* if $(1-e)Re = 0$. Two idempotents $e, f \in \text{idem}(R)$ are *isomorphic* (denoted by $e \simeq f$) if eR and fR are isomorphic submodules of R_R, iff $e = xy$, $f = yx$ for some $x, y \in R$. Moreover, we can always choose $y \in V(x)$. A stronger condition is that of similarity: e and f are *similar* if $f = ueu^{-1}$ for some unit $u \in \text{U}(R)$. Finally e and f are *left (resp. right) associates*, and we write $e \sim_\ell f$ (resp. $e \sim_r f$) if $Re = Rf$ or equivalently $ef = e$ and $fe = f$ (resp. $eR = fR$ or equivalently $ef = f$ and $fe = e$). Observe that these two relations do not commute in general. Also, of particular importance is the following fact [**6**], [**8**], [**9**], [**15**]: $e \sim_\ell f$ iff there exists some (unique) unit $u \in 1+(1-e)Re \subseteq \text{U}(R)$ such that $f = ue$. Thus the set $(1-e)Re$ parameterizes the left associates of e. As also $eu^{-1} = e$ in this case , $f = ueu^{-1}$ is similar to e, an important fact in consideration of the invariance of relations of left and right association under similarity [**15**, Lemma 2.1].

Following [**8**], [**14**] and [**15**], and using the terminology of [**10**], we recall the following definitions. Let $n \in \mathbb{N}$; a *left n-chain* from e to f consists of a sequence of $n+1$ idempotents $e_0; e_1; \ldots; e_n \in E(R)$ such that $e = e_0 \sim_\ell e_1 \sim_r e_2 \sim_\ell \cdots e_n = f$. We call the number n the length of the chain. When n is small, such as $n = 2$ or $n = 3$, we will write $e \sim_{\ell r} f$, respectively $e \sim_{\ell r \ell} f$. Right n-chains are defined dually (starting with a right equivalence).

We say that e and f are weakly (resp. strongly) n-chained if they are connected by either a left or a right (resp. both a left and a right) n-chain. We note $e \sim_n^w f$ (resp. $e \sim_n^s f$) if they are weakly (resp. strongly) n-chained, and $e \approx f$ if they are connected by a left (or right) n-chain, for some $n \in \mathbb{N}$. A ring R is weakly (resp. strongly) n-chained if any two isomorphic idempotents are weakly (resp. strongly) n-chained.

The following equivalences subsume the main connections between the previous notions ([**6**], [**8**], [**9**], [**10**], [**13**], [**14**], [**15**]).

(1) $e \in \text{idem}(R)$ is left semicentral iff $(\forall f \in \text{idem}(R))\ e \approx f \Rightarrow e \sim_r f$;
(2) $e \in \text{idem}(R)$ is central modulo the Jacobson radical $J(R)$ iff $(\forall f \in \text{idem}(R))\ e \approx f \Rightarrow e \sim_2^s f$ iff $(\forall f \in \text{idem}(R))\ e \approx f \Rightarrow e \sim_2^w f$;
(3) $(\forall f \in \text{idem}(R))\ e \approx f \Rightarrow eR \sim_\oplus fR$ iff $(\forall f \in \text{idem}(R))\ e \approx f \Rightarrow e \sim_3^s f$ iff $(\forall f \in \text{idem}(R))\ e \approx f \Rightarrow e \sim_3^w f$;
(4) the ring R is strongly 1-chained iff it is abelian;
(5) the ring R is weakly 1-chained iff idempotents are either left or right semicentral;

(6) the ring R is strongly 2-chained iff it is weakly 2-chained iff idempotents are central modulo $J(R)$ iff regular elements of R are strongly regular;
(7) the ring R is strongly 3-chained iff it is weakly 3-chained iff R perspective iff regular elements of R are perspective;
(8) the ring R is strongly 4-chained iff regular elements are special clean.

We list below some examples of rings with one of the above properties.

EXAMPLE 1.1.
(1) Let D be a division ring. Then for any integer $k \geq 1$, D^k is abelian hence strongly 1-chained;
(2) Let F be a field. Then the ring of 2-by-2 upper triangular matrices $\mathcal{U}_2(F) = \begin{pmatrix} F & F \\ 0 & F \end{pmatrix}$ is weakly 1-chained, but not strongly 1-chained;
(3) Let R be any quasi-duo ring. Then a simple argument due to Lam and Dugas [**11**] proves that regular elements of R are strongly regular, so that R is strongly 2-chained. It follows for instance that the quasi-duo ring $R = \mathcal{U}_3(F)$, F a field, is strongly 2-chained. It is not weakly 1-chained. Actually, very distinct considerations prove that we do not require F to be a field for upper triangular matrix rings to be strongly 2-chained. Indeed, by [**14**, Proposition 5.12], an upper triangular matrix ring $R = \mathcal{U}_k(S), k \geq 1$ is strongly n-chained iff S is strongly n-chained, for any $n \geq 2$. In particular, $R = \mathcal{U}_k(S)$ is strongly 2-chained for any commutative or abelian ring S (and any integer $k \geq 1$);
(4) Any matrix ring $\mathcal{M}_k(F)$ over a field F is unit-regular hence perspective [**4**]. If $k \geq 2$, it is not strongly regular (it contains nilpotents) hence not strongly 2-chained;
(5) By [**8**, Theorem 4.3 and Corollary 4.4], $\mathcal{M}_3(\mathbb{Z})$ is strongly 4-chained but not strongly 3-chained.

2. Transitivity of chaining

While unit-regular rings are known to be perspective (and in particular have perspectivity transitive), there are examples due to Bergman of regular rings that have perspectivity transitive but are not perspective. Recall that a ring R satisfies *internal cancellation* (IC) if isomorphic direct summands of R_R have isomorphic complementary direct summands. As a regular ring is IC iff it is unit-regular, it was questionable whether a (non-regular) IC ring with perspectivity transitive was perspective. In [**15**], we construct an IC ring R where perspectivity is transitive but still R is not perspective. In general, we only have the equivalence: $\mathcal{M}_2(R)$ has perspectivity transitive iff R has *stable range one* (forall $a, b \in R$, $aR + bR = R$ implies $a + bt \in \mathrm{U}(R)$ for some $t \in R$) by [**10**, Theorem 2.5] (and rings with stable range one are known to be perspective).

In any ring R, perspectivity is transitive iff relation $\sim_{r\ell r}$ is transitive on idem(R) from [**2**, Lemma 6.3]. And, still by [**2**, section 6], this holds iff any left 3-chain is a right 3-chain. This raises the following questions:

(1) is there a relationship between strongly 2-chained rings and rings where weak (or strong) 2-chaining is transitive?
(2) is there a relationship between strongly (resp. weakly) 1-chained rings and rings where strong (resp. weak) 1-chaining is transitive?

2.1. Transitivity of 2-chaining. Let R be a ring and $e \in \text{idem}(R)$. Following [**9**], we provide a general formula for idempotents $f \in \text{idem}(R)$ such that $e \sim_{r\ell r} f$. Let $f \in \text{idem}(R)$ be such that $e \sim_{r\ell r} f$. Then $e \sim_r g \sim_{\ell r} f$ for some $g \in \text{idem}(R)$. We decompose the ring R as

$$R = eRe \oplus eR(1-e) \oplus (1-e)Re \oplus (1-e)R(1-e)$$

and consider its Peirce matrix form

$$R = \begin{pmatrix} eRe & eR(1-e) \\ (1-e)Re & (1-e)R(1-e) \end{pmatrix}.$$

As $e \sim_r g$ then $g = e(1+t) = \begin{pmatrix} 1 & t \\ 0 & 0 \end{pmatrix} = (1+t)^{-1}e(1+t) = \begin{pmatrix} 1 & -t \\ 0 & 1 \end{pmatrix} e \begin{pmatrix} 1 & t \\ 0 & 1 \end{pmatrix}$ for some $t \in eR(1-e)$ (this is the parameterization of right associates of e by $eR(1-e)$, dual to the one explained in Section 1). By invariance of association under similarity, as $\begin{pmatrix} 1 & -t \\ 0 & 1 \end{pmatrix} e \begin{pmatrix} 1 & t \\ 0 & 1 \end{pmatrix} \sim_{\ell r} f$ then $e \sim_{\ell r} \begin{pmatrix} 1 & t \\ 0 & 1 \end{pmatrix} f \begin{pmatrix} 1 & -t \\ 0 & 1 \end{pmatrix}$ and there exists $h \in \text{idem}(R)$ such that $e \sim_\ell h \sim_r \begin{pmatrix} 1 & t \\ 0 & 1 \end{pmatrix} f \begin{pmatrix} 1 & -t \\ 0 & 1 \end{pmatrix}$. But as above there exists $r \in (1-e)Re$ such that $h = (1+r)e(1+r)^{-1} = \begin{pmatrix} 1 & 0 \\ r & 1 \end{pmatrix} e \begin{pmatrix} 1 & 0 \\ -r & 1 \end{pmatrix}$, so that once again by invariance of association under similarity

$$e \sim_r \begin{pmatrix} 1 & 0 \\ -r & 1 \end{pmatrix} \begin{pmatrix} 1 & t \\ 0 & 1 \end{pmatrix} f \begin{pmatrix} 1 & -t \\ 0 & 1 \end{pmatrix} \begin{pmatrix} 1 & 0 \\ r & 1 \end{pmatrix}.$$

Using the parameterization one last time we obtain

$$\begin{pmatrix} 1 & 0 \\ -r & 1 \end{pmatrix} \begin{pmatrix} 1 & t \\ 0 & 1 \end{pmatrix} f \begin{pmatrix} 1 & -t \\ 0 & 1 \end{pmatrix} \begin{pmatrix} 1 & 0 \\ r & 1 \end{pmatrix} = \begin{pmatrix} 1 & -s \\ 0 & 1 \end{pmatrix} e \begin{pmatrix} 1 & s \\ 0 & 1 \end{pmatrix}$$

for some $s \in eR(1-e)$. Finally as

$$\begin{pmatrix} 1 & 0 \\ r & 1 \end{pmatrix} \begin{pmatrix} 1 & -s \\ 0 & 1 \end{pmatrix} e \begin{pmatrix} 1 & s \\ 0 & 1 \end{pmatrix} \begin{pmatrix} 1 & 0 \\ -r & 1 \end{pmatrix} = \begin{pmatrix} 1 & 0 \\ r & 0 \end{pmatrix} \begin{pmatrix} 1-sr & s \\ 0 & 0 \end{pmatrix}$$

then

$$f = \begin{pmatrix} 1 & -t \\ 0 & 1 \end{pmatrix} \begin{pmatrix} 1 & 0 \\ r & 0 \end{pmatrix} \begin{pmatrix} 1-sr & s \\ 0 & 0 \end{pmatrix} \begin{pmatrix} 1 & t \\ 0 & 1 \end{pmatrix} = \begin{pmatrix} 1-tr & 0 \\ r & 0 \end{pmatrix} \begin{pmatrix} 1-sr & (1-sr)t+s \\ 0 & 0 \end{pmatrix}$$

for some $r \in (1-e)Re$ and $s, t \in eR(1-e)$. Conversely, for any $r \in (1-e)Re$ and $s, t \in eR(1-e)$, the previous formula

$$f = (1+t)^{-1}(1+r)(1+s)^{-1}e(1+s)(1+r)^{-1}(1+t) = \begin{pmatrix} 1-tr & 0 \\ r & 0 \end{pmatrix} \begin{pmatrix} 1-sr & (1-sr)t+s \\ 0 & 0 \end{pmatrix}$$

defines an idempotent f right 3-chained to e since

$$e \sim_r (1+t)^{-1}e(1+t) \sim_\ell (1+t)^{-1}(1+r)e(1+r)^{-1}(1+t) \sim_r f$$

(this also follows for instance from [**15**, Theorem 2.5], since $(e-tr)+zr$ is a unit for $z = t \in eR(1-e)$).

Let $r \in (1-e)Re$, $s, t \in eR(1-e)$ and pose $y = \begin{pmatrix} 1-tr & 0 \\ r & 0 \end{pmatrix}$, $x = \begin{pmatrix} 1-sr & (1-sr)t+s \\ 0 & 0 \end{pmatrix}$ and $f = yx$. Straightforward calculations prove that $e = xy$ and that $xyx =$

$x, yxy = y$ (so that x and y are reflexive inverses). Now, assume that the equivalence relation $\sim_2^w$ is transitive. As e and f are weakly 4-chained then they are weakly 2-chained by transitivity and either $e \sim_{r\ell} f$ or $e \sim_{\ell r} f$. But by [**15**, Proposition 2.2], this happens iff $exe = e - sr \in \mathrm{U}(eRe)$ or $eye = e - tr \in \mathrm{U}(eRe)$ respectively. Letting $s = t$ we deduce that in any case, $e - sr$ is a unit in eRe. We now conclude as in [**9**]. Let any $p \in R$ and $r = (1-e)pe$. Then by the previous arguments

$$1 - sp = e - s(1-e)pe + s(1-e)p(1-e) + (1-e) = \begin{pmatrix} 1 - sr & sp(1-e) \\ 0 & 1 \end{pmatrix}$$

is a triangular matrix with each diagonal entry invertible hence is a unit in R. This proves that $s \in J(R)$ and since s is arbitrary, $eR(1-e) \subseteq J(R)$. Working with the complementary idempotent $1-e$ instead of e we obtain the dual inclusion $(1-e)Re \subseteq J(R)$, and e is central modulo $J(R)$ (for any $a \in R, ea - ae = ea(1-e) - (1-e)ae \in J(R)$).

We have proved the following theorem.

THEOREM 2.1. *Let R be a ring such that weak 2-chaining is transitive. Then idempotents are central modulo the Jacobson radical.*

What about strong 2-chaining? The arguments above do not work since $e \sim_{r\ell r} f$ implies that e and f are strongly 4-chained, but not necessarily that $e \sim_2^s g \sim_2^s f$ for some $g \in \mathrm{idem}(R)$.

So let R be a ring where strong 2-chaining is transitive and let $e, f, g \in \mathrm{idem}(R)$ be such that $e \sim_2^s g \sim_2^s f$. First, as above as $e \sim_{\ell r} g$ then $g = \begin{pmatrix} 1 & 0 \\ r & 0 \end{pmatrix}\begin{pmatrix} 1-sr & s \\ 0 & 0 \end{pmatrix}$ for some $r \in (1-e)Re$ and $s \in eR(1-e)$. And since also $e \sim_{r\ell} g$ then $u = e - sr \in \mathrm{U}(eRe)$. But also $g = \begin{pmatrix} 1 & 0 \\ r & 1 \end{pmatrix}\begin{pmatrix} 1 & -s \\ 0 & 1 \end{pmatrix} e \begin{pmatrix} 1 & s \\ 0 & 1 \end{pmatrix}\begin{pmatrix} 1 & 0 \\ -r & 1 \end{pmatrix}$. Thus

$$f = \begin{pmatrix} 1 & 0 \\ r & 1 \end{pmatrix}\begin{pmatrix} 1 & -s \\ 0 & 1 \end{pmatrix}\begin{pmatrix} 1 & 0 \\ v & 0 \end{pmatrix}\begin{pmatrix} 1-wv & w \\ 0 & 0 \end{pmatrix}\begin{pmatrix} 1 & s \\ 0 & 1 \end{pmatrix}\begin{pmatrix} 1 & 0 \\ -r & 1 \end{pmatrix}$$

for some $v \in (1-e)Re$ and $w \in eR(1-e)$ such that $z = e - wv \in \mathrm{U}(eRe)$. Once again, the converse is also true. Any matrix f of the previous form with $r, v \in (1-e)Re$, $s, w \in eR(1-e)$ such that and $u = e - sr \in \mathrm{U}(eRe)$ and $z = e - wv \in \mathrm{U}(eRe)$ satisfies that $e \sim_2^s g \sim_2^s f$ for some $g \in \mathrm{idem}(R)$. In particular, we may let $r = w = 0$ (that is $e \sim_r g \sim_\ell f$). In this case $v \in (1-e)Re$ and $s \in eR(1-e)$ may be chosen arbitrarily and f takes the simpler form $f = \begin{pmatrix} 1-sv & 0 \\ v & 0 \end{pmatrix}\begin{pmatrix} 1 & s \\ 0 & 1 \end{pmatrix}$. By [**15**, Proposition 2.2], $e \sim_{\ell r} f$ iff $e - sv \in \mathrm{U}(eRe)$. By the same arguments as above this implies that $s \in J(R)$ and since s is arbitrary, $eR(1-e) \subseteq J(R)$.

THEOREM 2.2. *Let R be a ring such that strong 2-chaining is transitive. Then idempotents are central modulo the Jacobson radical.*

Using [**9**, Theorem 3.13], we obtain the following corollary.

COROLLARY 2.3. *Let R be a ring. Then the following statements are equivalent.*

(1) *Idempotents of R are central modulo $J(R)$;*
(2) *Isomorphic idempotents of R are strongly 2-chained;*
(3) *Idempotents of R in the same $\approx$-class are weakly 2-chained;*
(4) *Strong 2-chaining is transitive in R (equiv. in each $\approx$-class);*

(5) *Weak 2-chaining is transitive in R (equiv. in each $\approx$-class).*

2.2. Transitivity of 1-chaining. The case of 1-chain is simpler but still interesting since it exhibits a different behavior. Let R be a ring and $e, f \in \text{idem}(R)$. As $e \sim_1^s f$ iff $e = f$, then strong 1-chaining is always transitive. So is right 1-chaining and left 1-chaining. Thus, we are left with transitivity of weak 1-chaining. So let R be a ring with weak 1-chaining transitive and let $e \in \text{idem}(R)$. Let also $r \in (1-e)Re$ and $s \in eR(1-e)$. As in Section 2.1, $e \sim_{\ell r} f = \begin{pmatrix} 1-sr & s \\ r-rsr & rs \end{pmatrix} = \begin{pmatrix} 1 & 0 \\ r & 0 \end{pmatrix}\begin{pmatrix} 1-sr & s \\ 0 & 0 \end{pmatrix}$. By weak 1-chaining, either $e \sim_\ell f$ and $s = 0$ or $e \sim_r f$ and $r = 0$. Thus e is either right semicentral or left semicentral. As conversely, a ring with idempotents either right semicentral or left semicentral is weakly 1-chained, we obtain the following equivalences.

THEOREM 2.4. *Let R be a ring. Then the following statements are equivalent.*

(1) *Idempotents of R are either right semicentral or left semicentral;*
(2) *Isomorphic idempotents (equiv. in the same $\approx$-class) are weakly 1-chained;*
(3) *Weak 1-chaining is transitive in R (equiv. in each $\approx$-class).*

3. Acknowledgments

Some of the material discussed here was part of a presentation given at NCRA,VII, in honor of Prof. Tariq Rizvi. I wish him continued success. Also, this article benefited from stimulating exchanges with Dinesh Khurana and Pace P. Nielsen. Let me thank them. Finally, I am very grateful to the referee for his careful reading and insightful comments.

References

[1] Garrett Birkhoff, *Lattice Theory*, American Mathematical Society, New York, 1940. MR0001959

[2] Alexander J. Diesl, Samuel J. Dittmer, and Pace P. Nielsen, *Idempotent lifting and ring extensions*, J. Algebra Appl. **15** (2016), no. 6, 1650112, 16, DOI 10.1142/S0219498816501127. MR3479816

[3] L. Fuchs, *On a substitution property of modules*, Monatsh. Math. **75** (1971), 198–204, DOI 10.1007/BF01299099. MR296096

[4] Shelly Garg, Harpreet K. Grover, and Dinesh Khurana, *Perspective rings*, J. Algebra **415** (2014), 1–12, DOI 10.1016/j.jalgebra.2013.09.055. MR3229504

[5] David Handelman, *Perspectivity and cancellation in regular rings*, J. Algebra **48** (1977), no. 1, 1–16, DOI 10.1016/0021-8693(77)90289-7. MR447329

[6] Dinesh Khurana, T. Y. Lam, and Pace P. Nielsen, *An ensemble of idempotent lifting hypotheses*, J. Pure Appl. Algebra **222** (2018), no. 6, 1489–1511, DOI 10.1016/j.jpaa.2017.07.008. MR3754435

[7] Dinesh Khurana, T. Y. Lam, Pace P. Nielsen, and Janez Šter, *Special clean elements in rings*, J. Algebra Appl. **19** (2020), no. 11, 2050208, 27, DOI 10.1142/S0219498820502084. MR4141677

[8] Dinesh Khurana, Xavier Mary, and Pace P. Nielsen, *Idempotent chains and bounded generation of SL_2*, submitted.

[9] Dinesh Khurana and Pace P. Nielsen, *Perspectivity and von Neumann regularity*, Comm. Algebra **49** (2021), no. 12, 5483–5499, DOI 10.1080/00927872.2021.1948555. MR4328552

[10] ———, *Transitivity of perspectivity*, Algebr. Represent. Theory (2021), 1–7.

[11] T. Y. Lam and Alex S. Dugas, *Quasi-duo rings and stable range descent*, J. Pure Appl. Algebra **195** (2005), no. 3, 243–259, DOI 10.1016/j.jpaa.2004.08.011. MR2114274

[12] Xavier Mary, *Characterizations of clean elements by means of outer inverses in rings and applications*, J. Algebra Appl. **19** (2020), no. 7, 2050134, 23, DOI 10.1142/S0219498820501340. MR4129181

[13] ———, *Special clean elements, perspective elements and perspective rings*, J. Algebra Appl. (2021).

[14] Xavier Mary, *N-chained semigroups and $n/2$-perspective modules and rings*, Comm. Algebra **50** (2022), no. 1, 163–181, DOI 10.1080/00927872.2021.1955902. MR4370420

[15] Xavier Mary and Pace P. Nielsen, *IC rings and transitivity of perspectivity*, J. Algebra Appl. **21** (2022), no. 8, Paper No. 2250151, 8, DOI 10.1142/S0219498822501511. MR4469329

[16] Xavier Mary and Pedro Patrício, *Characterizations of special clean elements and applications*, Filomat **34** (2020), no. 14, 4847–4860, DOI 10.2298/fil2014847m. MR4290894

[17] John von Neumann, *Continuous geometry*, Princeton Mathematical Series, No. 25, Princeton University Press, Princeton, N.J., 1960. MR0120174

Laboratoire Modal'X, Université Paris Nanterre, 92000 Nanterre, France
Email address, Corresponding author: `xavier.mary@parisnanterre.fr`

Contemporary Mathematics
Volume **785**, 2023
https://doi.org/10.1090/conm/785/15788

On Continuous Hulls of Rings and Modules

Jae Keol Park and S. Tariq Rizvi

Dedicated to Professor Syed Tariq Rizvi for his contribution to Ring and Module Theory

ABSTRACT. Every module has an injective hull, a quasi-injective hull, and a quasi-continuous hull. However, there exists a uniform nonsingular cyclic module over a noncommutative ring which has no continuous hull. (Recall that the continuous hull of a module M is the smallest continuous extension of M in a fixed injective hull $E(M)$ of M.) When the base ring is commutative, every uniform cyclic module over a commutative ring has a continuous hull. Further, every nonsingular cyclic module over a commutative ring has a continuous hull. Motivated by the example and these results on continuous hulls, it is interesting to study whether R_R has a continuous hull for any ring R. We show that for any integer n such that $n > 1$, if R is the $n \times n$ matrix ring over any given ring or the $n \times n$ upper triangular matrix ring over any given ring, then R_R has a continuous hull and such a continuous hull is explicitly described. Moreover, if R is an abelian regular ring, it is shown that every nonsingular cyclic right R-module has a continuous hull. As an application, we show that R_R has a continuous hull when R is an abelian regular ring. In this case, $R\mathcal{B}(Q(R))_R$ (where $R\mathcal{B}(Q(R))$ is the idempotent closure of R) is the continuous hull of R_R. As a byproduct, we show that when R is an abelian regular ring, $R\mathcal{B}(Q(R))$ is the smallest continuous regular ring of quotients of R. We discuss the condition $(\star)$ of a commutative ring R, which is precisely that the classical ring of quotients of every uniform factor ring of R is self-injective. Moreover, by using the condition $(\star)$, we provide a detailed proof that every module over a commutative noetherian ring has a continuous hull. Various examples which illustrate our results are provided.

1. Introduction

Since the discovery of injective hulls of modules by Shoda [**16**] and Eckmann and Schopf [**3**] independently, the existence of generalized injective hulls such as quasi-injective hulls, continuous hulls, and quasi-continuous hulls of modules has been investigated in [**4**], [**7**], [**9**], and [**14**].

For a given module M, Johnson and Wong [**7**] showed the existence of the quasi-injective hull of M (Theorem 2.9). On the other hand, Goel and Jain [**4**] established the existence of the quasi-continuous hull of M (Theorem 2.12).

In contrast to the existence of quasi-injective hulls and quasi-continuous hulls, for the case of continuous hulls, there exists a uniform nonsingular cyclic module

2020 *Mathematics Subject Classification.* 16D50, 16E50, 16S50, 16U60.

Key words and phrases. quasi-injective hull, continuous hull, quasi-continuous hull, regular ring, reduced ring, abelian ring, idempotent closure.

over a *noncommutative* ring which does not have a continuous hull by Müller and Rizvi [**9**] (Example 3.1). In contrast to Example 3.1, each uniform cyclic module over a *commutative* ring does have a continuous hull by Müller and Rizvi [**9**] (Theorem 3.2). Moreover, every nonsingular cyclic module over a *commutative* ring has a continuous hull by Müller and Rizvi [**9**] (Theorem 3.3).

Motivated by Example 3.1, Theorem 3.2, and Theorem 3.3, it is of interest to study whether R_R has a continuous hull for any ring R (see Question 3.4). We show that for any ring A and any integer n such that $n > 1$, if R is the $n \times n$ matrix ring over A or the $n \times n$ upper triangular matrix ring over A, then R_R has a continuous hull and such a continuous hull is explicitly described (Theorem 3.10 and Theorem 3.15).

Moreover, if R is an abelian regular ring, it is shown that every nonsingular cyclic right R-module has a continuous hull (Theorem 3.23). In particular, R_R has a continuous hull when R is an abelian regular ring. In this case, $R\mathcal{B}(Q(R))_R$ is the continuous hull of R_R (Theorem 3.24). As a byproduct, we show that when R is an abelian regular ring, the idempotent closure $R\mathcal{B}(Q(R))$ is the smallest continuous regular ring of quotients of R (Corollary 3.30).

By adopting the argument used in the proof of Theorem 3.23, when R is a ring with a regular right ring of quotients T, it is shown that T is right continuous if and only T_R is continuous (Theorem 3.34). Thereby if R is a regular ring, then the ring $R\mathcal{B}(Q(R))$ is right continuous if and only if $R\mathcal{B}(Q(R))_R$ is continuous (Corollary 3.35).

In Section 4, we discuss condition $(\star)$ of a commutative ring R, which is exactly that the classical ring of quotients of every uniform factor ring of R is self-injective. By using condition $(\star)$, we provide detailed proofs for early results of Rizvi [**14**] which show that every commutative noetherian ring satisfies condition $(\star)$ (Theorem 4.13); that every continuous module over a commutative noetherian ring is quasi-injective (Theorem 4.14); and that every module M over a commutative noetherian ring has a continuous hull which is exactly the quasi-injective hull of M (Theorem 4.15).

Various examples which illustrate our results are provided.

All rings are assumed to have identity and all modules are assumed to be unitary. We use $E(M_R)$ or $E(M)$ to denote a fixed injective hull of a module M_R. For a module M, $K \leq M$, $L \trianglelefteq M$, $N \leq^{\text{ess}} M$, $U \leq^{\text{den}} M$, and $V \leq^{\oplus} M$ denote that K is a submodule of M, L is a fully invariant submodule of M, N is an essential submodule of M, U is a dense submodule of M, and V is a direct summand of M, respectively.

Let M_R and N_R be modules. Then $\text{Hom}(M_R, N_R)$ or $\text{Hom}(M, N)$ denotes the additive abelian group of all R-module homomorphisms from M_R to N_R. Further, R-homomorphisms and R-isomorphisms mean R-module homomorphisms and R-module isomorphisms, respectively. We let $\text{End}(M_R)$ or $\text{End}(M)$ denote the endomorphism ring of a module M_R.

For a set X, we let $|X|$ be the cardinality of X. Let M be a module and Λ be a nonempty set. Then $M^{(\Lambda)}$ stands for the direct sum of $|\Lambda|$ copies of M. When $|\Lambda| = n$, we use $M^{(n)}$ to denote $M^{(\Lambda)}$. For a positive integer n, we let $\text{Mat}_n(A)$ and $T_n(A)$ denote the $n \times n$ matrix ring over a ring A and the $n \times n$ upper triangular matrix ring over a ring A, respectively.

The symbols $\mathbb{Q}$, $\mathbb{Z}$, and $\mathbb{Z}_n (n > 1)$ denote the field of rationals, the ring of integers, and the ring of integers modulo n, respectively. For a prime integer p, we use $\mathbb{Z}_{p^\infty}$ to denote the Prüfer p-group. We note that $\mathbb{Z}_{p^\infty}$ is the injective hull of $\mathbb{Z}_{p^n}$, for any positive integer n, as $\mathbb{Z}$-modules.

Ideals without the adjective right or left mean two-sided ideals. When R is a ring, $I \trianglelefteq R$ denotes I is an ideal of R. For a set X and a subset Y of X, we let $X \setminus Y = \{x \in X \mid x \notin Y\}$.

2. Preliminary Results

In this section, we discuss the notions of quasi-injective modules, continuous modules, and quasi-continuous modules. Some properties of these modules are also introduced. The existence of quasi-injective hulls and quasi-continuous hulls of modules are discussed. Results of this section, except Example 2.11, are known and they will be used in Section 3 and Section 4. See [**2**, Chapter 2] for more details on the results of this section.

Let M and N be modules. Then M is called N-*injective* if, for any $W \leq N$ and $f \in \text{Hom}(W, M)$, there exists $\phi \in \text{Hom}(N, M)$, which is an extension of f.

We recall that a module M is called *injective* if M is N-injective for all modules N. Note that a module M_R is injective if and only if M_R is R_R-injective by Baer's Criterion. If N is semisimple, then every module M is N-injective.

Definition 2.1. A module M is called *quasi-injective* if M is M-injective.

Proposition 2.2. Let M be a module. Then M is quasi-injective if and only if $M \trianglelefteq E(M)$.

Definition 2.3. Let M and N are modules. Then M and N are said to be *relatively injective* (to each other) if M is N-injective and N is M-injective.

Let M and N be modules. If $M \cong N$, then $E(M) \cong E(N)$. But the converse does not hold true. Indeed, let p be a prime integer, and let k, ℓ be distinct positive integers. Then $E(\mathbb{Z}_{p^k}) = \mathbb{Z}_{p^\infty} = E(\mathbb{Z}_{p^\ell})$. But $\mathbb{Z}_{p^k} \not\cong \mathbb{Z}_{p^\ell}$ as $\mathbb{Z}$-modules.

In contrast to the preceding example, the following holds true.

Theorem 2.4. Let M and N be modules. Then we have the following.

(i) M is N-injective if and only if $\varphi(N) \subseteq M$ for every $\varphi \in \text{Hom}(E(N), E(M))$.

(ii) let M and N be relatively injective modules. If $E(M) \cong E(N)$, then $M \cong N$.

Proposition 2.5. Let $M_1, M_2, \ldots, M_n$ be modules, where n is a positive integer. Then the following are equivalent.

(i) $M_1 \oplus M_2 \oplus \cdots \oplus M_n$ is quasi-injective.

(ii) M_i and M_j are relatively injective for $i, j = 1, 2, \ldots, n$.

Corollary 2.6. Let M be a module and n a positive integer. Then $M^{(n)}$ is quasi-injective if and only if M is quasi-injective.

Consider the following conditions for a module M:

(C_1) Every submodule of M is essential in a direct summand of M.

(C_2) If $V \leq M$ and $V \cong N \leq^{\oplus} M$, then $V \leq^{\oplus} M$.

(C_3) If M_1 and M_2 are direct summands of M with $M_1 \cap M_2 = 0$, then $M_1 \oplus M_2$ is a direct summand of M.

Definition 2.7. Let M be a module. Then:

(i) M is called *continuous* if it satisfies (C_1) and (C_2) conditions;

(ii) M is called *quasi-continuous* if it satisfies (C_1) and (C_3) conditions.

It is well-known that the following implications hold true for modules while the reverse implications do not hold true (see [**2**, Section 2.1]):

$$\text{injective} \Rightarrow \text{quasi-injective} \Rightarrow \text{continuous} \Rightarrow \text{quasi-continuous}.$$

Definition 2.8. Let M be a module. We fix an injective hull $E(M)$ of M.

(i) The smallest quasi-injective extension of M in $E(M)$ is called the *quasi-injective hull* of M.

(ii) The smallest continuous extension of M in $E(M)$ is called the *continuous hull* of M.

(iii) The smallest quasi-continuous extension of M in $E(M)$ is called the *quasi-continuous hull* of M.

The next result on the existence of quasi-injective hulls is due to Johnson and Wong [**7**].

Theorem 2.9. Let M be a right R-module and let $H = \text{End}(E(M))$. Then HM is the quasi-injective hull of M.

Let M be a module and $N \leq M$. By Zorn's Lemma, there exists a submodule C of M which is maximal with respect to $N \cap C = 0$. Then $N \oplus C \leq^{\text{ess}} M$. In this case, C is called a *complement* of N in M.

The equivalence of (i) through (iv) in the following theorem is due to Goel and Jain [**4**], while the equivalence of (i) and (v) is shown by Oshiro and Rizvi [**12**].

Theorem 2.10. The following are equivalent for a module M.

(i) M is quasi-continuous.

(ii) $M = X \oplus Y$ for any two submodules X and Y which are complements of each other in M.

(iii) $fM \subseteq M$ for every $f^2 = f \in \text{End}(E(M))$.

(iv) $E(M) = \oplus_{i \in \Lambda} E_i$ implies $M = \oplus_{i \in \Lambda}(M \cap E_i)$.

(v) Any essential extension V of M with a decomposition $V = \oplus_{\alpha \in \Gamma} V_\alpha$ implies $M = \oplus_{\alpha \in \Gamma}(M \cap V_\alpha)$.

In the following Example 2.11, condition (iv) of Theorem 2.10 is discussed. Indeed, we give an example of a module M such that:

(1) M is not quasi-continuous;

(2) There exist two sets $\{U_1, U_2\}$ and $\{V_1, V_2\}$ of submodules of $E(M)$ with $E(M) = U_1 \oplus U_2$ and $E(M) = V_1 \oplus V_2$; and

(3) $M = (M \cap U_1) \oplus (M \cap U_2)$, but $M \neq (M \cap V_1) \oplus (M \cap V_2)$.

Example 2.11. For $k \in \mathbb{Z}$, let k_n denotes the canonical image of k in $\mathbb{Z}_n$, where n is an integer such that $n > 1$. We let

$$M = \{0_2 + 0_4,\ 0_2 + 1_4,\ 0_2 + 2_4,\ 0_2 + 3_4,\ 1_2 + 0_4,\ 1_2 + 1_4,\ 1_2 + 2_4,\ 1_2 + 3_4\},$$

where $a_2 + b_4 = c_2 + d_4$ if and only if $a_2 = c_2$ and $b_4 = d_4$ for $a_2 + b_4$ and $c_2 + d_4$ in M. Next for $k_2 + \ell_4$ and $v_2 + w_4$ in M and $m \in \mathbb{Z}$, define

$$(k_2 + \ell_4) + (v_2 + w_4) = (k + v)_2 + (\ell + w)_4 \text{ and } m(k_2 + \ell_4) = (mk)_2 + (m\ell)_4.$$

Then M is a $\mathbb{Z}$-module. Note that $M \cong \mathbb{Z}_2 \oplus \mathbb{Z}_4$ as $\mathbb{Z}$-modules.

We show that M is not quasi-continuous. For this, let $N_1 = \{0_2 + 0_4,\, 1_2 + 0_4\}$ and $N_2 = \{0_2 + 0_4, 1_2 + 2_4\}$. Then $N_1 \leq M$ and $N_2 \leq M$. We put

$$A = \{0_2 + 0_4, 0_2 + 1_4, 0_2 + 2_4, 0_2 + 3_4\}.$$

Then $A \leq M$, $N_1 \oplus A = M$, and $N_2 \oplus A = M$. Thus $N_1 \leq^{\oplus} M$ and $N_2 \leq^{\oplus} M$. Further, $N_1 \cap N_2 = 0_2 + 0_4\, (= 0)$. But, $N_1 \oplus N_2$ is not a direct summand of M. In fact, by direct computation, we can check that $N_1 \oplus N_2 \leq^{\text{ess}} M$. So, if $N_1 \oplus N_2 \leq^{\oplus} M$, $M = N_1 \oplus N_2$.

Since $|M| = 8$ and $|N_1 \oplus N_2| = 4$, we get a contradiction. Therefore, $N_1 \oplus N_2$ is not a direct summand of M. So M is not quasi-continuous as a $\mathbb{Z}$-module. Hence condition (iv) of Theorem 2.10 does not hold true.

We illustrate condition (iv) of Theorem 2.10 further in detail for M as follows.

(i) Put $M_1 = N_1$ and $M_2 = A$. Then $M = M_1 \oplus M_2$. Hence we have that $E(M) = E(M_1 \oplus M_2) = E(M_1) \oplus E(M_2)$. Note that $E(M_1) \cap M_2 = 0$ since $M_1 \cap M_2 = 0$.

As $M = M_1 \oplus M_2$ and $M_1 \leq E(M_1) \cap M \leq M$, the modular law yields that

$$E(M_1) \cap M = M_1 + (E(M_1) \cap M \cap M_2) = M_1 \oplus (E(M_1) \cap M_2) = M_1.$$

Similarly, $E(M_2) \cap M = M_2$. So $M = (E(M_1) \cap M) \oplus (E(M_2) \cap M)$ because $M = M_1 \oplus M_2$.

(ii) On the other hand, from the preceding argument, $N_1 \oplus N_2 \leq^{\text{ess}} M$. Thus $E(M) = E(N_1 \oplus N_2) = E(N_1) \oplus E(N_2)$. Since $N_1 = M_1$, the argument used in part (i) yields that $E(N_1) \cap M = E(M_1) \cap M = M_1 = N_1$. Consider

$$\bigcup_{n=1}^{\infty} \mathbb{Z}(1_{2^n} + 2_{2^{n+1}}) = \mathbb{Z}\,(1_2 + 2_4) \cup \mathbb{Z}\,(1_4 + 2_8) \cup \mathbb{Z}\,(1_8 + 2_{16}) \cup \cdots .$$

Then $\cup_{n=1}^{\infty}\mathbb{Z}(1_{2^n} + 2_{2^{n+1}}) \cong \mathbb{Z}_{2^\infty}$ as $\mathbb{Z}$-modules. So $\cup_{n=1}^{\infty}\mathbb{Z}(1_{2^n} + 2_{2^{n+1}})$ is an injective $\mathbb{Z}$-module. Further, we can check that $\cup_{n=1}^{\infty}\mathbb{Z}(1_{2^n} + 2_{2^{n+1}})$ is an essential extension of $N_2 = \mathbb{Z}(1_2 + 2_4)$ as $\mathbb{Z}$-modules. Thus $E(N_2) = \cup_{n=1}^{\infty}\mathbb{Z}(1_{2^n} + 2_{2^{n+1}})$.

We recall that

$$M = \{0_2 + 0_4,\, 0_2 + 1_4,\, 0_2 + 2_4,\, 0_2 + 3_4,\, 1_2 + 0_4,\, 1_2 + 1_4,\, 1_2 + 2_4,\, 1_2 + 3_4\}.$$

Therefore $E(N_2) \cap M = \mathbb{Z}\,(1_2 + 2_4) = N_2$.

Note that $E(M) = E(N_1) \oplus E(N_2)$, $E(N_1) \cap M = N_1$, and $E(N_2) \cap M = N_2$. Hence $(E(N_1) \cap M) \oplus (E(N_2) \cap M) = N_1 \oplus N_2 \neq M$.

(iii) Consequently, for $\{E(M_1),\, E(M_2)\}$ in part (i),

$$E(M) = E(M_1) \oplus E(M_2) \quad \text{implies} \quad M = (M \cap E(M_1)) \oplus (M \cap E(M_2)),$$

while, for $\{E(N_1),\, E(N_2)\}$ in part (ii),

$$E(M) = E(N_1) \oplus E(N_2) \quad \text{does not imply} \quad M = (M \cap E(N_1)) \oplus (M \cap E(N_2)).$$

Therefore, condition (iv) of Theorem 2.10 which is, namely

"for *any* $\{E_i \leq E(M) \mid i \in \Lambda\}$ with $E(M) = \oplus_{i \in \Lambda} E_i$ implies $M = \oplus_{i \in \Lambda}(M \cap E_i)$"

does not hold true for the case when

$$M = \{0_2 + 0_4,\, 0_2 + 1_4,\, 0_2 + 2_4,\, 0_2 + 3_4,\, 1_2 + 0_4,\, 1_2 + 1_4,\, 1_2 + 2_4,\, 1_2 + 3_4\},$$

which is isomorphic to $\mathbb{Z}_2 \oplus \mathbb{Z}_4$ as $\mathbb{Z}$-modules.

As an application of Theorem 2.10, it is shown that the quasi-continuous hull of any module always exists (see Theorem 2.12). From Theorem 2.10, a module M is quasi-continuous if and only if $fM \subseteq M$ for any $f^2 = f \in \mathrm{End}(E(M))$.

Let Ω be the subring of $H = \mathrm{End}(E(M))$ generated by all idempotents in H. Then from Theorem 2.10, ΩM is a quasi-continuous intermediate module between M and $E(M)$. Further, in the following, ΩM is the quasi-continuous hull of M, as shown by Goel and Jain [**4**].

Theorem 2.12. Assume that M is a right R-module and $H = \mathrm{End}(E(M))$. Let Ω be the subring of H generated by the set of all idempotents of H. Then ΩM is the quasi-continuous hull of M.

Related to Theorem 2.10, the following theorem is shown by Müller and Rizvi [**10**] for direct sums of (quasi-)continuous modules.

Theorem 2.13. Let $M = \oplus_{i \in \Lambda} M_i$, and assume that $\oplus_{i \in \Lambda} E(M_i)$ is an injective right R-module (e.g., Λ is finite or R is right noetherian). Then $M = \oplus_{i \in \Lambda} M_i$ is (quasi-)continuous if and only if each M_i is (quasi-)continuous and M_j-injective for all $j \neq i$.

3. A Counterexample and Continuous Hulls of Rings

We begin this section with an example of a uniform nonsingular cyclic module over a *noncommutative* ring which has no continuous hull (Example 3.1). In spite of Example 3.1, each uniform cyclic module over a *commutative* ring R has a continuous hull (Theorem 3.2). Further, every nonsingular cyclic module over a *commutative* ring has a continuous hull (Theorem 3.3).

Motivated by Example 3.1, Theorem 3.2, and Theorem 3.3, one may raise the following question (Question 3.4):

Does there exist the continuous hull of R_R for any ring R?

For this question, it is shown that for any integer n such that $n > 1$, if R is the $n \times n$ matrix ring over any given ring or the $n \times n$ upper triangular matrix ring over any given ring, then R_R has a continuous hull and such a continuous hull is explicitly described (Theorem 3.10 and Theorem 3.15).

Moreover, if R is an abelian regular ring, it is shown that every nonsingular cyclic right R-module has a continuous hull (Theorem 3.23). As an application of Theorem 3.23, when R is an abelian regular ring, we prove that $R\mathcal{B}(Q(R))_R$ is the continuous hull of R_R, where $R\mathcal{B}(Q(R))$ is the idempotent closure of R (Theorem 3.24). We give an example of a regular ring R for which $R\mathcal{B}(Q(R))_R$ is not continuous (Example 3.25). Thereby, Theorem 3.24 does not hold true for regular rings in general.

As a byproduct, it is shown that when R is an abelian regular ring, $R\mathcal{B}(Q(R))$ is the smallest continuous regular ring of quotients of R (Corollary 3.30). To illustrate Corollary 3.30, we give an example of an abelian regular ring R for which the smallest continuous regular ring of quotients $R\mathcal{B}(Q(R))$ of R is a proper intermediate ring between R and $Q(R)$ (Example 3.31). Further, we also give an example of a regular ring R for which the regular ring $R\mathcal{B}(Q(R))$ is neither right nor left continuous (Example 3.33).

When R is a ring and T is a regular right ring of quotients of R, by adopting a similar method to the argument used in Example 3.33, we show that T is right continuous if and only if T_R is continuous (Theorem 3.34). Thereby, when R is a regular ring, the idempotent closure $R\mathcal{B}(Q(R))$ is a right continuous ring if and only if $R\mathcal{B}(Q(R))_R$ is continuous (Corollary 3.35).

Examples and remarks which illustrate our results, and applications of our results are provided. Some necessary known results are introduced.

We recall that a nonzero module M is said to be *uniform* if every nonzero submodule of M is essential in M. Thus M is uniform if and only if $N \cap K \neq 0$ for any nonzero submodules N and K of M.

For a module M_R, let

$$Z(M_R) = \{m \in M \mid mI = 0 \text{ for some essential right ideal } I \text{ of } R\}.$$

The submodule $Z(M_R)$ is called the *singular submodule* of M_R. Similarly, for a left R-module ${}_RM$, $Z({}_RM)$ is defined. We say that a module M_R is *nonsingular* if $Z(M_R) = 0$.

From Definition 2.8, recall that the *continuous hull* of a module M is the smallest continuous extension of M in a fixed injective hull $E(M)$ of M. The following example, which is constructed by Müller and Rizvi [**9**], shows that there exists a uniform nonsingular cyclic module over a *noncommutative* ring which has no continuous hull.

Example 3.1. Assume that V is a vector space over a field F with basis elements $v_m, w_k\,(m, k = 0, 1, 2, \dots)$ in a basis of V. We denote by V_n the subspace generated by the $v_m\,(m \geq n)$ and all the w_k. Also we denote by W_n the subspace generated by the $w_k\,(k \geq n)$. We write $\mathcal{S}$ for the shifting operator such that $\mathcal{S}(w_k) = w_{k+1}$ and $\mathcal{S}(v_i) = 0$ for all k, i.

Let R be the set of all $\rho \in \operatorname{End}_F(V)$ such that $\rho(v_m) \in V_m$, $\rho(w_0) \in W_0$ and $\rho(w_k) = \mathcal{S}^k\rho(w_0)$, for $m, k = 0, 1, 2, \dots$. We note that

$$\tau\rho(w_k) = \mathcal{S}^k\tau\rho(w_0), \quad \text{for} \quad \rho, \tau \in R, \quad \text{and so } \tau\rho \in R.$$

Thus it is routine to check that R is a subring of $\operatorname{End}_F(V)$. Further, we see that $V_n = Rv_n$, $W_n = Rw_n$, and $V_{n+1} \subseteq V_n$ for all n. (When $f \in R$ and $v \in V$, we also use fv for the image $f(v)$ of v under f.)

Consider the left R-module $M = W_0$. First, we show that $M = Rw_0$ is uniform. For this, take $fw_0 \neq 0, gw_0 \neq 0$ in M, where $f, g \in R$. We need to find $h_1, h_2 \in R$ such that $h_1fw_0 = h_2gw_0 \neq 0$. Let

$$fw_0 = b_0w_0 + b_1w_1 + \cdots + b_mw_m \in Rw_0$$

and

$$gw_0 = c_0w_0 + c_1w_1 + \cdots + c_mw_m \in Rw_0,$$

where $b_i, c_j \in F$, $i, j = 0, 1, \dots, m$, and some terms of b_i and c_j may be zero. Put

$$h_1w_0 = x_0w_0 + x_1w_1 + \cdots + x_\ell w_\ell$$

and

$$h_2w_0 = y_0w_0 + y_1w_1 + \cdots + y_\ell w_\ell,$$

where $x_i, y_i \in F$, $i = 0, 1, \dots, \ell$ (also some terms of x_i and y_j may be zero). Since $h_1(w_k) = \mathcal{S}^kh_1(w_0)$ and $h_2(w_k) = \mathcal{S}^kh_2(w_0)$ for $k = 0, 1, 2 \dots$, we need to find such

$x_i, y_i \in F$, $0 \le i \le \ell$ so that $h_1 f w_0 = h_2 g w_0 \neq 0$ from the following equations:

$$b_0x_0 = c_0y_0,\ b_0x_1 + b_1x_0 = c_0y_1 + c_1y_0,$$
$$b_0x_2 + b_1x_1 + b_2x_0 = c_0y_2 + c_1y_1 + c_2y_0,$$
$$b_0x_3 + b_1x_2 + b_2x_1 + b_3x_0 = c_0y_3 + c_1y_2 + c_2y_1 + c_3y_0,$$

and so on.

Now say $\alpha(t) = b_0 + \cdots + b_mt^m \neq 0$ and $\beta(t) = c_0 + \cdots + c_mt^m \neq 0$ in the polynomial ring $F[t]$. Then $\alpha(t)F[t] \cap \beta(t)F[t] \neq 0$.

We may note that finding such $x_0, x_1 \ldots, x_\ell, y_0, y_1 \ldots, y_\ell$ in F above is the same job for finding $x_0, x_1 \ldots, x_\ell, y_0, y_1 \ldots, y_\ell$ such that

$$\alpha(t)(x_0 + x_1t + \cdots + x_\ell t^\ell) = \beta(t)(y_0 + y_1t + \cdots + y_\ell t^\ell) \neq 0$$

in the polynomial ring $F[t]$.

Observing that $0 \neq \alpha(t)\beta(t) \in \alpha(t)F[t] \cap \beta(t)F[t]$, take

$$h_1w_0 = c_0w_0 + c_1w_1 + \cdots + c_mw_m$$

by putting $\ell = m, x_i = c_i$ for $0 \le i \le m$, and

$$h_2w_0 = b_0w_0 + b_1w_1 + \cdots + b_mw_m$$

by putting $\ell = m, y_i = b_i$ for $0 \le i \le m$. Because $\alpha(t)\beta(t) \neq 0$, we have that $0 \neq h_1fw_0 = h_2gw_0 \in Rfw_0 \cap Rgw_0$. So M is uniform.

Next, we show that each V_n is an essential extension of M (hence each V_i is uniform). Indeed, let $0 \neq \mu v_n \in Rv_n = V_n$, where $\mu \in R$. Say

$$\mu v_n = a_{n+k}v_{n+k} + \cdots + a_{n+k+\ell}v_{n+k+\ell} + b_sw_s + \cdots + b_{s+m}w_{k+m}.$$

If $a_{n+k} = \cdots = a_{n+k+\ell} = 0$, then $\mu v_n \in W_0$. Otherwise, we may assume that $a_{n+k} \neq 0$. Let $\omega \in R$ such that $\omega(v_{n+k}) = w_0$ and $\omega(v_i) = 0$ for $i \neq n+k$ and $\omega(w_j) = 0$ for all j. Then $0 \neq \omega\mu v_n = a_{n+k}w_0 \in W_0$. Thus $M = W_0$ is essential in V_n. Since M is uniform, V_n is also uniform for all n.

We prove that ${}_RM$ is nonsingular. For this, assume that $u \in Z({}_RM)$ and let $K = \{\alpha \in R \mid \alpha u = 0\}$. Then K is an essential left ideal of R. So $K \cap R\mathcal{S}^2 \neq 0$. Thus there exists $\rho \in R$ such that $\rho\mathcal{S}^2 \neq 0$ and $\rho\mathcal{S}^2(u) = 0$. Say $u = a_kw_k + a_{k+1}w_{k+1} + \cdots + a_nw_n$ with $a_k, a_{k+1}, \ldots, a_n \in F$. Assume on the contrary that $u \neq 0$. Then we may suppose that $a_k \neq 0$. Because $\rho(w_n) = \mathcal{S}^n\rho(w_0)$ for $n = 0, 1, 2, \ldots$,

$$\begin{aligned} 0 &= a_k\rho\mathcal{S}^2(w_k) + a_{k+1}\rho\mathcal{S}^2(w_{k+1}) + \cdots + a_n\rho\mathcal{S}^2(w_n) \\ &= a_k\mathcal{S}^{k+2}\rho(w_0) + a_{k+1}\mathcal{S}^{k+3}\rho(w_0) + \cdots + a_n\mathcal{S}^{n+2}\rho(w_0). \end{aligned}$$

Here we put $\rho(w_0) = b_\ell w_\ell + b_{\ell+1}w_{\ell+1} + \cdots + b_tw_t$. If $\rho(w_0) = 0$, then we see that

$$\rho\mathcal{S}^2(w_0) = \rho(w_2) = \mathcal{S}^2\rho(w_0) = 0.$$

Also, $\rho\mathcal{S}^2(w_m) = 0$ for all $m = 1, 2, \ldots$, and $\rho\mathcal{S}^2(v_i) = 0$ for all $i = 0, 1, \ldots$. Thus $\rho\mathcal{S}^2 = 0$, a contradiction. Hence $\rho(w_0) \neq 0$, and so we may assume that $b_\ell \neq 0$. Note that

$$\mathcal{S}^{k+2}\rho(w_0) = b_\ell w_{\ell+k+2} + b_{\ell+1}w_{\ell+k+3} + \cdots + b_tw_{t+k+2},$$
$$\mathcal{S}^{k+3}\rho(w_0) = b_\ell w_{\ell+k+3} + b_{\ell+1}w_{\ell+k+4} + \cdots + b_tw_{t+k+3},$$

and so on. Thus $0 = \rho\mathcal{S}^2(u) = a_kb_\ell w_{\ell+k+2} + (a_kb_{\ell+1} + a_{k+1}b_\ell)w_{\ell+k+3} + \cdots$, and hence $a_kb_\ell = 0$, which is a contradiction because $a_k \neq 0$ and $b_\ell \neq 0$. Therefore $u = 0$, and so M is nonsingular.

We show now that V_n is continuous. Note that V_n is uniform. So clearly, V_n has (C_1) condition. Thus, to show that V_n is continuous, it suffices to prove that every R-monomorphism of V_n is onto for V_n to satisfy (C_2) condition. Let $\varphi : V_n \to V_n$ be an R-monomorphism.

Now we put

$$\varphi(v_n) = \rho v_n \in Rv_n = V_n, \text{ where } \rho \in R.$$

We claim that $\rho v_n \notin V_{n+1}$. For this, assume on the contrary that $\rho v_n \in V_{n+1}$. Let $\lambda \in R$ such that $\lambda v_n = v_n$, $\lambda v_k = 0$ for $k \neq n$, and $\lambda w_m = 0$ for all m. Then $\varphi(\lambda v_n) = \lambda(\rho v_n) = 0$ since $\rho(v_n) \in V_{n+1}$. But $\lambda v_n = v_n \neq 0$. Thus φ is not one-to-one, a contradiction. Therefore $\rho v_n \notin V_{n+1}$.

As $\rho v_n \in V_n$, write

$$\rho v_n = a_n v_n + a_{n+1} v_{n+1} + \cdots + a_{n+\ell} v_{n+\ell} + b_0 w_0 + \cdots + b_h w_h,$$

where $a_n, a_{n+1}, \ldots, a_{n+\ell}, b_0, b_1, \ldots, b_h \in F$, and $a_n \neq 0$. Take $\nu \in R$ such that $\nu v_n = a_n^{-1} v_n$, $\nu v_k = 0$ for $k \neq n$ and $\nu w_m = 0$ for all m. Then we see that $v_n = \nu \rho v_n \in R\rho v_n$. So $Rv_n \subseteq R\rho v_n$, hence $V_n = Rv_n = R\rho v_n$. Thus $\varphi(Rv_n) = R\varphi(v_n) = R\rho v_n = V_n$, so φ is onto. Therefore each V_n is continuous.

Finally, note that the uniform nonsingular module $M = Rw_0$ is not continuous, since the shifting operator $\mathcal{S}$ provides an R-monomorphism which is not onto. Hence, M has no continuous hull (in $E(M) = E(V)$), because such a hull would have to be contained in each V_n, and hence in $M = \bigcap_n V_n$, a contradiction.

Assume that A is a commutative ring. An element $s \in A$ is called a non-zero-divisor if $bs = 0$ with $b \in A$ implies $b = 0$. Let

$$Q_{c\ell}(A) = \{as^{-1} \mid a,\, s \in A \text{ and } s \text{ is a non-zero-divisor in } A\},$$

where addition and multiplication are defined as follows:

$$a_1 s_1^{-1} + a_2 s_2^{-1} = (a_1 s_2 + a_2 s_1)(s_1 s_2)^{-1}$$

and

$$(a_1 s_1)^{-1})(a_2 s_2^{-1}) = (a_1 a_2)(s_1 s_2)^{-1}$$

for $a_1 s_1^{-1}$, $a_2 s_2^{-1} \in Q_{c\ell}(A)$. Then $Q_{c\ell}(A)$ is an overring of A, which is called the *classical ring of quotients* (or the *total ring of fractions*) of A.

In contrast to Example 3.1, the following theorem is shown by Rizvi [**14**].

Theorem 3.2. Let R be a commutative ring. Then every uniform cyclic module has a continuous hull.

Further, we have the following which is due to Müller and Rizvi [**9**].

Theorem 3.3. Let R be a commutative ring. Then every nonsingular cyclic module has a continuous hull.

Motivated by Example 3.1, Theorem 3.2, and Theorem 3.3, we raise the following interesting question. According to our best knowledge, the question is still open.

Question 3.4. *Let R be a ring. Then does there exist the continuous hull of R_R?*

For Question 3.4, we study rings R for which R_R has a continuous hull in Theorem 3.10, Theorem 3.15, and Theorem 3.24. For this, we need some preparations.

Recall that a ring R is called *regular* (in the sense of von Neumann) if for each $a \in R$ there exists $b \in R$ such that $a = aba$. A ring R is said to be *right nonsingular* if $Z(R_R) = 0$. Similarly, a ring R is called *left nonsingular* if $Z({}_RR) = 0$. We note that any regular ring is right and left nonsingular.

An extension module B_R of a module A_R is called a *dense* (or *rational*) extension if for $b_1, b_2 \in B$ with $b_2 \neq 0$, there exists $r \in R$ such that $b_1 r \in B$ and $b_2 r \neq 0$. In this case, we write $A \leq^{\text{den}} B$.

We note that every dense submodule of a module M is essential in M. But the converse does not hold in general. For example, take $M = \mathbb{Z}/p^{n+1}\mathbb{Z}$ and $N = p\mathbb{Z}/p^{n+1}\mathbb{Z}$ as $\mathbb{Z}$-modules, where p is a prime integer and n is an integer such that $n > 1$. Then N is an essential submodule of M, but N is not dense in M.

Each module has a maximal dense extension. For a module M, we use $\widetilde{E}(M)$ to denote the maximal dense extension of M. For a ring R, $\widetilde{E}(R_R)$ has an overring structure of R. We let $Q(R) = \widetilde{E}(R_R)$, which is called the *maximal right ring of quotients* of R. Therefore $R_R \leq^{\text{den}} Q(R)_R$.

Similarly, the *maximal left ring of quotients* of R is defined. We use $Q^{\ell}(R)$ to denote the maximal left ring of quotients of R. For more details for dense extensions and the maximal right ring of quotients, see [**2**, Section 1.3].

The following result is due to Utumi [**17**].

Theorem 3.5. (i) Assume that R is a ring and n is a positive integer. Then $Q(\text{Mat}_n(R)) = \text{Mat}_n(Q(R))$.

(ii) Let $\{R_i \mid i \in \Lambda\}$ be a set of rings. Then $Q(\prod_{i\in\Lambda} R_i) = \prod_{i\in\Lambda} Q(R_i)$.

Theorem 3.6 and Theorem 3.7 are known results, which will be used for our results and examples.

Theorem 3.6. Let R be a ring. Then R is right nonsingular if and only if $Q(R)$ is a right self-injective regular ring.

Theorem 3.7. Assume that R is a right nonsingular ring. Then $R_R \leq^{\text{den}} E(R_R)$, and so $Q(R) = \widetilde{E}(R_R) = E(R_R)$.

An overring T of a ring R is called a *right ring of quotients* of R if $R_R \leq^{\text{den}} T_R$. Thus an overring T of a ring R is a right ring of quotients of R if and only if T is an intermediate ring between R and $Q(R)$. Similarly, a *left ring of quotients* of a ring is defined.

The following will be used for the study of the endomorphism ring $\text{End}(T_R)$ of T_R, when T is a right ring of quotients of a ring R. For the proof, see [**2**, Proposition 2.1.32].

Proposition 3.8. Let R be a ring and T a right ring of quotients of R. Then $\text{End}(T_R) = \text{End}(T_T) \cong T$.

The following result provides information on a right ring of quotients which is right self-injective. See [**2**, Theorem 7.1.3] for the proof.

Theorem 3.9. Let T be a right ring of quotients of a ring R. Then T is right self-injective if and only if $T = E(R_R)$.

By the preceding preparation, now we consider Question 3.4. Let A be any ring. For any integer n such that $n > 1$, let R be $\mathrm{Mat}_n(A)$, the $n \times n$ matrix ring over A or let R be $T_n(A)$, the $n \times n$ upper triangular matrix ring over A. Then we show that R_R has a continuous hull in Theorem 3.10 and Theorem 3.15.

Theorem 3.10. Let R be the $n \times n$ matrix ring over any given ring, where n is any given integer such that $n > 1$. Then we have the following.

(i) R_R has a continuous hull.

(ii) The continuous hull of R_R coincides with the quasi-continuous and the quasi-injective hull of R_R.

(iii) Addtionally, if R is right nonsingular, then the continuous hull of R_R is $Q(R)_R$, the injective hull of R_R.

Proof. (i) Put $R = \mathrm{Mat}_n(A)$ for any given ring A and any given integer n such that $n > 1$. Let $e_i \in R$ be the matrix with 1 in the (i,i)-position and 0 elsewhere, where $i = 1, 2, \ldots, n$. Then $e_iR_R \cong e_jR_R$, and hence $E(e_iR_R) \cong E(e_jR_R)$ for all i, $j = 1, 2, \ldots, n$. Note that

$$R_R = e_1R_R \oplus e_2R_R \oplus \cdots \oplus e_nR_R.$$

We show that R_R has a continuous hull. For this, first note that R_R has a quasi-continuous hull from Theorem 2.12. Say V is the quasi-continuous hull of R_R. Now $E(R_R) = E(e_1R) \oplus \cdots \oplus E(e_nR)$. Since $E(V) = E(R_R)$ and V is quasi-continuous, Theorem 2.10 yields that

$$V = (V \cap E(e_1R)) \oplus \cdots \oplus (V \cap E(e_nR)).$$

From Theorem 2.13, each $V \cap E(e_iR)$ is quasi-continuous and $V \cap E(e_jR)$-injective for all j, where $j \neq i$. As $e_iR = V \cap e_iR \leq^{\mathrm{ess}} V \cap E(e_iR)$, we have that $E(e_iR) = E(V \cap E(e_iR))$ for each i. Note that $e_iR \cong e_jR$ for all i, j. Hence

$$E(V \cap E(e_iR)) = E(e_iR) \cong E(e_jR) = E(V \cap E(e_jR))$$

for all i, j. Thus Theorem 2.4 yields that $V \cap E(e_iR) \cong V \cap E(e_jR)$ for all i, j, where $i \neq j$ because $V \cap E(e_iR)$ and $V \cap E(e_jR)$ are relatively injective for all i, j, where $i \neq j$. Consequently, $V \cap E(e_iR) \cong V \cap E(e_jR)$ for all i, j.

Put $B_R = V \cap E(e_1R)$. Then $V_R \cong B_R^{(n)}$. Further, let $C_R = V \cap E(e_2R)$. Then B_R is C_R-injective. Since $B_R \cong C_R$, B_R is B_R-injective. So $B_R^{(n)}$ is quasi-injective by Proposition 2.5 or Corollary 2.6. Hence V_R is continuous.

Now let W be a continuous module such that $R_R \leq W \leq E(R)$. Then W is quasi-continuous. Since V is the quasi-continuous hull of R_R, $V \leq W$. Consequently V_R is the continuous hull of R_R.

(ii) Let V be the quasi-continuous hull of R_R. Then V is quasi-injective by the argument as in (i). Let Y be a quasi-injective module such that $R_R \leq Y \leq E(R)$. Since Y is quasi-injective and V is the quasi-continuous hull of R_R, $V \leq Y$. Hence V is the quasi-injective hull of R_R. Further, note that V is the continuous hull of R_R from (i).

(iii) Additionally, assume that R is right nonsingular. Then Theorem 3.7 yields that $E(R_R) = Q(R)$. On the other hand, $\mathrm{End}(Q(R)_R) = \mathrm{End}(Q(R)_{Q(R)}) \cong Q(R)$

by Proposition 3.8. Hence

$$\mathrm{End}(E(R_R))R_R = \mathrm{End}(Q(R)_R)R_R = Q(R)R_R = Q(R)_R$$

is the quasi-injective hull of R_R by Theorem 2.9. From part (ii), $Q(R)_R$ is the continuous hull, the quasi-continuous hull, and the (quasi-)injective hull of R_R. □

The following example illustrates Theorem 3.10.

Example 3.11. Assume that F is a field which has a proper subfield K. Let

$$A = \left\{ (x_k)_{k=1}^{\infty} \in \prod_{k=1}^{\infty} F \mid x_k \in K \text{ eventually} \right\},$$

which is a subring of $\prod_{k=1}^{\infty} F$. Let n be a positive integer. Put $R = \mathrm{Mat}_n(A)$. We show that the continuous hull of R_R exists as follows.

(i) When $n = 1$. Note that $R = A$. In this case, we claim that $Q(R)_R$ is the (quasi-)injective hull of R_R. Further, R_R itself is the (quasi-)continuous hull of R_R.

First, we show that $E(R_R) = Q(R) = \prod_{k=1}^{\infty} F$. For this, put $V = \prod_{k=1}^{\infty} F$. As R is a regular ring, R is right nonsingular. Hence Theorem 3.7 yields that $R_R \leq^{\mathrm{den}} E(R_R)$. Note that $R_R \leq^{\mathrm{ess}} V_R$ and so $R_R \leq^{\mathrm{den}} V_R$. Therefore V is a right ring of quotients of R.

From Theorem 3.5, $Q(V) = \prod_{k=1}^{\infty} Q(F) = \prod_{k=1}^{\infty} F = V$. Since the ring V is regular, V is a right nonsingular ring. Therefore $Q(V)\,(= V)$ is right self-injective from Theorem 3.6. Now as V is a right ring of quotients of R and further V is right self-injective, Theorem 3.9 yields that $V = E(R_R)$. Again since R is right nonsingular, $Q(R) = E(R_R)$ by Theorem 3.7. Hence $V = Q(R)$ and so

$$E(R_R) = Q(R) = \prod_{k=1}^{\infty} F.$$

Note that the set of all idempotents of $Q(R)$ and the set of all idempotents of R coincide. By Proposition 3.8

$$\mathrm{End}(E(R_R)_R) = \mathrm{End}(Q(R)_R) = \mathrm{End}(Q(R)_{Q(R)}) \cong Q(R).$$

Thus from Theorem 2.12, R_R is quasi-continuous. Since R is regular, R_R is continuous. Thus R_R itself is the continuous hull of R_R.

By Theorem 2.9, $Q(R)_R$ is the quasi-injective hull of R_R. As $E(R_R) = Q(R)$, $Q(R)_R$ is the injective hull of R_R. However, $Q(R)_R$ is not the continuous hull of R (note that $Q(R)_R \neq R_R$ and R_R itself is the continuous hull of R_R by the preceding argument). Further, as R_R is quasi-continuous from the preceding argument, R_R itself is the quasi-continuous hull of R_R.

(ii) When $n > 1$. In this case, the (quasi-)continuous hull and the (quasi-)injective hull of R_R coincide. Indeed, R_R has a continuous hull by Theorem 3.10. Furthermore, since R is right nonsingular, $Q(R)_R$ is the (quasi-)continuous hull, and the (quasi-)injective hull of R_R by Theorem 3.10.

A ring R (not necessarily with identity) is said to satisfy a polynomial identity (or simply PI), if there is a polynomial $f(x_1, x_2, \ldots, x_n)$ with integer coefficients and not necessarily commuting indeterminates $x_1, x_2, \ldots, x_n$ such that the coefficient of one of the monomials in $f(x_1, x_2, \ldots, x_n)$ of maximal degree is 1 and $f(a_1, a_2, \ldots, a_n) = 0$ for all $a_1, a_2, \ldots, a_n$ in R.

Recall that a module M is called *extending* if M satisfies (C_1) condition, i.e., every submodule of M is essential in a direct summand of M. A ring R is called *right extending* if R_R is extending. Similarly, a ring R is called *left extending* if ${}_RR$ is extending.

The following result appears in [**2**, Corollary 2.12]. For details on this result, see [**2**, Section 3.3].

Proposition 3.12. Let R be a semiprime PI-ring. Then R is right extending if and only if R is left extending.

A ring R is said to be *right continuous* if R_R is continuous. Similarly a ring R is called *left continuous* if ${}_RR$ is continuous.

In [**5**, Example 13.8], it is shown that the ring $R = \mathrm{Mat}_2(A)$ of Example 3.13 is a regular ring which is neither right nor left continuous. However, as an application of our results, we exhibit Example 3.13 in a way different from that of [**5**, Example 13.8] as follows.

Example 3.13. Let A be the ring as in Example 3.11 and let $R = \mathrm{Mat}_n(A)$, where n is an integer such that $n > 1$. By Theorem 3.10, $Q(R)_R$ is the continuous hull of R_R since R is right nonsingular. Now

$$Q(R) = Q(\mathrm{Mat}_n(A)) = \mathrm{Mat}_n(Q(A))$$

by Theorem 3.5. Since $Q(A) = \prod_{k=1}^{\infty} F$ (see Example 3.11(i)), $Q(A) \neq A$. Therefore we have that

$$R = \mathrm{Mat}_n(A) \neq \mathrm{Mat}_n(Q(A)) = Q(R).$$

So R_R is not continuous as $Q(R)_R$ is the continuous hull of R_R (Example 3.11(ii)) and $R \neq Q(R)$. Because R is a regular ring and R is not right continuous, R is not right extending.

Note that R is a PI-ring as A is commutative. Hence by Proposition 3.12, R is not left extending as R is a semiprime ring. Therefore R is a regular ring which is neither right nor left continuous.

Remark 3.14. In Example 3.13, the ring A is right continuous (see Example 3.11(i)), but the ring $\mathrm{Mat}_n(A)$ is not right extending for any given integer n such that $n > 1$. Therefore we have the following as a byproduct:

(1) The property for a ring being right continuous is not Morita invariant;
(2) The property for a ring being right quasi-continuous is not Morita invariant;
(3) The property for a ring being right extending is not Morita invariant.

Note that a ring is right self-injective if and only if it is right quasi-injective by Baer's Criterion. Since the property for a ring being right self-injective is a Morita invariant property, the property for a ring being right quasi-injective is also a Morita invariant property.

Related to Question 3.4, the following result shows the existence of a continuous hull of the $n \times n$ upper triangular matrix ring over any ring, where n is any integer such that $n > 1$.

Theorem 3.15. Let A be any given ring and n be any given integer such that $n > 1$. We put $R = T_n(A)$, the $n \times n$ upper triangular matrix ring over A. Then we have the following.

(i) R_R has a continuous hull.

(ii) The continuous hull of R_R coincides with the quasi-continuous and the quasi-injective hull of R_R.

(iii) If A is right nonsingular, then the continuous hull of R_R is $\mathrm{Mat}_n(Q(A))_R$, which is the injective hull of R_R.

Proof. (i) and (ii) Let $e_i \in R$ be the matrix with 1 in the (i,i)-position and 0 elsewhere, for $i = 1, 2, \ldots, n$. It can be checked that $e_iR_R \leq^{\text{ess}} e_i\mathrm{Mat}_n(A)_R$ for all i. Hence $E(e_iR_R) = E(e_i\mathrm{Mat}_n(A)_R)$ for all i.

We note that $e_i\mathrm{Mat}_n(A)_R \cong e_j\mathrm{Mat}_n(A)_R$ for all i, j. Hence

$$E(e_i\mathrm{Mat}_n(A)_R) \cong E(e_j\mathrm{Mat}_n(A)_R)$$

for all i, j. Therefore

$$E(e_iR_R) = E(e_i\mathrm{Mat}_n(A)_R) \cong E(e_j\mathrm{Mat}_n(A)_R) = E(e_jR_R)$$

for all i, j. Because $R_R = e_1R_R \oplus e_2R_R \oplus \cdots \oplus e_nR_R$, the argument used in the proof of Theorem 3.10 yields that R_R has a continuous hull, and furthermore the continuous hull, the quasi-continuous hull, and the quasi-injective hull of R_R coincide.

(iii) Additionally, we assume that A is right nonsingular. We show that R is right nonsingular. For this, let $s \in Z(R_R)$, the singular submodule of R_R. Then there exists $I_R \leq^{\text{ess}} R_R$ such that $sI = 0$.

We now first show that $I\mathrm{Mat}_n(A)_{\mathrm{Mat}_n(A)} \leq^{\text{ess}} \mathrm{Mat}_n(A)_{\mathrm{Mat}_n(A)}$. For this, take $0 \neq t \in \mathrm{Mat}_n(A)$. Because $e_iR_R \leq^{\text{ess}} e_i\mathrm{Mat}_n(A)_R$ for all i from the proof of parts (i) and (ii),

$$\begin{aligned} R_R &= e_1R_R \oplus e_2R_R \oplus \cdots \oplus e_nR_R \\ &\leq^{\text{ess}} e_1\mathrm{Mat}_n(A)_R \oplus e_2\mathrm{Mat}_n(A)_R \oplus \cdots \oplus e_n\mathrm{Mat}_n(A)_R \\ &= \mathrm{Mat}_n(A)_R. \end{aligned}$$

Hence there exists $w \in R$ such that $0 \neq tw \in R$. Because $I_R \leq^{\text{ess}} R_R$, there exists $w_1 \in R$ such that $0 \neq tww_1 \in I \subseteq I\mathrm{Mat}_n(A)$. Thus

$$I\mathrm{Mat}_n(A)_{\mathrm{Mat}_n(A)} \leq^{\text{ess}} \mathrm{Mat}_n(A)_{\mathrm{Mat}_n(A)}.$$

Now since $sI = 0$, $sI\mathrm{Mat}_n(A) = 0$. Because $\mathrm{Mat}_n(A)$ is right nonsingular, $s = 0$ and thus R is right nonsingular.

From Theorem 3.7, $E(R_R) = Q(R)$. So by Proposition 3.8,

$$\mathrm{End}(E(R_R)_R) = \mathrm{End}(Q(R)_R) = \mathrm{End}(Q(R)_{Q(R)}) \cong Q(R).$$

Therefore $Q(R)R_R = Q(R)_R$ is the quasi-injective hull of R_R from Theorem 2.9. From parts (i) and (ii), $Q(R)_R$ is the continuous hull and the quasi-continuous hull of R_R. Further, since $E(R_R) = Q(R)$, $Q(R)_R$ is the injective hull of R_R.

We describe $Q(R)$ explicitly. First, we show that $R_R \leq^{\text{ess}} \mathrm{Mat}_n(Q(A))_R$. For this, take a nonzero matrix $[q_{ij}] \in \mathrm{Mat}_n(Q(A))$. Say $q_{11} \neq 0$. Since $A_A \leq^{\text{ess}} Q(A)_A$, there exists $a_1 \in A$ such that $0 \neq q_{11}a_1 \in A$. If all $q_{ij}a_1 = 0$ for $i \neq 1$ and $j \neq 1$, then we are done. Otherwise, say $q_{12}a_1 \neq 0$. Then there exists $a_2 \in A$ such that $0 \neq q_{12}a_1a_2 \in R$. If $q_{ij}a_1a_2 = 0$ for all $i \neq 1$ and $j \neq 2$, then we are done.

Otherwise, say $q_{13}a_1a_2 \neq 0$, then there exists $a_3 \in A$ such that $0 \neq q_{13}a_1a_2a_3 \in A$, and so on.

Continuing this procedure, there exists $a \in A$ such that $q_{ij}a \in A$ for all i, j, and $q_{k\ell}a \neq 0$ for some k and ℓ.

Let $\alpha \in R$ be the diagonal matrix with a on the diagonal. Then $0 \neq [q_{ij}a] = [q_{ij}]\alpha \in \mathrm{Mat}_n(A)$. So $\mathrm{Mat}_n(A)_R \leq^{\mathrm{ess}} \mathrm{Mat}(Q(A))_R$. As $R_R \leq^{\mathrm{ess}} \mathrm{Mat}_n(A)_R$ from the preceding argument, we have that $R_R \leq^{\mathrm{ess}} \mathrm{Mat}_n(Q(A))_R$.

We note that R_R is nonsingular by the preceding argument, so $\mathrm{Mat}_n(Q(R))_R$ is nonsingular because $R_R \leq^{\mathrm{ess}} \mathrm{Mat}_n(Q(A))_R$. Hence $R_R \leq^{\mathrm{den}} \mathrm{Mat}_n(Q(A))_R$ from Theorem 3.7. Therefore $\mathrm{Mat}_n(Q(A))$ is a right ring of quotients of R.

On the other hand, $\mathrm{Mat}_n(Q(A)) = Q(\mathrm{Mat}_n(A))$ from Theorem 3.5. Since A is right nonsingular, so is $\mathrm{Mat}_n(A)$. Hence $Q(\mathrm{Mat}_n(A))$ is right self-injective by Theorem 3.6. So $\mathrm{Mat}_n(Q(A))$ is right self-injective. Thus Theorem 3.9 yields that $\mathrm{Mat}_n(Q(A)) = E(R_R)$. As $E(R_R) = Q(R)$, $Q(R) = \mathrm{Mat}_n(Q(A))$. □

We have the following remark from Example 3.11 and Theorem 3.15.

Remark 3.16. (i) From the arguments used for the proof of Theorem 3.15(iii), we note that if a ring A is right nonsingular and n is a positive integer, then $T_n(A)$ is right nonsingular and $Q(T_n(A)) = \mathrm{Mat}_n(Q(A))$. Hence from Theorem 3.5, we have that $Q(T_n(A)) = \mathrm{Mat}_n(Q(A)) = Q(\mathrm{Mat}_n(A))$.

(ii) Let A be the ring as in Example 3.11 and n be an integer such that $n > 1$. Put $R = T_n(A)$. Note that $Q(A) = \prod_{k=1}^{\infty} F$ (see Example 3.11). As A is right nonsingular, $Q(R) = Q(T_n(A)) = \mathrm{Mat}_n(Q(A)) = Q(\mathrm{Mat}_n(A))$ from part (i). Hence Theorem 3.15 yields that $\mathrm{Mat}_n(Q(A))_R = \mathrm{Mat}_n(\prod_{k=1}^{\infty} F)_R$ is the continuous hull of R_R.

(iii) Theorem 3.15(iii) does not hold true for the case when $n = 1$. Let A be the ring as in Example 3.11. From Example 3.11(i), A_A itself is the (quasi-)continuous hull of A_A. Also recall from Example 3.11(i), $Q(A) \neq A$ because $Q(A) = \prod_{k=1}^{\infty} F$. Hence $Q(A)_A$ is not the (quasi-)continuous hull of A_A.

We use $\mathcal{B}(-)$ to denote the set of all central idempotents of a ring.

Definition 3.17. Let R be a ring and $R\mathcal{B}(Q(R))$ be the subring of $Q(R)$ generated by R and $\mathcal{B}(Q(R))$. The ring $R\mathcal{B}(Q(R))$ is called the *idempotent closure* of R (see [**1**] for more details).

The following is due to Oshiro [**11**]. It is used for the study of the idempotent closure of a ring (see also [**2**, Lemma 8.3.38]).

Proposition 3.18. Let T be an overring of a ring R and $\{f_1, \ldots, f_n\} \subseteq \mathcal{B}(T)$. Then there exists a set of orthogonal idempotents $\{e_1, \ldots, e_m\} \subseteq \mathcal{B}(T)$ such that $\sum_{i=1}^{n} f_iR \subseteq \sum_{j=1}^{m} e_iR$.

Proposition 3.18 is useful for describing elements of $R\mathcal{B}(Q(R))$ in a convenient form. Let R be a ring and let $v \in R\mathcal{B}(Q(R))$. Then $v = r_1f_1 + r_2f_2 + \cdots + r_kf_k$ for some $r_i \in R$ and $f_i \in \mathcal{B}(Q(R))$, $i = 1, 2, \ldots, k$. By Proposition 3.18, there exist $a_1, a_2, \ldots, a_n \in R$ and orthogonal idempotents $e_1, e_2, \ldots, e_n$ in $\mathcal{B}(Q(R))$ such that $v = a_1e_1 + a_2e_2 + \cdots + a_ne_n$.

The next result follows from [**2**, Corollary 8.3.38]. But, we give its proof for the convenience of the reader.

Proposition 3.19. Let R be a regular ring. Then so is $R\mathcal{B}(Q(R))$.

Proof. Take $q \in R\mathcal{B}(Q(R))$. By Proposition 3.18, $q = a_1e_1 + a_2e_2 + \cdots + a_ne_n$, where $a_i \in R$, $e_i \in \mathcal{B}(Q(R))$ for some n, and $e_1, e_2, \ldots, e_n$ are orthogonal. Since R is regular, there exists $b_i \in R$ such that $a_i = a_ib_ia_i$ for each i, $1 \leq i \leq n$.

We put $v = b_1e_1 + \cdots + b_ne_n \in R\mathcal{B}(Q(R))$. Then $q = qvq$. Hence $R\mathcal{B}(Q(R))$ is a regular ring. $\square$

Recall that a ring R is said to be *reduced* if R has no nonzero nilpotent elements. A ring R is called *abelian* if every idempotent of R is central. Any reduced ring is abelian, but not conversely because there is a commutative ring (hence an abelian ring) which is not reduced.

See [**5**, Corollary 7.4] for the next proposition (note that $R\mathcal{B}(Q(R))$ is a regular right ring of quotients of R when R is a regular ring from Proposition 3.19).

Proposition 3.20. Let R be an abelian regular ring and T a regular right ring of quotients of R. Then T is abelian, in particular $R\mathcal{B}(Q(R))$ and $Q(R)$ are abelian.

The following lemma is [**5**, Theorem 3.2].

Lemma 3.21. Let R be a regular ring. Then the following are equivalent.

(i) R is abelian.

(ii) R/P is a division ring for every prime ideal P of R.

(iii) R is reduced.

(iv) All right (left) ideals of R are ideals.

(v) Every nonzero right (left) ideal of R contains a nonzero central idempotent.

Lemma 3.22. Let R be a semiprime ring. Assume that I and J are ideals of R. Then $IJ = 0$ if and only if $JI = 0$.

Proof. Assume that $IJ = 0$. Then $(JI)^2 = 0$. Since R is semiprime, $JI = 0$. Conversely if $JI = 0$, then similarly $IJ = 0$. $\square$

For Question 3.4, we have the following which shows that $R\mathcal{B}(Q(R))_R$ is the continuous hull of R_R when R is an abelian regular ring.

We use $\mathbf{I}(-)$ to denote the set of all idempotents of a ring. It is of interest to compare the following theorem with Theorem 3.3.

Theorem 3.23. Let R be an abelian regular ring. Then every nonsingular cyclic right R-module has a continuous hull.

Proof. Let $M = xR$ be a nonsingular cyclic right R-module. Put $I = r_R(x)$, the right annihilator of x in R. Since R is an abelian regular ring, I is an ideal of R from Lemma 3.21. So the ring R/I is a regular ring. To show that R/I is abelian, let B be a right ideal of R/I. Then there exists a right ideal C of R such that $B = C/I$. As R is abelian regular, C is an ideal of R from Lemma 3.21, and hence $B = C/I$ is an ideal of the ring R/I. By Lemma 3.21, the regular ring $A := R/I$ is abelian.

Since $M \cong (R/I)_R$ is nonsingular, $E := E((R/I)_R)$ is nonsingular. We show that $EI = 0$. For this, take $z \in E$ and let

$$J = \{r \in R \mid zr \in R/I\}.$$

Then J is a right ideal of R, and so J is an ideal of R by Lemma 3.21 as R is abelian regular.

We show that $J_R \leq^{\text{ess}} R_R$. For this, say $0 \neq b \in R$. If $zb = 0 \in E$, then $zb = 0 \in R/I$ and so $0 \neq b \in J$. If $0 \neq zb \in E$, then there exists $r \in R$ such that $0 \neq zbr \in R/I$ because $(R/I)_R \leq^{\text{ess}} E_R$. Hence $0 \neq br \in J$. Therefore $J_R \leq^{\text{ess}} R_R$. Further, $zJ \subseteq R/I$. Thus $zJI = 0$.

Put $K = r_R(z)$, the right annihilator of z in R. Then K is a right ideal of R. As R is abelian regular, Lemma 3.21 yields that K is an ideal of R. From $zJI = 0$, $JI \subseteq r_R(z)$, so $JI \subseteq K$. Therefore

$$[(J+K)/K][(I+K)/K] = 0$$

in the ring R/K. From the preceding argument, the factor ring R/K is abelian regular (so R/K is semiprime). Hence from Lemma 3.22, it follows that

$$[(I+K)/K][(J+K)/K] = 0.$$

Thus $IJ \subseteq K$, so $IJ \subseteq r_R(z)$. Therefore $zIJ = 0$. Since $J_R \leq^{\text{ess}} R_R$ and E_R is nonsingular, $zI = 0$. So $EI = 0$.

Now for $v \in E$ and $r + I \in R/I$, define $v(r+I) = vr$. Then this scalar multiplication of E over R/I is well-defined since $EI = 0$. Thus E is a right (R/I)-module. Therefore $E = E((R/I)_R)$ has a right (R/I)-module structure induced from the right R-module structure of E.

We prove that $E((R/I)_R)$ is the injective hull of $(R/I)_{(R/I)}$. For this, first take $0 \neq y \in E((R/I)_R)$. Then there exists $r \in R$ such that $0 \neq yr \in R/I$ since $(R/I)_R \leq^{\text{ess}} E((R/I)_R)$. But we note that $yr = y(r+I)$, so we have that $(R/I)_{(R/I)} \leq^{\text{ess}} E((R/I)_R)_{(R/I)}$.

Next we let $(K/I)_{(R/I)} \leq (R/I)_{(R/I)}$ and $f \in \text{Hom}((K/I)_{(R/I)}, E_{(R/I)})$. Then $f \in \text{Hom}((K/I)_R, E_R)$. As E_R is injective, there exists

$$g \in \text{Hom}((R/I)_R, E_R), \quad \text{which is an extension of } f.$$

Now $g \in \text{Hom}((R/I)_{(R/I)}, E_{(R/I)})$ because $E = E((R/I)_R)$ has a right (R/I)-module structure induced from the right R-module structure of E as is noted previously. Hence $E_{(R/I)}$ is injective as a right (R/I)-module by Baer's Criterion. So $E_{(R/I)}$ is the injective hull of $(R/I)_{(R/I)}$. Since R/I is an abelian regular ring, it is right nonsingular. Hence

$$E_{(R/I)} = E((R/I)_{(R/I)}) = Q(R/I)$$

by Theorem 3.7.

From the preceding argument, we have that $A = R/I$ is an abelian regular ring. Theorem 3.6 and Proposition 3.20 yield that $Q(A)$ is abelian regular. Hence all idempotents of $Q(A)$ are central. Thus $\mathbf{I}(Q(A)) = \mathcal{B}(Q(A))$.

We claim that $A\mathcal{B}(Q(A))_R$ is the quasi-continuous hull of A_R. For this, note first that $E(A_R)I = 0$ from the preceding argument, and hence $E(A_R)$ is a right (R/I)-module and $\text{Hom}(E(A_R)_R) = \text{Hom}(E(A_R)_{(R/I)})$. Also from the preceding argument, $E_{(R/I)} = E((R/I)_{(R/I)}) = Q(R/I)$, and so $E(A_R)_A = E(A_A) = Q(A)$. Therefore

$$\begin{aligned}\mathrm{End}(E(A_R)_R) &= \mathrm{End}(E(A_R)_{(R/I)}) = \mathrm{End}(Q(A)_A)\\ &= \mathrm{End}(Q(A)_{Q(A)}) \cong Q(A)\end{aligned}$$

as $\mathrm{End}(Q(A)_A) = \mathrm{End}(Q(A)_{Q(A)})$ (see Proposition 3.8).

Now let Ω be the subring of $\mathrm{End}(E(A_R)_R)$ generated by $\mathbf{I}(\mathrm{End}(E(A_R)_R)$. Since $\mathrm{End}(E(A_R)_R) \cong Q(A)$ and $\mathbf{I}(Q(A)) = \mathcal{B}(Q(A))$ from preceding arguments, we have that $\Omega A_R = \mathcal{B}(Q(A))A_R$. Therefore $A\mathcal{B}(Q(A))_R$ is the quasi-continuous hull of A_R by Theorem 2.12.

To show that $A\mathcal{B}(Q(A))_R$ is continuous, first we note from Proposition 3.19 that the ring $A\mathcal{B}(Q(A))$ is regular. Next, we let $N_R \leq^{\oplus} A\mathcal{B}(Q(A))_R$ and assume that $W_R \leq A\mathcal{B}(Q(A))_R$ such that $W_R \cong N_R$. Since $N_R \leq^{\oplus} A\mathcal{B}(Q(A))_R$, there exists $h^2 = h \in \mathrm{End}(A\mathcal{B}(Q(A))_R)$ such that $N = hA\mathcal{B}(Q(A))_R$. Furthermore, $NI \subseteq A\mathcal{B}(Q(A))I = 0$, so N is a right A-module induced from the right R-module structure of N_R and $N_A \leq A\mathcal{B}(Q(A))_A$.

Because $N = hA\mathcal{B}(Q(A))$, N is a right $A\mathcal{B}(Q(A))$-module which is compatible with the right A-module structure of N_A. Note that from Proposition 3.8,

$$\begin{aligned}h \in \mathrm{End}(A\mathcal{B}(Q(A))_R) &= \mathrm{End}(A\mathcal{B}(Q(A))_A)\\ &= \mathrm{End}(A\mathcal{B}(Q(A))_{A\mathcal{B}(Q(A))}) \cong A\mathcal{B}(Q(A)).\end{aligned}$$

Hence $N_{A\mathcal{B}(Q(A))} \leq^{\oplus} A\mathcal{B}(Q(A))_{A\mathcal{B}(Q(A))}$.

Let $\theta : W_R \to N_R$ be an R-isomorphism. Since $W_R \leq A\mathcal{B}(Q(A))_R$ and $A\mathcal{B}(Q(A))I = 0$, $WI = 0$ and hence W is a right A-module induced from the right R-module structure from W_R. Therefore we have that $\mathrm{Hom}(W_R, N_R) = \mathrm{Hom}(W_A, N_A)$. Hence $\theta : W_A \to N_A$ is an A-isomorphism.

We establish a right $A\mathcal{B}(Q(A))$-module structure on W which is compatible with the right A-module structure of W. For this, let $y \in W$ and $\alpha \in A\mathcal{B}(Q(A))$. Define

$$y \cdot \alpha = \theta^{-1}(\theta(y)\alpha).$$

Then for $y \in W$ and $\alpha, \beta \in A\mathcal{B}(Q(A))$, we have that

$$\begin{aligned}y \cdot (\alpha + \beta) = \theta^{-1}(\theta(y)(\alpha + \beta)) &= \theta^{-1}(\theta(y)\alpha + \theta(y)\beta)\\ &= \theta^{-1}(\theta(y)\alpha) + \theta^{-1}(\theta(y)\beta)\\ &= y \cdot \alpha + y \cdot \beta.\end{aligned}$$

For any $y \in W$ and $\alpha \in A\mathcal{B}(Q(A))$, as $y \cdot \alpha = \theta^{-1}(\theta(y)\alpha)$, $\theta(y \cdot \alpha) = \theta(y)\alpha$.

Also, for $\alpha, \beta \in A\mathcal{B}(Q(A))$,

$$\theta(y \cdot (\alpha\beta)) = \theta(y)(\alpha\beta) \ \text{ and } \ \theta((y \cdot \alpha) \cdot \beta) = \theta(y \cdot \alpha)\beta = (\theta(y)\alpha)\beta = \theta(y)(\alpha\beta).$$

Thus $\theta(y \cdot (\alpha\beta)) = \theta((y \cdot \alpha) \cdot \beta)$. As θ is one-to-one, $y \cdot (\alpha\beta) = (y \cdot \alpha) \cdot \beta$. Therefore W becomes a right $A\mathcal{B}(Q(A))$-module under the scalar multiplication "$\cdot$" .

Next for $y \in W$ and $a \in A$,

$$y \cdot a = \theta^{-1}(\theta(y)a) = \theta^{-1}(\theta(ya)) = ya.$$

Hence the right $A\mathcal{B}(Q(A))$-module structure of W is compatible with the right A-module structure of W. Furthermore, since $\theta(y \cdot \alpha) = \theta(y)\alpha$ for $y \in W$ and $\alpha \in A\mathcal{B}(Q(A))$, θ is an $A\mathcal{B}(Q(A)$-isomorphism from $W_{A\mathcal{B}(Q(A))}$ to $N_{A\mathcal{B}(Q(A))} =$

$h\mathcal{AB}(Q(A))_{\mathcal{AB}(Q(A))}$. Thus $W_{\mathcal{AB}(Q(A))}$ is a principal right ideal of the regular ring $\mathcal{AB}(Q(A))$. Hence

$$\mathcal{AB}(Q(A))_{\mathcal{AB}(Q(A))} = W_{\mathcal{AB}(Q(A))} \oplus U_{\mathcal{AB}(Q(A))}$$

for some $U_{\mathcal{AB}(Q(A))} \leq \mathcal{AB}(Q(A))_{\mathcal{AB}(Q(A))}$. Therefore $\mathcal{AB}(Q(A))_A = W_A \oplus U_A$. So $\mathcal{AB}(Q(A))_R = W_R \oplus U_R$. Hence $\mathcal{AB}(Q(A))_R$ satisfies (C_2) condition. Since $\mathcal{AB}(Q(A))_R$ is quasi-continuous, $\mathcal{AB}(Q(A))_R$ satisfies (C_1) condition. Consequently, $\mathcal{AB}(Q(A))_R$ is a continuous module.

Let V_R be a continuous R-module such that $A_R \leq V_R \leq E(A_R) = Q(A)_R$. Then V_R is quasi-continuous and $\mathcal{AB}(Q(A))_R$ is the quasi-continuous hull of A_R. Hence $\mathcal{AB}(Q(A))_R \leq V_R$. As $\mathcal{AB}(Q(A))_R$ is continuous, $\mathcal{AB}(Q(A))_R$ is the continuous hull of A_R. Thus $M = xR_R$ has a continuous hull since $M \cong A_R$. □

As an application of Theorem 3.23, we obtain the following theorem related to Question 3.4.

Theorem 3.24. Let R be an abelian regular ring. Then $eR\mathcal{B}(Q(R))_R$ is the continuous hull of eR_R for $e^2 = e \in R$. In particular, $R\mathcal{B}(Q(R))_R$ is the continuous hull of R_R.

Proof. Let $B_R = eR_R$ with $e^2 = e \in R$. Then B_R is a nonsingular cyclic right R-module. Note that e is central, so $B = eRe$ is a ring with e identity. Since $Q(R)$ is abelian regular by Proposition 3.20, all idempotents of $Q(R)$ are central.

To show that $\mathbf{I}(eQ(R)e) = e\mathbf{I}(Q(R))e$, let $y \in \mathbf{I}(eQ(R)e)$. Then $y^2 = y$ and $y \in Q(R)$. Hence $y \in \mathbf{I}(Q(R))$, and so $y = eye \in e\mathbf{I}(Q(R))e$. Thus we have that $\mathbf{I}(eQ(R)e) \subseteq e\mathbf{I}(Q(R))e$.

Conversely, take $v \in e\mathbf{I}(Q(R))e$. Then $v = eqe$ for some $q \in \mathbf{I}(Q(R))$. So $v = eqe \in eQ(R)e$ and $v^2 = (eqe)(eqe) = eq^2e = eqe = v$. Hence $v \in \mathbf{I}(eQ(R)e)$. Thus $e\mathbf{I}(Q(R))e \subseteq \mathbf{I}(eQ(R)e)$. Consequently, $\mathbf{I}(eQ(R)e) = e\mathbf{I}(Q(R))e$.

As $\mathbf{I}(eQ(R)e) = \mathcal{B}(eQ(R)e)$ and $\mathbf{I}(Q(R)) = \mathcal{B}(Q(R))$, $e\mathbf{I}(Q(R))e = e\mathcal{B}(Q(R))e$ and thus $\mathcal{B}(eQ(R)e) = e\mathcal{B}(Q(R))e$.

We show that $Q(eRe) = eQ(R)e$. Since eRe is a regular ring, eRe is right nonsingular and so $Q(eRe) = E(eRe_{eRe})$ from Theorem 3.7. Now we note that $eRe_{eRe} \leq^{\text{ess}} eQ(R)e_{eRe}$. Indeed, take $0 \neq eqe = eq \in eQ(R)e = eQ(R)$ with $q \in Q(R)$. Since $E(eR_R) = eQ(R)_R$, there exists $r \in R$ such that $0 \neq eqr \in eR$. Thus $0 \neq eqr = eqere \in eR = eRe$ and $ere \in eRe$ because e is central in $Q(R)$. Thus $eRe_{eRe} \leq^{\text{ess}} eQ(R)e_{eRe}$.

Next, to prove that $eQ(R)e_{eRe}$ is injective, we now take $I_{eRe} \leq eRe_{eRe}$ and $f \in \text{Hom}(I_{eRe}, eQ(R)e_{eRe})$. Since $I = eIe$, $IR = eIeR = IeRe \subseteq I$ and so $I_R \leq eRe_R = eR_R$. For $a \in I$ and $s \in R$,

$$f(as) = f(aes) = f(aese) = f(a)ese = f(ae)s = f(a)s$$

because e is a central idempotent in $Q(R)$. Hence $f \in \text{Hom}(I_R, eQ(R)_R)$. Because $E(eR_R) = eQ(R)_R$, $eQ(R)_R$ is injective. So there exists $g \in \text{Hom}(eR_R, eQ(R)_R)$ which is an extension of f. In this case, we note that $g \in \text{Hom}(eRe_{eRe}, eQ(R)e_{eRe})$. Thus $eQ(R)e_{eRe}$ is injective by Baer's Criterion. Hence, $E(eRe_{eRe}) = eQ(R)e_{eRe}$ because $eRe_{eRe} \leq^{\text{ess}} eQ(R)e_{eRe}$. From Theorem 3.7, $Q(eRe) = E(eRe_{eRe})$ since eRe is regular (so eRe is right nonsingular). Thus $Q(eRe) = eQ(R)e$.

As a consequence,

$$B\mathcal{B}(Q(B)) = (eR)\mathcal{B}(Q(eRe)) = eR\mathcal{B}(eQ(R)e) = eRe\mathcal{B}(Q(R))e = eR\mathcal{B}(Q(R)).$$

So by Theorem 3.23, $eR\mathcal{B}(Q(R))_R$ is the continuous hull of eR_R. □

There exists a regular ring R for which $R\mathcal{B}(Q(R))_R$ is not the continuous hull of R_R (Example 3.25(i)). Thereby, Theorem 3.24 does not hold true for the case when R is a regular ring in general.

Furthermore, Example 3.25(ii) shows that Theorem 3.24 does not hold true for the case when R is a commutative domain (hence R is abelian) which is not a field.

Example 3.25. (i) There exists a regular ring R such that $Q(R)_R$ is the continuous hull of R_R, but $R\mathcal{B}(Q(R))_R$ is not the continuous hull of R_R.

Indeed, let

$$R = \left\{ (A_n)_{n=1}^{\infty} \in \prod_{n=1}^{\infty} \mathrm{Mat}_2(\mathbb{Q}) \mid A_n \text{ is constant eventually} \right\},$$

which is a subring of $\prod_{n=1}^{\infty} \mathrm{Mat}_2(\mathbb{Q})$. Then R is regular. By applying the argument used in Example 3.11(i), $E(R_R) = Q(R) = \prod_{n=1}^{\infty} \mathrm{Mat}_2(\mathbb{Q})$.

We show that $Q(R)_R$ is the continuous hull of R_R. For this, take $\alpha \in Q(R)$. Say

$$\alpha = \left(\begin{bmatrix} v_1 & s_1 \\ t_1 & u_1 \end{bmatrix}, \begin{bmatrix} v_2 & s_2 \\ t_2 & u_2 \end{bmatrix}, \cdots \right),$$

where $v_i, s_i, t_i, u_i \in \mathbb{Q}$, $i = 1, 2, \cdots$. Take e_1, e_2, e_3, and e_4 in $Q(R)$ such that

$$e_1 = \left(\begin{bmatrix} 1 & v_1 \\ 0 & 0 \end{bmatrix}, \begin{bmatrix} 1 & v_2 \\ 0 & 0 \end{bmatrix}, \cdots \right), \; e_2 = \left(\begin{bmatrix} 1 & s_1 \\ 0 & 0 \end{bmatrix}, \begin{bmatrix} 1 & s_2 \\ 0 & 0 \end{bmatrix}, \cdots \right),$$

and

$$e_3 = \left(\begin{bmatrix} 0 & 0 \\ t_1 & 1 \end{bmatrix}, \begin{bmatrix} 0 & 0 \\ t_2 & 1 \end{bmatrix}, \cdots \right), \; e_4 = \left(\begin{bmatrix} 0 & 0 \\ u_1 & 1 \end{bmatrix}, \begin{bmatrix} 0 & 0 \\ u_2 & 1 \end{bmatrix}, \cdots \right).$$

Then $e_i^2 = e_i$ for $i = 1, 2, 3, 4$.

Next take r_1, r_2, r_3, and r_4 in R such that

$$r_1 = \left(\begin{bmatrix} 0 & 0 \\ 1 & 0 \end{bmatrix}, \begin{bmatrix} 0 & 0 \\ 1 & 0 \end{bmatrix}, \cdots \right), \; r_2 = \left(\begin{bmatrix} 0 & 0 \\ 0 & 1 \end{bmatrix}, \begin{bmatrix} 0 & 0 \\ 0 & 1 \end{bmatrix}, \cdots \right)$$

and

$$r_3 = \left(\begin{bmatrix} 1 & 0 \\ 0 & 0 \end{bmatrix}, \begin{bmatrix} 1 & 0 \\ 0 & 0 \end{bmatrix}, \cdots \right), \; r_4 = \left(\begin{bmatrix} 0 & 1 \\ 0 & 0 \end{bmatrix}, \begin{bmatrix} 0 & 1 \\ 0 & 0 \end{bmatrix}, \cdots \right).$$

Then we see that

$$\alpha = e_1 r_1 + e_2 r_2 + e_3 r_3 + e_4 r_4.$$

On the other hand, Proposition 3.8 yields that

$$\mathrm{End}(E(R_R)) = \mathrm{End}(Q(R)_R) = \mathrm{End}(Q(R)_{Q(R)}) \cong Q(R).$$

Let Ω be the subring of $Q(R) \cong \mathrm{End}(E(R_R))$ generated by $\mathbf{I}(Q(R))$ (recall that $\mathbf{I}(Q(R))$ denotes the set of all idempotents of $Q(R)$). Then ΩR_R is the quasi-continuous hull of R_R from Theorem 2.12.

Now we note that

$$\alpha = e_1 r_1 + e_2 r_2 + e_3 r_3 + e_4 r_4 \in \Omega R.$$

Hence it follows that $E(R_R) = Q(R) \subseteq \Omega R$, and so $Q(R) = \Omega R$. Thus $Q(R)_R$ is the quasi-continuous hull of R_R. Therefore the quasi-continuous hull, the continuous hull, and the injective hull of R_R all coincide, and so $Q(R)_R$ is the continuous hull of R_R.

We take

$$f = \left(\begin{bmatrix} 1 & 1 \\ 0 & 0 \end{bmatrix}, \begin{bmatrix} 1 & 2 \\ 0 & 0 \end{bmatrix}, \dots, \begin{bmatrix} 1 & k \\ 0 & 0 \end{bmatrix}, \dots \right) \in Q(R).$$

Then $f \notin R\mathcal{B}(Q(R))$. Thus $R\mathcal{B}(Q(R)) \neq Q(R)$, so $R\mathcal{B}(Q(R))_R$ is not the continuous hull of R_R. Further, note that $R \neq R\mathcal{B}(Q(R))$ because

$$\beta = \left(\begin{bmatrix} 1 & 0 \\ 0 & 1 \end{bmatrix}, \begin{bmatrix} 0 & 0 \\ 0 & 0 \end{bmatrix}, \begin{bmatrix} 1 & 0 \\ 0 & 1 \end{bmatrix}, \begin{bmatrix} 0 & 0 \\ 0 & 0 \end{bmatrix}, \dots \right) \in \mathcal{B}(Q(R)) \subseteq R\mathcal{B}(Q(R)),$$

but $\beta \notin R$.

(ii) Assume that R is a commutative domain (hence R is reduced). Because $R = R\mathcal{B}(Q(R))$, it follows that $R\mathcal{B}(Q(R))_R$ is the continuous hull of R_R if and only if R is a field. Therefore Theorem 3.24 does not hold for the case when R is a commutative domain which is not a field.

By Theorem 3.2, R_R has a continuous hull. In this case, the continuous hull of R_R is $Q(R)_R$ (note that $Q(R)$ is the field of fractions of F) from [**9**] (see also [**2**, Theorem 8.4.11]).

The following, related to Theorem 3.3, follows from [**9**, Lemma 2 and Corollary 3]. See also [**13**, Corollary 4.5.5].

Proposition 3.26. Assume that R is a commutative semiprime ring and P is the subring of $Q(R)$ generated by all idempotents of $Q(R)$. Let V be the intersection of all continuous regular rings of quotients of R. Then V_R is the continuous hull of R_R.

Let A be a ring and $R = \mathrm{Mat}_n(A)$ or $R = T_n(A)$, where $n > 1$. Then R_R has a continuous hull. Further, the continuous hull, the quasi-injective hull, and the quasi-continuous hull of R_R coincide (see Theorem 3.10 and Theorem 3.15).

When a ring R is abelian regular, $R\mathcal{B}(Q(R))_R$ is the continuous hull of R_R, and further $R\mathcal{B}(Q(R))_R$ is the quasi-continuous hull of R_R (see Theorem 3.24). Thereby, when R is an abelian regular ring, the continuous hull of R_R and the quasi-continuous hull of R_R coincide.

However, in the next example from [**9**], there exists a commutative semiprime ring R such that the quasi-continuous hull of R_R is distinct from the continuous hull which is, in turn, distinct from the (quasi-)injective hull.

Example 3.27. Let R the subring of $\prod_{n=1}^{\infty} \mathbb{R}$ generated by $\oplus_{n=1}^{\infty}\mathbb{R}$ and $1_{\prod_{n=1}^{\infty} \mathbb{R}}$. Since R is commutative semiprime, R is right nonsingular. In this case, note that

$$R = \left\{ (a_n)_{n=1}^{\infty} \in \prod_{n=1}^{\infty} \mathbb{R} \mid a_n \in \mathbb{Z} \text{ eventually and constant} \right\}$$

By applying the argument used in Example 3.11(i), $E(R_R) = Q(R) = \prod_{n=1}^{\infty} F_n$.

(i) $Q(R)_R$ is the (quasi-)injective hull of R_R. By Thorem 3.7 and Proposition 3.8,

$$\mathrm{End}(E(R_R)) = \mathrm{End}(Q(R)_R) = \mathrm{End}(Q(R)_{Q(R)}) \cong Q(R) = \prod_{n=1}^{\infty} \mathbb{R}$$

Thus $Q(R)_R$ is the (quasi-)injective hull of R_R from Theorem 2.9.

(ii) We note that

$$A = \left\{ (a_n)_{n=1}^{\infty} \in \prod_{n=1}^{\infty} \mathbb{R} \mid a_n \in \mathbb{Z} \text{ eventually} \right\}$$

is the quasi-continuous hull of R_R (see Theorem 2.15).

(iii) From Proposition 3.26, it follows that

$$B = \left\{ (a_n)_{n=1}^{\infty} \in \prod_{n=1}^{\infty} \mathbb{R} \mid a_n \in \mathbb{Q} \text{ eventually} \right\}$$

is the continuous hull of R_R because B is the smallest continuous regular intermediate ring between R and $Q(R)$.

We recall that $Q^{\ell}(R)$ denotes the maximal left ring of quotients of a ring R.

Proposition 3.28. Let R be an abelian regular ring. Then we have the following.

(i) ([**5**, Theorem 3.8]) $Q(R) = Q^{\ell}(R)$.

(ii) (see [**2**, Corollary 3.3.3]) R is right continuous if and only if R is left continuous.

When $Q(R) = Q^{\ell}(R)$, we say that $Q(R)$ is the *maximal ring of quotients* of R. Hence the maximal ring of quotients of a ring R is both the maximal right ring of quotients of R and the maximal left ring of quotients of R.

Recall that an overring T of a ring R is called a left ring of quotients of R if ${}_RR \leq^{\text{den}} {}_RT$. Thus an overring T of a ring R is a left ring of quotients of R if and only if T is an intermediate ring between R and $Q^{\ell}(R)$.

An overring T of a ring R is said to be a *ring of quotients* of R if T is both a right ring of quotients of R and a left ring of quotients of R.

A ring R is said to be *continuous* if R is both right and left continuous. Proposition 3.28 yields that an abelian regular ring R is continuous if and only if R is right (or left) continuous

Recall that a ring R is called *unit-regular* if for each $a \in R$, there exists a unit (i.e., an invertible element) u in R such that $a = aua$. Abelian regular rings are unit-regular (see [**5**, Corollary 4.2]).

Example 3.29. ([**5**, Example 13.35]) There exists a unit-regular ring which cannot be embedded as a subring in any continuous regular ring.

In contrast to Example 3.29, we get the following corollary from Theorem 3.24 as a byproduct for the case of abelian rings (note that if R is an abelian regular ring, then $Q(R)$ is right and left self-injective, hence the maximal ring of quotients $Q(R)$ of R is a continuous abelian regular ring by Theorem 3.6, Proposition 3.20, and Proposition 3.28).

Corollary 3.30. Assume that R is an abelian regular ring. Then $R\mathcal{B}(Q(R))$ is the smallest continuous regular ring of quotients of R.

Proof. From Proposition 3.20 and Proposition 3.28, it is enough to prove that $R\mathcal{B}(Q(R))$ is the smallest right continuous regular right ring of quotients of R. For this, we show first that for an abelian regular ring A, A is right continuous if and only if $\mathcal{B}(Q(A)) \subseteq A$.

So assume that A is an abelian regular ring which is right continuous. Then A_A itself is the continuous hull of A_A. Thus by Theorem 3.24, $A = A\mathcal{B}(Q(A))$ and so $\mathcal{B}(Q(A)) \subseteq A$. Conversely, assume that A is an abelian regular ring and $\mathcal{B}(Q(A)) \subseteq A$. Then $A = A\mathcal{B}(Q(A))$ and so A_A is the continuous hull of A_A by Theorem 3.24. Consequently, if A is an abelian regular ring, then A is right continuous if and only if $\mathcal{B}(Q(A)) \subseteq A$.

Now we note that $R\mathcal{B}(Q(R))$ is a right ring of quotients of R. Because R is abelian regular, Proposition 3.20 yields that $R\mathcal{B}(Q(R))$ is abelian regular. Let $T = R\mathcal{B}(Q(R))$. Because $Q(T) = Q(R)$, $\mathcal{B}(Q(T)) = \mathcal{B}(Q(R))$. Therefore $\mathcal{B}(Q(T)) = \mathcal{B}(Q(R)) \subseteq T$. By the preceding argument, the ring T is right continuous since T is abelian regular.

Next U be a right continuous regular right ring of quotients of R. Then U is abelian since $Q(U) = Q(R)$ is abelian from Proposition 3.20. So $\mathcal{B}(Q(U)) \subseteq U$ from the preceding argument. Since $Q(U) = Q(R)$, $\mathcal{B}(Q(U)) = \mathcal{B}(Q(R))$. Therefore $T = R\mathcal{B}(Q(R)) \subseteq U$. As a consequence, $R\mathcal{B}(Q(R))$ is the smallest right continuous regular right ring of quotients of R. □

We illustrate Corollary 3.30. In the following example, there exists an abelian regular ring R for which the smallest continuous regular ring of quotients $R\mathcal{B}(Q(R))$ of R is a *proper intermediate ring* between R and $Q(R)$.

As is mentioned before Corollary 3.30, we note that when R is an abelian regular ring, $Q(R)$ is continuous abelian regular ring.

Example 3.31. Assume that D is a division ring such that $|D| = \infty$. Consider the following ring

$$R = \left\{ (a_k)_{k=1}^{\infty} \in \prod_{k=1}^{\infty} D \mid a_k \text{ is constant eventually} \right\},$$

which is a subring of $\prod_{k=1}^{\infty} D$.

The ring R is abelian regular. Furthermore, by the argument used in Example 3.11(i), $E(R_R) = Q(R) = \prod_{k=1}^{\infty} D$. Let $\{d_k\}$ for $k = 1, 2, 3, \ldots$ be a set of mutually distinct elements in D. Then $(d_k)_{k=1}^{\infty} \in Q(R)$, but we have that $(d_k)_{k=1}^{\infty} \notin R\mathcal{B}(Q(R))$.

On the other hand, take $\alpha = (1, 0, 1, 0, \ldots) \in \mathcal{B}(Q(R))$. Then $\alpha \in R\mathcal{B}(Q(R))$. But $\alpha \notin R$. So $R\mathcal{B}(Q(R))$ is a proper intermediate ring between R and $Q(R)$.

The next result is [**2**, Proposition 8.1.4] which is used for Example 3.33.

Proposition 3.32. Let T be a right ring of quotients of a ring R. Then T is right extending if and only if T_R is extending.

When R is a regular ring, recall that $R\mathcal{B}(Q(R))$ is a regular ring from Proposition 3.19. As an application of our results, the ring $R\mathcal{B}(Q(R))$ of Example 3.33 is another example of a regular ring which is neither right nor left continuous (cf. Example 3.13). Therefore Corollary 3.30 does not hold true for the case when R is a regular ring in general.

Example 3.33. Let R be the ring as in Example 3.25(i). Then R is a regular ring and hence $R\mathcal{B}(Q(R))$ is a regular ring by Proposition 3.19.

First, we claim that the ring $R\mathcal{B}(Q(R))$ is not right continuous. For this, assume on the contrary that $R\mathcal{B}(Q(R))$ is right continuous. So $R\mathcal{B}(Q(R))$ is right extending. Hence Proposition 3.32 yields that $R\mathcal{B}(Q(R))_R$ is extending.

Now we show that $R\mathcal{B}(Q(R))_R$ satisfies (C_2). For this, let $V_R \leq R\mathcal{B}(Q(R))_R$ and

$$V_R \cong N_R \leq^{\oplus} R\mathcal{B}(Q(R))_R.$$

Then there exists $h^2 = h \in \mathrm{End}(R\mathcal{B}(Q(R))_R)$ such that $N = hR\mathcal{B}(Q(R))$. By Proposition 3.8,

$$\mathrm{End}(R\mathcal{B}(Q(R))_R) = \mathrm{End}(R\mathcal{B}(Q(R))_{R\mathcal{B}(Q(R))}) \cong R\mathcal{B}(Q(R)).$$

Hence $h \in R\mathcal{B}(Q(R))$ such that $N = hR\mathcal{B}(Q(R))$ and so

$$N = hR\mathcal{B}(Q(R))_{R\mathcal{B}(Q(R))} \leq^{\oplus} R\mathcal{B}(Q(R))_{R\mathcal{B}(Q(R))}.$$

By a similar method to the argument used in the proof of Theorem 3.23, V becomes a right $R\mathcal{B}(Q(R))$-module for which the $R\mathcal{B}(Q(R))$-module scalar multiplication is compatible with the right R-module scalar multiplication. Furthermore,

$$V_{R\mathcal{B}(Q(R))} \cong N_{R\mathcal{B}(Q(R))} = hR\mathcal{B}(Q(R)).$$

Since $R\mathcal{B}(Q(R))$ is regular, $R\mathcal{B}(Q(R))_{R\mathcal{B}(Q(R))}$ satisfies (C_2), and hence

$$V_{R\mathcal{B}(Q(R))} \leq^{\oplus} R\mathcal{B}(Q(R))_{R\mathcal{B}(Q(R))}.$$

Therefore $V_R \leq^{\oplus} R\mathcal{B}(Q(R))_R$. So $R\mathcal{B}(Q(R))_R$ satisfies (C_2).

As a consequence, $R\mathcal{B}(Q(R))_R$ is a continuous module. However from Example 3.25(i), $R\mathcal{B}(Q(R))_R$ is not continuous. Thus we get a contradiction. Consequently, the ring $R\mathcal{B}(Q(R))$ is not right continuous. Since $R\mathcal{B}(Q(R))$ is regular and not right continuous, $R\mathcal{B}(Q(R))$ is not right extending.

From Example 3.25(i), we have that $Q(R) = \prod_{n=1}^{\infty} \mathrm{Mat}_2(\mathbb{Q})$. Threfore $Q(R)$ is a PI-ring and hence the ring $R\mathcal{B}(Q(R))$ is a PI-ring. Hence $R\mathcal{B}(Q(R))$ is a regular PI-ring. Because $R\mathcal{B}(Q(R))$ is not right extending, Proposition 3.12 yields that $R\mathcal{B}(Q(R))$ is not left extending. Thus $R\mathcal{B}(Q(R))$ is not left continuous. Consequently, the ring $R\mathcal{B}(Q(R))$ is neither right nor left continuous.

Motivated from the argument used for Example 3.33, we obtain the following consequence by adopting a similar method to the argument used in the proof of Theorem 3.24.

Theorem 3.34. Let R be a ring and T be a regular right ring of quotients of R. Then the following are equivalent.

(i) T is right continuous.

(ii) T_R is continuous.

Proof. (i)⇒(ii) The proof of this implication is a slight modification of the argument used for Example 3.33. However, we give the proof in details for convenience.

Assume that T is right continuous. Then since T is right extending, T_R is extending from Proposition 3.32. Now we show that T_R satisfies (C_2). For this, let $V_R \leq T_R$ and $V_R \cong N_R \leq^{\oplus} T_R$. Then there exists $h^2 = h \in \mathrm{End}(T_R)$ such that $N = hT$. By Proposition 3.8, $\mathrm{End}(T_R) = \mathrm{End}(T_T) \cong T$ as T is a right ring of quotients of R. Hence $h \in T$ such that $N = hT$ and so $N = hT_T \leq^{\oplus} T_T$.

By a similar method to the argument used in the proof of Theorem 3.23, V becomes a right T-module for which the T-module scalar multiplication is compatible with the right R-module scalar multiplication. Furthermore, $V_T \cong N_T = hT$.

Since T is regular, T_T satisfies (C_2), and hence $V_T \leq^{\oplus} T_T$. Therefore $V_R \leq^{\oplus} T_R$. So T_R satisfies (C_2). As a consequence T_R is a continuous module.

(ii)$\Rightarrow$(i) Assume that T_R is continuous. Then T_R is extending. By Proposition 3.32, T is right extending. Because T is a regular ring, T_T satisfies (C_2). Consequently, the ring T is right continuous. □

When R is a regular ring, $R\mathcal{B}(Q(R))$ is a regular ring from Proposition 3.19. Obviously, $R\mathcal{B}(Q(R))$ is a right ring of quotients of R. Therefore the following corollary follows immediately from Theorem 3.34.

Corollary 3.35. Let R be a regular ring. Then the following are equivalent.
(i) $R\mathcal{B}(Q(R))$ is right continuous.
(ii) $R\mathcal{B}(Q(R))_R$ is continuous.

We get the following from the argument used in the proof of Theorem 3.10 as a byproduct.

Proposition 3.36. Let R be a ring. Then, for any R-module M_R and any set Λ such that $|\Lambda| > 1$, we have the following.
(i) $M_R^{(\Lambda)}$ has a continuous hull.
(ii) The continuous hull, the quasi-injective hull, and the quasi-continuous hull of $M_R^{(\Lambda)}$ coincide.

Proof. Assume first that Λ is finite such that $|\Lambda| = n > 1$. Then $M_R^{(\Lambda)} = M_R^{(n)}$ has a continuous hull and further, the continuous hull of $M_R^{(n)}$ is the quasi-injective hull of $M_R^{(n)}$ and the quasi-continuous hull of $M_R^{(n)}$ by the argument used in the proof of Theorem 3.10.

Next we assume that Λ is infinite. Then there exist Λ_1 and Λ_2 such that $\Lambda = \Lambda_1 \cup \Lambda_2$ (disjoint union) and $|\Lambda_1| = |\Lambda_2|$. We put $M_1 = M_R^{(\Lambda_1)}$ and $M_2 = M_R^{(\Lambda_2)}$. Then $M_R^{(\Lambda)} = M_1 \oplus M_2$ and $M_1 \cong M_2$. From the argument used in the proof of Theorem 3.10, $M_R^{(\Lambda)}$ has a continuous hull and further, the continuous hull, the quasi-injective hull, and the quasi-continuous hull of $M_R^{(\Lambda)}$ coincide. □

We obtain following corollaries from Theorem 3.3 and Proposition 3.36.

Corollary 3.37. Let R be a commutative semiprime ring. Then any free R-module $R_R^{(\Lambda)}$ has a continuous hull. Moreover, if $|\Lambda| > 1$, then the continuous hull, the quasi-injective hull, and the quasi-continuous hull of $R_R^{(\Lambda)}$ coincide.

Proof. Since R is commutative semiprime, R_R is nonsingular. By Theorem 3.3, R_R has a continuous hull. If $|\Lambda| > 1$, then the proof follows from Proposition 3.36. □

A module is called *purely infinite* if it is isomorphic to the direct sum of two copies of itself.

Corollary 3.38. Let M be a purely infinite module. Then we have the following.
(i) M has a continuous hull.

(ii) The continuous hull, the quasi-injective hull, and the quasi-continuous hull of M coincide.

(iii) If M is continuous, then M is quasi-injective.

Proof. Since M is purely infinite, $M \cong M \oplus M$. Thus parts (i) and (ii) follows from Proposition 3.36. Part (iii) follows immediately from part (ii). □

The next result follows immediately from Theorem 3.24 and Proposition 3.36.

Corollary 3.39. Let R be an abelian regular ring. Then any free right R-module has a continuous hull.

Finally, we have te following remark for Proposition 3.36.

Remark 3.40. Let n be an integer such that $n > 1$. Then for any ring R, $R_R^{(n)}$ has a continuous hull and the quasi-injective hull of $R^{(n)}$ is the continuous hull of $R_R^{(n)}$ by Proposition 3.36.

Now we have that $\text{End}(E(R_R^{(n)})) = \text{End}(E(R_R)^{(n)}) = \text{Mat}_n(\text{End}(E(R_R))$ because $E(R_R^{(n)}) = E(R_R)^{(n)}$. Therefore $\text{End}(E(R_R^{(n)}))R_R^{(n)} = (\text{End}(E(R_R))R)_R^{(n)}$ is the quasi-injective hull of $R_R^{(n)}$ by Theorem 2.9. Hence $(\text{End}(E(R_R))R)_R^{(n)}$ is the continuous hull of $R_R^{(n)}$ by Proposition 3.36.

Additionally, if R is right nonsingular, then $Q(R) = E(R_R)$ by Theorem 3.7. Hence $\text{End}(E(R_R)) = \text{End}(Q(R)_R) = \text{End}(Q(R)_{Q(R)}) \cong Q(R)$ from Proposition 3.8. So

$$(\text{End}(E(R_R))R)_R^{(n)} = (Q(R)R)_R^{(n)} = Q(R)_R^{(n)} = E(R_R)_R^{(n)}$$

is the continuous hull of $R_R^{(n)}$, which is the injective hull of $R_R^{(n)}$.

4. Condition $(\star)$ and Continuous Hulls of Modules

In this section, we discuss the existence and description of continuous hulls of modules over a commutative ring. For the discussion, we consider a commutative ring R with the following condition $(\star)$:

$Q_{c\ell}(R/I)$ *is self-injective for every uniform factor ring* R/I *of* R.

In this case, the ring R is said to satisfy condition $(\star)$. We study a necessary and sufficient condition, over an arbitrary commutative ring, for every indecomposable (uniform) continuous module to be quasi-injective by using condition $(\star)$.

Also we study necessary and sufficient conditions for special classes of commutative rings such as finitary rings to have the property that every continuous module is quasi-injective. Thereby, in contrast to Example 3.1, when R is a commutative noetherian ring, every module M over R has a continuous hull and the continuous hull of M is exactly the quasi-injective hull of M.

Except Proposition 4.10, Theorem 4.11, Lemma 4.12, and Example 4.18, all results of this section are early results due to Rizvi [**14**]. However, we provide proofs of these results in detail for the sake of completeness and convenience.

We begin with the following definition.

Definition 4.1. An element $a \in R$ is said to *act regularly* on an R-module M_R, if $ma = 0$ implies $m = 0$, for $m \in M$. The set of all such elements of R is denoted by $\mathfrak{C}(M)$, that is $\mathfrak{C}(M) = \{a \in R \mid ma = 0 \text{ implies } m = 0 \text{ for } m \in M\}$.

The next lemma is necessary for the study of continuous hulls of modules over a commutative ring.

Lemma 4.2. Let R be a commutative ring and M_R be a uniform R-module. Then we have the following.

(i) $\mathfrak{C}(M) = \mathfrak{C}(E(M))$.

(ii) For any $m \in E(M)$ and $c \in \mathfrak{C}(M)$, there exists a unique $y \in E(M)$ such that $yc = m$.

In this case, we write $y = mc^{-1}$.

(iii) $E(M) = \{xc^{-1} \mid x \in E(M) \text{ and } c \in \mathfrak{C}(M)\}$.

(iv) For $x \in E(M)$ and $c \in \mathfrak{C}(M)$, $xc^{-1}c = x$ and $xcc^{-1} = x$.

(v) For $x \in E(M)$, $r \in R$, and $c \in \mathfrak{C}(M)$, $xrc^{-1} = xc^{-1}r$.

(vi) For $x \in E(M)$ and $c, d \in \mathfrak{C}(M)$, $x(cd)^{-1} = xd^{-1}c^{-1} = xc^{-1}d^{-1}$.

(vii) $M\mathfrak{C}(M)^{-1} := \{mc^{-1} \mid m \in M \text{ and } c \in \mathfrak{C}(M)\}$ is an intermediate R-module between M and $E(M)$.

Proof. (i) Let $E = E(M)$. Take $c \in \mathfrak{C}(M)$. Then c acts regularly on E. Indeed, suppose $vc = 0$ with $v \in E$. If $v \neq 0$, then there exists $a \in R$ with $0 \neq va \in M$. So $(va)c = vca = 0$, and hence $va = 0$ because c acts regularly on M and $0 \neq va \in M$, which is a contradiction. Therefore $v = 0$. Hence c acts regularly on E. Thus $\mathfrak{C}(M) \subseteq \mathfrak{C}(E)$. Obviously, $\mathfrak{C}(E) \subseteq \mathfrak{C}(M)$. Hence $\mathfrak{C}(M) = \mathfrak{C}(E)$.

(ii) Since M is uniform, $E := E(M)$ is indecomposable (and injective). Now for $c \in \mathfrak{C}(M)$, $E \cong Ec \leq E$ as c acts regularly on E because $\mathfrak{C}(M) = \mathfrak{C}(E)$. Thus Ec is injective, so Ec is a direct summand of E. As E is indecomposable, we have that $E = Ec$. Take $m \in E$. Then $m \in E = Ec$, so there exists $y \in E$ such that $m = yc$. If $y_1 \in E$ such that $m = yc = y_1c$. Then $(y - y_1)c = 0$. As $c \in \mathfrak{C}(E)$, $y - y_1 = 0$, so $y = y_1$.

(iii) For any $y \in E(M)$, $y1 = y$ and $1 \in \mathfrak{C}(M)$. Thus $y = y1^{-1}$. So the proof follows from part (ii).

(iv) First, we show that $xc^{-1}c = x$. For this, let $u = xc^{-1}$. Then $uc = x$ and hence $x = uc = xc^{-1}c$. Next for $xcc^{-1} = x$, put $w = xcc^{-1}$. Then $wc = xc$, so $(w - x)c = 0$ and hence $w - x = 0$. So $w = x$. Thus $xcc^{-1} = x$.

(v) Let $x \in E(M)$, $r \in R$, and $c \in \mathfrak{C}(M)$. Then $xc^{-1}rc = xc^{-1}cr = xr$ because $xc^{-1}c = x$ by part (iv) and $rc = cr$. Therefore $xc^{-1}r = xrc^{-1}$.

(vi) Put $y = x(cd)^{-1}$. Then $x = y(cd) = (yc)d$ and so $xd^{-1} = yc$. Hence $xd^{-1}c^{-1} = y$, and therefore $xd^{-1}c^{-1} = x(cd)^{-1}$. Further, from $y = x(cd)^{-1}$, $y = x(dc)^{-1}$ and hence $y = xc^{-1}d^{-1}$ from the preceding argument. Thus we have that $x(cd)^{-1} = xc^{-1}d^{-1}$.

(vii) First $M\mathfrak{C}(M)^{-1} \subseteq E(M)$ from part (ii). Take $mc^{-1} \in M\mathfrak{C}(M)^{-1}$ with $m \in M$, $c \in \mathfrak{C}(M)$, and take $r \in R$. By part (v), $(mc^{-1})r = (mr)c^{-1} \in M\mathfrak{C}(M)^{-1}$.

Next, let $m_1c_1^{-1}, m_2c_2^{-1} \in M\mathfrak{C}(M)^{-1}$, where $m_i \in M$ and $c_i \in \mathfrak{C}(M), 1 \leq i \leq 2$. Then by part (iv)

$$\begin{aligned}(m_1c_1^{-1} + m_2c_2^{-1})(c_1c_2) &= (m_1c_1^{-1})(c_1c_2) + (m_2c_2^{-1})(c_1c_2)\\ &= (m_1c_1^{-1}c_1)c_2 + (m_2c_2^{-1}c_2)c_1\\ &= m_1c_2 + m_2c_1.\end{aligned}$$

Because $c_1c_2 \in C$, $m_1c_1^{-1} + m_2c_2^{-1} = (m_1c_2 + m_2c_1)(c_1c_2)^{-1}$ and thus we have that $m_1c_1^{-1} + m_2c_2^{-1} \in M\mathfrak{C}(M)^{-1}$. Other conditions for $M\mathfrak{C}(M)^{-1}$ to be an R-submodule of $E(M)$ can be checked routinely.

For $m \in M$, let $y = m1^{-1}$. Then $y = y1 = m$, so $m = m1^{-1}$. Since $1 \in C$, $m = m1^{-1} \in M\mathfrak{C}(M)^{-1}$. Thus $M = \{m1^{-1} \mid m \in M\}$ is an R-submodule of $M\mathfrak{C}(M)^{-1}$. So $M\mathfrak{C}(M)^{-1}$ is an intermediate R-module between M and $E(M)$. □

The following theorem describes Theorem 3.2 explictly. Some parts of its proof are routine. However, we give the proof in detail for convenience.

Theorem 4.3. Assume that R is a commutative ring, and M_R is a uniform cyclic R-module. Then $M\mathfrak{C}(M)_R^{-1}$ is the continuous hull of M_R.

Proof. Let $M = mR$ and put $C = \mathfrak{C}(mR)$. By Lemma 4.2(vii), $mR \leq (mR)C^{-1} \leq E(mR)$ as R-modules. Since mR_R is uniform and $mR_R \leq^{\text{ess}} (mR)C^{-1}$, $(mR)C^{-1}$ is uniform. So $(mR)C^{-1}$ is an extending R-module.

First, we show that $(mR)C^{-1}$ is a continuous R-module. For this, let $\varphi \in \text{End}_R((mR)C^{-1})$ be a monomorphism. Write $\varphi(m) = (mr_0)c_0^{-1}$, where $r_0 \in R$ and $c_0 \in C$. To prove that r_0 acts regularly on mR, take $v \in mR$ and assume that $vr_0 = 0$. Say $v = ma$ with $a \in R$. By Lemma 4.2(v), $mr_0c_0^{-1}a = mr_0ac_0^{-1}$. Therefore

$$\begin{aligned}\varphi(v) = \varphi(ma) = \varphi(m)a = mr_0c_0^{-1}a = mr_0ac_0^{-1}\\ = mar_0c_0^{-1} = vr_0c_0^{-1} = 0.\end{aligned}$$

Since φ is a monomorphism, $v = 0$, and so r_0 acts regularly on mR. Hence $r_0 \in C$. As $C = \mathfrak{C}(E(mR))$ from Lemma 4.2(i), r_0 also acts regularly on $E(mR)$.

We show that $\varphi(mRC^{-1}) = \varphi(m)RC^{-1}$. Indeed, for $r \in R$ and $c \in C$,

$$\varphi(mrc^{-1})c = \varphi(mrc^{-1}c) = \varphi(mr) = \varphi(m)r \quad \text{and} \quad \varphi(m)rc^{-1}c = \varphi(m)r.$$

So $\varphi(mrc^{-1})c = \varphi(m)r = \varphi(m)rc^{-1}c$. As $c \in C$, $\varphi(mrc^{-1}) = \varphi(m)rc^{-1}$. Hence $\varphi(mRC^{-1}) = \varphi(m)RC^{-1}$.

Recall that $\varphi(m) = mr_0c_0^{-1}$ with $r_0 \in R$ and $c_0 \in C$. Clearly, we see that $(mR)(c_0C)^{-1} \subseteq (mR)C^{-1}$. Conversely, let $mbd^{-1} \in (mR)C^{-1}$ with $b \in R$ and $d \in C$. Then from Lemma 4.2 (iv) and (vi),

$$mbd^{-1} = mbc_0c_0^{-1}d^{-1} = mbc_0(c_0d)^{-1},$$

so $mbd^{-1} \in (mR)(c_0C)^{-1}$. Hence $(mR)C^{-1} \subseteq (mR)(c_0C)^{-1}$. Therefore, we have that $(mR)(c_0C)^{-1} = (mR)C^{-1}$.

We claim that $(mR)C^{-1}r_0 = (mR)C^{-1}$. From Lemma 4.2(v), $(mR)C^{-1}r_0 = (mR)r_0C^{-1} \subseteq (mR)C^{-1}$. Next, take $mbd^{-1} \in (mR)C^{-1}$ with $b \in R$ and $d \in C$.

Note that $r_0 \in C$ by the preceding argument. From Lemma 4.2(iv), (v), and (vi),

$$\begin{aligned} mbd^{-1} &= (mb)r_0r_0^{-1}d^{-1} = (mbr_0)r_0^{-1}d^{-1} = (mbr_0)(r_0d)^{-1} \\ &= (mb)r_0(r_0d)^{-1} = (mb)(r_0d)^{-1}r_0. \end{aligned}$$

So $mbd^{-1} \in (mR)C^{-1}r_0$. Thus $(mR)C^{-1} \subseteq (mR)C^{-1}r_0$. Therefore, it follows that $(mR)C^{-1}r_0 = (mR)C^{-1}$. (In general, $(mR)C^{-1}r = (mR)C^{-1}$ for any $r \in C$.)

By Lemma 4.2(v) and (vi), it follows that

$$\begin{aligned} \varphi(mRC^{-1}) &= \varphi(m)RC^{-1} = mr_0c_0^{-1}RC^{-1} = mr_0Rc_0^{-1}C^{-1} \\ &= mr_0R(c_0C)^{-1} = mRr_0(c_0C)^{-1} = mR(c_0C)^{-1}r_0 \\ &= (mR)C^{-1}r_0 = mRC^{-1} \end{aligned}$$

because $(mR)(c_0C)^{-1} = mRC^{-1}$ from the preceding argument. Hence φ is onto. So $(mR)C^{-1}$ is a continuous R-module.

Finally, assume that V is a continuous R-module such that $mR \leq V \leq E(fR)$. Take $d \in C$. Since $C = \mathfrak{C}(E(mR))$ from Lemma 4.2(i), d acts regularly on $E(mR)$, so does on V. Hence we have that $V \cong Vd \leq V$. As V is uniform and continuous, $Vd = V$ and so $Vd^{-1} = V$. Thus $VC^{-1} = V$. Because $mR \leq V$, $(mR)C^{-1} \leq V$. As a consequence, $(mR)C^{-1}$ is the continuous hull of mR. □

Let R be a commutative ring and I be an ideal of R. To define R-module scalar multiplication on $Q_{c\ell}(R)$, we take $(a+I)(c+I)^{-1} \in Q_{c\ell}(R/I)$ and $r \in R$ with $a+I$, $c+I \in R/I$ and $c+I$ a non-zero-divisor in R/I. Define

$$(a+I)(c+I)^{-1}r = (a+I)(c+I)^{-1}(r+I).$$

Then $Q_{c\ell}(R/I)$ is a right R-module.

Lemma 4.4. Let R be a commutative ring and mR be a uniform cyclic R-module. Put $I = \mathrm{Ann}_R(m)$. Then $(mR)\mathfrak{C}(mR)^{-1} \cong Q_{c\ell}(R/I)$ as right R-modules.

Proof. Put $C = \mathfrak{C}(mR)$. First, we prove that $c \in C$ if and only if $c+I$ is a non-zero-divisor in the ring R/I. For this, assume that $c \in C$. Then c acts regularly on mR. Take $a+I \in R/I$. If $(a+I)(c+I) = 0$, then $ac+I = 0$ and so $ac \in I = \mathrm{Ann}_R(m)$. Hence $0 = m(ac) = (ma)c$, so $ma = 0$. Thus $a \in \mathrm{Ann}_R(m) = I$. Therefore $a+I = 0$, and hence $c+I$ is a non-zero-divisor in R/I.

Conversely, suppose that $c+I$ is a non-zero-divisor in R/I. Let $ma \in mR$ with $a \in R$ and $(ma)c = 0$. Then $ac \in \mathrm{Ann}_R(m) = I$, and thus $ac + I = 0$. So $(a+I)(c+I) = 0$. As $c+I$ is a non-zero-divisor in R/I, $a+I = 0$ and hence $a \in I$. Therefore $ma = 0$, and so $c \in C$.

Define $\theta : (mR)C^{-1} \to Q_{c\ell}(R/I)$ by

$$\theta(mac^{-1}) = (a+I)(c+I)^{-1}, \quad \text{where} \quad a \in R \text{ and } c \in C.$$

If $ma_1c_1^{-1} = ma_2c_2^{-1}$ with a_1, $a_2 \in R$ and c_1, $c_2 \in C$, then $ma_1 = ma_2c_2^{-1}c_1 = ma_2c_1c_2^{-1}$ by Lemma 4.2(v). Hence $ma_1c_2 = ma_2c_1$, so $m(a_1c_2 - a_2c_1) = 0$. Thus $a_1c_2 - a_2c_1 \in \mathrm{Ann}_R(m) = I$, and hence $(a_1+I)(c_2+I) = (a_2+I)(c_1+I)$. So $(a_1+I)(c_1+I)^{-1} = (a_2+I)(c_2+I)^{-1}$. Therefore $\theta(ma_1c_1^{-1}) = \theta(ma_2c_2^{-1})$, and thus θ is well-defined.

To show that θ is additive, let $ma_ic_i^{-1} \in (mR)C^{-1}$ with $a_i \in R$ and $c_i \in C$ for i, $1 \leq i \leq 2$. Then from the proof of Lemma 4.2(vii),

$$ma_1c_1^{-1} + ma_2c_2^{-1} = (ma_1c_2 + ma_2c_1)(c_1c_2)^{-1} = [m(a_1c_2 + a_2c_1)](c_1c_2)^{-1}.$$

Hence $\theta(ma_1c_1^{-1} + ma_2c_2^{-1}) = ((a_1c_2 + a_2c_1) + I)(c_1c_2 + I)^{-1}$. On the other hand,

$$\begin{aligned}\theta(ma_1c_1^{-1}) + \theta(ma_2c_2^{-1}) &= (a_1 + I)(c_1 + I)^{-1} + (a_2 + I)(c_2 + I)^{-1}\\ &= ((a_1c_2 + a_2c_1) + I)(c_1c_2 + I)^{-1}.\end{aligned}$$

Therefore $\theta(ma_1c_1^{-1} + ma_2c_2^{-1}) = \theta(ma_1c_1^{-1}) + \theta(ma_2c_2^{-1})$. Hence θ is additive.

Now if $\theta(mac^{-1}) = (a + I)(c + I)^{-1} = 0$, then $a + I = 0$ and hence $a \in I = \mathrm{Ann}_R(m)$. Thus $ma = 0$, so $mac^{-1} = 0$. Therefore θ is one-to-one. On the other hand, we can easily check that θ is onto. Therefore θ is one-to-one and onto.

Take $mac^{-1} \in mRC^{-1}$ with $a \in R$ and $c \in C$. From Lemma 4.2(v), we note that $mac^{-1}r = marc^{-1}$ for $r \in R$. Hence

$$\begin{aligned}\theta(mac^{-1}r) &= \theta(marc^{-1}) = (ar + I)(c + I)^{-1} = (a + I)(r + I)(c + I)^{-1}\\ &= (a + I)(c + I)^{-1}(r + I) = (a + I)(c + I)^{-1}r\\ &= \theta(mac^{-1})r.\end{aligned}$$

Hence θ is an R-isomorphism. So $mRC^{-1} \cong Q_{c\ell}(R/I)$ as right R-modules. □

The next theorem follows from Theorem 4.3 and Lemma 4.4.

Theorem 4.5. Let R be a commutative ring, and mR_R a uniform cyclic module. Put $I = \mathrm{Ann}_R(m)$. Then we have the following.

(i) $mR_R \cong (R/I)_R$.

(ii) $Q_{c\ell}(R/I)_R$ is the continuous hull of $(R/I)_R$.

Proof. (i) is evident, and (ii) follows from Theorem 4.3 and Lemma 4.4. □

Theorem 4.6. Let R be a commutative ring. Then the following are equivalent.

(i) Every uniform continuous right R-module is quasi-injective.

(ii) $Q_{c\ell}(R/I)$ is a self-injective ring for every uniform factor ring R/I of R.

Proof. (i)⇒(ii) Assume that every uniform continuous R-module is quasi-injective. Let $I \trianglelefteq R$ such that $(R/I)_R$ is a uniform R-module. By Theorem 4.5 (or Theorem 3.2(ii)), $Q_{c\ell}(R/I)_R$ is the continuous hull of $(R/I)_R$ because $(R/I)_R$ is uniform and cyclic. We note that $Q_{c\ell}(R/I)_R$ is also uniform. By assumption, $Q_{c\ell}(R/I)_R$ is quasi-injective.

To show that $Q_{c\ell}(R/I)$ is self-injective, we assume that $K \trianglelefteq Q_{c\ell}(R/I)$ and $\varphi \in \mathrm{Hom}_{Q_{c\ell}(R/I)}(K, Q_{c\ell}(R/I))$. Thus $\varphi \in \mathrm{Hom}_{(R/I)}(K, Q_{c\ell}(R/I))$. Note that $Q_{c\ell}(R/I)I = 0$ because $(R/I)I = 0$. Thus $KI = 0$. Hence both (R/I)-module structures of K and $Q_{c\ell}(R/I)$ are compatible with R-module structures of K and $Q_{c\ell}(R/I)$, respectively. Because $\varphi \in \mathrm{Hom}_{(R/I)}(K, Q_{c\ell}(R/I))$, it follows that $\varphi \in \mathrm{Hom}_R(K, Q_{c\ell}(R/I))$.

Since $Q_{c\ell}(R/I)_R$ is quasi-injective by assumption and $K_R \leq Q_{c\ell}(R/I)_R$, there exists $\phi \in \mathrm{End}_R(Q_{c\ell}(R/I))$, an extension of φ. Thus $\phi \in \mathrm{End}_{(R/I)}(Q_{c\ell}(R/I))$ because $Q_{c\ell}(R/I)I = 0$.

For $x \in Q_{c\ell}(R/I)$ and a non-zero-divisor v in R/I, $\phi(xv^{-1})v = \phi(xv^{-1}v) = \phi(x)$. So $\phi(xv^{-1}) = \phi(x)v^{-1}$. Hence $\phi \in \mathrm{End}_{Q_{c\ell}(R/I)}(Q_{c\ell}(R/I))$. (This also follows from $\mathrm{End}_{(R/I)}(Q_{c\ell}(R/I)) = \mathrm{End}_{Q_{c\ell}(R/I)}(Q_{c\ell}(R/I))$ by Proposition 2.8 because $Q_{c\ell}(R/I)$ is a right ring of quotients of the ring R/I.) Thus $Q_{c\ell}(R/I)$ is self-injective by Baer's Criterion.

(ii)$\Rightarrow$(i) **Step 1.** Let R/I be a uniform factor ring of R. We show that $Q_{c\ell}(R/I)_R$ is quasi-injective. By hypothesis, $Q_{c\ell}(R/I)$ is self-injective. We note that $Q_{c\ell}(R/I)_{(R/I)}$ is a dense extension of $(R/I)_{(R/I)}$, so $Q_{c\ell}(R/I)$ is a right ring of quotients of R/I. By Theorem 3.9, $Q_{c\ell}(R/I) = E((R/I)_{(R/I)})$. Therefore $Q_{c\ell}(R/I)_{(R/I)}$ is injective.

To show that $Q_{c\ell}(R/I)_R$ is quasi-injective, we assume that $N_R \leq Q_{c\ell}(R/I)_R$ and $f \in \mathrm{Hom}_R(N, Q_{c\ell}(R/I))$. Since $(R/I)\,I = 0$, $Q_{c\ell}(R/I)I = 0$ and hence $NI = 0$. Therefore N is an (R/I)-module, and furthermore $f \in \mathrm{Hom}_{(R/I)}(N, Q_{c\ell}(R/I))$.

As $Q_{c\ell}(R/I)_{(R/I)}$ is injective by the preceding argument, there exists $g \in \mathrm{End}_{(R/I)}(Q_{c\ell}(R/I))$, which is an extension of f. Hence $g \in \mathrm{End}_R(Q_{c\ell}(R/I))$ because $Q_{c\ell}(R/I)I = 0$. Therefore $Q_{c\ell}(R/I)_R$ is quasi-injective.

Step 2. Assume that A_R is a uniform continuous R-module. Let $0 \neq x \in A$. We put $J = \mathrm{Ann}_R(x)$. Then $xR_R \cong (R/J)_R$. By Lemma 4.4,

$$(xR)\mathfrak{C}(xR)_R^{-1} \cong Q_{c\ell}(R/J)_R,$$

and by Theorem 4.3, $(xR)\mathfrak{C}(xR)_R^{-1}$ is the continuous hull of xR_R.

Since A_R is continuous, $xR_R \leq A_R$, and $(xR)\mathfrak{C}(xR)_R^{-1}$ is the continuous hull of xR_R, we have that $(xR)\mathfrak{C}(xR)_R^{-1} \leq A_R$. By Step 1, $Q_{c\ell}(R/J)_R$ is quasi-injective because the ring R/J is uniform. As $(xR)\mathfrak{C}(xR)_R^{-1} \cong Q_{c\ell}(R/J)_R$, it follows that $(xR)\mathfrak{C}(xR)_R^{-1}$ is quasi-injective.

We observe that $E(A) = E((xR)\mathfrak{C}(xR)^{-1})$ as $(xR)\mathfrak{C}(xR)_R^{-1} \leq^{\text{ess}} A_R$. We take $\theta \in \mathrm{End}_R(E(A))$. Since $(xR)\mathfrak{C}(xR)_R^{-1}$ is quasi-injective, $\theta((xR)\mathfrak{C}(xR)^{-1}) \subseteq (xR)\mathfrak{C}(xR)^{-1}$. So $\theta(x) \in (xR)\mathfrak{C}(xR)^{-1} \subseteq A$. Thus $\theta(x) \in A$. Consequently, A_R is quasi-injective. □

Motivated from Theorem 4.6, we shall now be interested in the following condition on a commutative ring.

4.7 Condition ($\star$). We say that a commutative ring R satisfies *condition* ($\star$) if $Q_{c\ell}(R/I)$ is self-injective for every uniform factor ring R/I of R.

Definition 4.8. A commutative ring R is called *finitary* if every cyclic R-module is of finite uniform dimension. Examples of such rings include commutative noetherian rings, and more generally, commutative rings with Krull dimension in the sense of Rentschler and Gabriel (see [**6**] for Krull dimension).

While every quasi-injective module is continuous, the converse is not true (e.g., [**2**, Example 2.1.36]). The following provides a class of commutative rings over which every continuous module is quasi-injective.

For a right R-module M, $\mathrm{udim}(M)$ denotes the uniform dimension of M.

Theorem 4.9. Assume that R is a commutative finitary ring. Then the following are equivalent.

(i) Every continuous right R-module is quasi-injective.
(ii) R satisfies condition ($\star$).

Proof. (i)⇒(ii) The proof follows immediately from Theorem 4.6.

(ii)⇒(i) Let R satisfy condition $(\star)$, and let M_R be a continuous R-module. We put $M = M_R$. To prove that M is quasi-injective, we need to show that $M \trianglelefteq E(M)$ from Proposition 2.2. For this, take $0 \neq x \in M$ and $\phi \in \text{End}(E(M))$. We now claim that $\phi(x) \in M$.

Put $xR = xR_R$. Since R is finitary, xR is of finite uniform dimension, say $n = \text{udim}(xR)$. There exist nonzero uniform submodules $U_1, \dots, U_n$ of xR such that $(U_1 \oplus \cdots \oplus U_n)_R \leq^{\text{ess}} xR_R$. Hence

$$E(xR) = E(U_1 \oplus \cdots \oplus U_n) = E(U_1) \oplus \cdots \oplus E(U_n).$$

We note that each $E(U_i)$ is uniform.

Let $E_i = E(U_i)$ for $i = 1, 2, \dots, n$. Then $E(xR) = \oplus_{i=1}^n E_i$, where E_i is injective and uniform. Because $E(xR) \leq E(M)$, $E(M) = E_0 \oplus (\oplus_{i=1}^n E_i)$ for some injective module E_0.

Now we simply assume that $n = 3$. Then $E(M) = E_0 \oplus E_1 \oplus E_2 \oplus E_3$. Since M is continuous, Theorem 2.10 yields that

$$M = (M \cap E_0) \oplus (M \cap E_1) \oplus (M \cap E_2) \oplus (M \cap E_3).$$

Put $M_i = M \cap E_i$ for $i = 0, 1, 2, 3$. Then $M = M_0 \oplus M_1 \oplus M_2 \oplus M_3$.

As M is continuous, each M_i, $0 \leq i \leq 3$, is continuous, and each M_i is M_j-injective for all $j \neq i$ from Theorem 2.13. Note that M_1, M_2, and M_3 are uniform and continuous. Hence by Theorem 4.6, M_1, M_2, and M_3 are quasi-injective.

Because $\phi \in \text{End}_R(E(M)) = \text{End}_R(E_0 \oplus E_1 \oplus E_2 \oplus E_3)$, we have that

$$\phi = \begin{bmatrix} \phi_{0\,0} & \phi_{0\,1} & \phi_{0\,2} & \phi_{0\,3} \\ \phi_{1\,0} & \phi_{1\,1} & \phi_{1\,2} & \phi_{1\,3} \\ \phi_{2\,0} & \phi_{2\,1} & \phi_{2\,2} & \phi_{2\,3} \\ \phi_{3\,0} & \phi_{3\,1} & \phi_{3\,2} & \phi_{3\,3} \end{bmatrix},$$

where $\phi_{i\,j} \in \text{Hom}_R(E_j, E_i)$, $0 \leq i, j \leq 3$.

Write $x = m_0 + m_1 + m_2 + m_3$ with $m_i \in M_i$, $0 \leq i \leq 3$. Note that

$$x \in xR \subseteq E(xR) = E_1 \oplus E_2 \oplus E_3.$$

So $m_0 = 0$ and hence $x = m_1 + m_2 + m_3$. Therefore

$$\phi(x) = \begin{bmatrix} \phi_{0\,0} & \phi_{0\,1} & \phi_{0\,2} & \phi_{0\,3} \\ \phi_{1\,0} & \phi_{1\,1} & \phi_{1\,2} & \phi_{1\,3} \\ \phi_{2\,0} & \phi_{2\,1} & \phi_{2\,2} & \phi_{2\,3} \\ \phi_{3\,0} & \phi_{3\,1} & \phi_{3\,2} & \phi_{3\,3} \end{bmatrix} \begin{bmatrix} 0 \\ m_1 \\ m_2 \\ m_3 \end{bmatrix} = \begin{bmatrix} \phi_{0\,1}(m_1) + \phi_{0\,2}(m_2) + \phi_{0\,3}(m_3) \\ \phi_{1\,1}(m_1) + \phi_{1\,2}(m_2) + \phi_{1\,3}(m_3) \\ \phi_{2\,1}(m_1) + \phi_{2\,2}(m_2) + \phi_{2\,3}(m_3) \\ \phi_{3\,1}(m_1) + \phi_{3\,2}(m_2) + \phi_{3\,3}(m_3) \end{bmatrix}$$

and

$$\phi(x) \in \begin{bmatrix} E_0 \\ E_1 \\ E_2 \\ E_3 \end{bmatrix} = E.$$

Note that $E_i = E(M_i)$ for $0 \le i \le 3$. As M_1, M_2, M_3 are quasi-injective and $\phi_{i\,i} \in \mathrm{End}_R(E_i)$, $\phi_{i\,i}(m_i) \in M_i$ for $1 \le i \le 3$ (see Proposition 2.2).

For $j \ne i$ where $0 \le i, j \le 3$, $\phi_{i\,j} \in \mathrm{Hom}_R(E_j, E_i)$ and M_i is M_j-injective, hence $\phi_{i\,j}(M_j) \subseteq M_i$ by Theorem 2.4 and so $j \ne i$, $\phi_{j\,i}(m_j) \in M_i$. Therefore

$$\phi(x) = \begin{bmatrix} \phi_{0\,1}(m_1) + \phi_{0\,2}(m_2) + \phi_{0\,3}(m_3) \\ \phi_{1\,1}(m_1) + \phi_{1\,2}(m_2) + \phi_{1\,3}(m_3) \\ \phi_{2\,1}(m_1) + \phi_{2\,2}(m_2) + \phi_{2\,3}(m_3) \\ \phi_{3\,1}(m_1) + \phi_{3\,2}(m_2) + \phi_{3\,3}(m_3) \end{bmatrix} \in \begin{bmatrix} M_0 \\ M_1 \\ M_2 \\ M_3 \end{bmatrix} = M.$$

Hence $\phi(x) \in M$, and so $M \trianglelefteq E(M)$. Therefore M is quasi-injective from Proposition 2.2. The preceding method also can be applied to general case for n. □

The next result follows from [**15**, Proposition 4.23, p.107 and Proposition 5.12, p.134].

Proposition 4.10. Assume that R is a commutative local noetherian ring with the maximal ideal M. We let $E = E((R/M)_R)$ and let $A_n = \mathrm{Ann}_E(M^n)$, where $n = 1, 2, \ldots$. Then we have the following.

(i) $E = \bigcup_{n=1}^{\infty} A_n$.

(ii) For $f \in \mathrm{End}_R(E)$ and a positive integer k, there exists $r_k \in R$ such that $f(v) = vr_k$ for all $v \in A_k$.

We obtain the next result by using Proposition 4.10.

Theorem 4.11. Assume that R is a commutative local noetherian ring with the maximal ideal M. Let $E = E((R/M)_R)$. Then every submodule of E is quasi-injective.

Proof. Let $V_R \le E_R$. We may assume that $V \ne 0$. So $E(V)$ is a nonzero direct summand of E. Therefore $E(V) = E$ since E is indecomposable. To show that V is quasi-injective, take $f \in \mathrm{End}_R(E(V)) = \mathrm{End}_R(E)$ and $w \in V$. We show that $f(w) \in V$. By Proposition 4.10, there exists a positive integer k such that $w \in A_k$ since $E = \bigcup_{n=1}^{\infty} A_n$.

On the other hand, there exists $r_k \in R$ such that $f(v) = vr_k$ for all $v \in A_k$ from Proposition 4.10. In particular, $f(w) = wr_k \in V$. So $f(V) \subseteq V$ for all $f \in \mathrm{End}_R(E(V))$. Hence V is quasi-injective by Proposition 2.2. □

The next lemma is in [**8**] (see also [**15**, Lemma 2.30, p.50]).

Lemma 4.12. Let R be a commutative noetherian ring, and E be an indecomposable injective R-module. Then there exists a unique prime ideal P of R such that $E \cong E((R/P)_R)$ as R-modules.

Proof. Take $0 \ne v \in E$, and put $M = vR$. Then M is a nonzero noetherian R-module. Let

$$\Omega = \{\mathrm{Ann}_R(x) \mid 0 \ne x \in M\},$$

where $\mathrm{Ann}_R(x) = \{r \in R \mid xr = 0\}$, the annihilator of x in R. Then Ω is nonempty because $0 \ne v \in M$ and so $\mathrm{Ann}_R(v) \in \Omega$.

We show that there exists a maximal element in Ω. If not, then there exists nonzero $x_1, x_2, x_3, \ldots$ in M such that $\mathrm{Ann}_R(x_1) \subset \mathrm{Ann}_R(x_2) \subset \mathrm{Ann}_R(x_3) \subset \cdots$, where "$\subset$" means proper inclusion. So we have the following ascending chain of submodule of R-module $R/\mathrm{Ann}_R(x_1)$, which is

$$0 \neq \mathrm{Ann}_R(x_2)/\mathrm{Ann}_R(x_1) \subset \mathrm{Ann}_R(x_3)/\mathrm{Ann}_R(x_1) \subset \cdots ,$$

where "$\subset$" also means proper inclusion.

Note that $R/\mathrm{Ann}_R(x_1) \cong x_1R$ and x_1R is a noetherian R-module. So there is a positive integer n such that $\mathrm{Ann}_R(x_n)/\mathrm{Ann}_R(x_1) = \mathrm{Ann}_R(x_{n+1})/\mathrm{Ann}_R(x_1) = \cdots$. Hence $\mathrm{Ann}_R(x_n) = \mathrm{Ann}_R(x_{n+1}) = \cdots$, which is a contradiction. Hence there exists a maximal element in Ω.

Let $\mathrm{Ann}_R(m)$, with $0 \neq m \in M$, be a maximal element in Ω. We put

$$P = \mathrm{Ann}_R(m).$$

We claim that P is a prime ideal of R. For this, first note that P is a proper ideal because $m \neq 0$. Let $a, b \in R$ such that $ab \in P$. Assume that $b \notin P$. Then $mab = 0$, but $mb \neq 0$. Thus $\mathrm{Ann}(mb) \in \Omega$ and $\mathrm{Ann}_R(m) \subseteq \mathrm{Ann}_R(mb)$. As $\mathrm{Ann}_R(m)$ is a maximal element in Ω, it follows that $\mathrm{Ann}_R(m) = \mathrm{Ann}_R(mb)$. Note that $a \in \mathrm{Ann}_R(mb)$, so $a \in \mathrm{Ann}_R(m) = P$. Therefore P is a prime ideal of R.

Now $R/P = R/\mathrm{Ann}_R(m) \cong mR \leq E$ as R-modules. So $E(mR_R) \leq^{\oplus} E$. Since E is an indecomposable injective R-module, $E = E(mR_R) \cong E((R/P)_R)$ as R-modules.

Let P_1 be a prime ideal of R such that $E \cong E((R/P_1)_R)$ as R-modules. Then, from the preceding argument, it follows that $E \cong E((R/P)_R) \cong E((R/P_1)_R)$ as R-modules. Thus $\mathfrak{C}(E((R/P)_R)) = \mathfrak{C}(E((R/P_1)_R))$. Note that

$$\mathfrak{C}(E((R/P)_R)) = \mathfrak{C}((R/P)_R) \text{ and } \mathfrak{C}(E((R/P_1)_R)) = \mathfrak{C}((R/P_1)_R)$$

from Lemma 4.2(i). Hence $\mathfrak{C}((R/P)_R) = \mathfrak{C}((R/P_1)_R)$.

We show that $\mathfrak{C}((R/P)_R) = R \setminus P$, where $R \setminus P = \{r \in R \mid r \notin P\}$. Obviously, we see that $R \setminus P \subseteq \mathfrak{C}((R/P)_R)$. Conversely, let $r \in \mathfrak{C}((R/P)_R)$. If $r \in P$, then $(1+P)r = 0$ but $1 + P \neq 0$. Thus $r \notin \mathfrak{C}((R/P)_R)$, a contradiction. So $r \in R \setminus P$. Hence $\mathfrak{C}((R/P)_R) = R \setminus P$. Similarly, $\mathfrak{C}((R/P_1)_R) = R \setminus P_1$.

As $\mathfrak{C}((R/P)_R) = \mathfrak{C}((R/P_1)_R)$, $\mathfrak{C}((R/P)_R) = R \setminus P$, and $\mathfrak{C}((R/P_1)_R) = R \setminus P_1$, we have that $R \setminus P = R \setminus P_1$, and so $P = P_1$. Hence the prime ideal P is a unique prime ideal of R such that $E \cong E((R/P)_R)$ as R-modules. □

Let R be a commutative ring and P a prime ideal of R. We let R_P be the localization of R at P. Note that R_P is a local ring with the maximal ideal PR_P.

Some parts of the proof of the following theorem are routine. However, we provide the proof in detail for convenience and completeness.

Theorem 4.13. Any commutative noetherian ring satisfies condition $(\star)$.

Proof. For the proof, first let R be a uniform, commutative, and noetherian ring. We show that $Q_{c\ell}(R)$ is self-injective. Since R is uniform, $E(R_R)$ is an indecomposable injective R-module. By Lemma 4.12, there exists a prime ideal P of R uniquely such that $E(R_R) \cong E((R/P)_R)$ and as R-modules. From Lemma 4.2(i),

$$\mathfrak{C}(R_R) = \mathfrak{C}(E(R_R)) \text{ and } \mathfrak{C}(E((R/P)_R) = \mathfrak{C}((R/P)_R).$$

Hence

$$\mathfrak{C}(R_R) = \mathfrak{C}(E(R_R)) = \mathfrak{C}(E(R/P)_R) = \mathfrak{C}((R/P)_R).$$

Note that $\mathfrak{C}((R/P)_R) = R \setminus P$ from the proof of Lemma 4.12. As $\mathfrak{C}(R_R)$ is the set of all non-zero-divisors of R. Now since $R \setminus P = \mathfrak{C}(R_R)$, $R \setminus P$ is the set of all non-zero-divisors of R.

Note that $R_P = Q_{c\ell}(R)$ because $R \setminus P$ is the set of all non-zero-divisors of R. Put $C = \mathfrak{C}(R_R) = \mathfrak{C}(E((R/P)_R) = R \setminus P$.

Step 1. We claim that $E((R/P)_R)$ becomes an R_P-module with the induced R_P-module scalar multiplication of $E((R/P)_R)$ from the R-module scalar multiplication of $E((R/P)_R)$. Further, the induced R_P-module scalar multiplication extends the R-module scalar multiplication of $E((R/P)_R)$.

For our claim, we adopt similar methods used in the proof of Lemma 4.2. Now take $v \in E((R/P)_R)$, and $rc^{-1} \in R_P$ with $r \in R$ and $c \in C = R \setminus P$. Because $E((R/P)_R)$ is indecomposable injective and c acts regularly on $E((R/P)_R)$, we have that $E((R/P)_R) = E((R/P)_R)c$ and thus there exists $y \in E((R/P)_R)$ uniquely such that $v = yc$ as in the proof of Lemma 4.2(ii).

Define

$$v \cdot (rc^{-1}) = yr \in E((R/P)_R).$$

To show that "$\cdot$" is well-defined, suppose that $r_1c_1^{-1} = r_2c_2^{-1}$ with $r_1, r_2 \in R$ and $c_1, c_2 \in C$. Then $r_1c_2 = r_2c_1$. There exist $y_1, y_2 \in E((R/P)_R)$ such that $v = y_1c_1$ and $v = y_2c_2$ by the preceding argument. Therefore $vr_1 = y_1c_1r_1 = y_1r_1c_1$ and $vr_2 = y_2c_2r_2 = y_2r_2c_2$. Hence $vr_1c_2 = y_1r_1c_1c_2$ and $vr_2c_1 = y_2r_2c_2c_1$.

Because $r_1c_2 = r_2c_1$, $vr_1c_2 = vr_2c_1$ and thus $y_1r_1c_1c_2 = y_2r_2c_2c_1 = y_2r_2c_1c_2$. Hence $(y_1r_1 - y_2r_2)c_1c_2 = 0$. As $c_1c_2 \in C$, c_1c_2 acts regularly on $E((R/P)_R)$ and so $y_1r_1 - y_2r_2 = 0$. Thus $y_1r_1 = y_2r_2$ and therefore $v \cdot (r_1c_1^{-1}) = v \cdot (r_2c_2^{-1})$. So the scalar multiplication "$\cdot$" is well-defined.

Further, note that $v \cdot (rc^{-1}) = yr$, where $v = yc$. Hence $y = v \cdot c^{-1}$. So we have that

$$v \cdot (rc^{-1}) = (v \cdot c^{-1})r.$$

Next, let $v \in E((R/P)_R)$, and $r_1c_1^{-1}$, $r_2c_2^{-1} \in R_P$ with r_1, $r_2 \in R$ and c_1, $c_2 \in C$. We show that

$$v \cdot (r_1c_1^{-1} + r_2c_2^{-1}) = v \cdot (r_1c_1^{-1}) + v \cdot (r_2c_2^{-1}).$$

Indeed, as c_1 acts regularly on $E((R/P)_R)$, there exists uniquely $v_1 \in E((R/P)_R)$ such that $v = v_1c_1$ by the preceding argument. Thus $v \cdot c_1^{-1} = v_1$. Similarly, there exists uniquely $v_2 \in E((R/P)_R)$ such that $v = v_2c_2$. So $v \cdot c_2^{-1} = v_2$. Since $v \cdot (r_1c_1^{-1}) = (v \cdot c_1^{-1})r_1$ and $v \cdot (r_2c_2^{-1}) = (v \cdot c_2^{-1})r_2$ from the preceding argument,

$$v \cdot (r_1c_1^{-1}) = (v \cdot c_1^{-1})r_1 = v_1r_1 \quad \text{and} \quad v \cdot (r_2c_2^{-1}) = (v \cdot c_2^{-1})r_2 = v_2r_2.$$

Thus $v \cdot (r_1c_1^{-1}) + v \cdot (r_2c_2^{-1}) = v_1r_1 + v_2r_2$.

On the other hand, note that c_1c_2 acts regularly on $E((R/P)_R)$. Thus there exists uniquely $u \in E(R/P)_R)$ such that $v = uc_1c_2$. Hence $u = v \cdot (c_1c_2)^{-1}$.

Recall that $v = v_1c_1$ and $v = v_2c_2$. Thus $uc_2c_1 = uc_1c_2 = v = v_1c_1$. Since c_1 acts regularly on $E((R/P)_R)$, we have that $uc_2 = v_1$. Also $uc_1c_2 = v = v_2c_2$ and c_2 acts regularly on $E((R/P)_R)$, and hence $uc_1 = v_2$. Further, also recall from our preceding argument that $w \cdot (ad^{-1}) = (w \cdot d^{-1})a$ for any $w \in E((R/P)_R)$, $a \in R$, and $d \in C$. Therefore

$$\begin{aligned} v \cdot (r_1 c_1^{-1} + r_2 c_2^{-1}) &= v \cdot ((r_1 c_2 + r_2 c_1)(c_1 c_2)^{-1}) = (v \cdot (c_1 c_2)^{-1})(r_1 c_2 + r_2 c_1) \\ &= u(r_1 c_2 + r_2 c_1) = u r_1 c_2 + u r_2 c_1 = (u c_2) r_1 + (u c_1) r_2 \\ &= v_1 r_1 + v_2 r_2 = v \cdot (r_1 c_1^{-1}) + v \cdot (r_2 c_2^{-1}). \end{aligned}$$

Other axioms for the scalar multiplication for $E((R/P)_R)$ to be an R_P-module can be routinely checked. As a consequence, $E((R/P)_R)$ is an R_P-module.

Take $w \in E((R/P)_R)$ and $a \in R$. As $w = w1$ and $1 \in C$, $w \cdot 1^{-1} = w$. Therefore

$$w \cdot a = w \cdot (a 1^{-1}) = (w \cdot 1^{-1}) a = wa.$$

Therefore the R_P-module scalar multiplication "$\cdot$" of $E((R/P)_R)$ extends the R-module scalar multiplication of $E((R/P)_R)$.

Now, we may write $v \cdot (rc^{-1}) = v(rc^{-1})$ for $v \in E((R/P)_R)$, $r \in R$, and $c \in C$.

Step 2. We show that $Q_{c\ell}(R)$ is self-injective. For this, we first note that $\mathrm{End}_{R_P}(E((R/P)_R)) \subseteq \mathrm{End}_R(E((R/P)_R))$. We take $\varphi \in \mathrm{End}_R(E((R/P)_R))$. Then for $v \in E((R/P)_R)$ and $c \in C$, $\varphi(vc^{-1})c = \varphi(vc^{-1}c) = \varphi(v)$ and so $\varphi(vc^{-1}) = \varphi(v)c^{-1}$ because $c \in C$. So it follows that $\varphi \in \mathrm{End}_{R_P}(E((R/P)_R))$. Therefore

$$\mathrm{End}_R(E((R/P)_R)) = \mathrm{End}_{R_P}(E((R/P)_R)).$$

Now to show that $E((R/P)_R)$ is injective as an R_P-module, let A be an ideal of R_P and we take $f \in \mathrm{Hom}_{R_P}(A, E((R/P)_R))$. As $E((R/P)_R)$ is injective as an R-module, there is $g \in \mathrm{Hom}_R(R_P, E((R/P)_R))$, which is an extension of f. Since $\mathrm{Hom}_R(R_P, E((R/P)_R)) = \mathrm{Hom}_{R_P}(R_P, E((R/P)_R))$ (similarly from the preceding argument),

$$g \in \mathrm{Hom}_{R_P}(R_P, E((R/P)_R))$$

and hence $E((R/P)_R)$ is injective as an R_P-module by Baer's Criterion.

We now take $0 \neq x \in R/P$. Then obviously $PR_P \subseteq \mathrm{Ann}_{R_P}(x)$. Conversely, let $ac^{-1} \in \mathrm{Ann}_{R_P}(x)$ with $a \in R$ and $c \in C$. Then $xac^{-1} = 0$, so $xa = 0$. Put $x = r + P$ with $r \in R$. Then $r \notin P$ because $x \neq 0$. Now $0 = xa = (r+P)a = ra + P$. Thus $ra \in P$. As $r \notin P$, $a \in P$ and hence $ac^{-1} \in PR_P$. Therefore $\mathrm{Ann}_{R_P}(x) \subseteq PR_P$. Consequently, $\mathrm{Ann}_{R_P}(x) = PR_P$. So $0 \neq xR_P \cong R_P/\mathrm{Ann}_{R_P}(x) = R_P/PR_P$. Thus

$$E((xR_P)_{R_P}) \cong E((R_P/PR_P)_{R_P})$$

as R_P-modules.

On the other hand, by the preceding argument, $E((R/P)_R)$ is injective as an R_P-module. Now we note that $E((R/P)_R)$ is indecomposable as an R_P-module because $E((R/P)_R)$ is indecomposable as an R-module. Therefore $E((xR_P)_{R_P}) = E((R/P)_R)$ as R_P-modules because $0 \neq xR_P \leq E((R/P)_R)$ as R_P-modules.

Since $E((xR_P)_{R_P}) \cong E((R_P/PR_P)_{R_P})$ and $E((xR_P)_{R_P}) = E((R/P)_R)$ as R_P-modules, we have that $E((R/P)_R) \cong E((R_P/PR_P)_{R_P})$ as R_P-modules. Thus there exists an R_P-isomorphism

$$\theta : E((R/P)_R) \to E((R_P/PR_P)_{R_P}).$$

We note that $R_P \leq E(R_R) \cong E((R/P)_R)$ as R-modules. Let

$$\lambda : E(R_R) \to E((R/P)_R)$$

be an R-isomorphism. Take $y \in E(R_R)$ and $s \in R_P$, by adopting the idea of the proof of Theorem 3.23, define $y \circ s$ as follows:

$$y \circ s = \lambda^{-1}(\lambda(y)s).$$

Then $y \circ s \in E(R_R)$ and $\lambda(y \circ s) = \lambda(y)s$. Thus, we can check that $\circ$ is an R_P-module scalar multiplication of $E(R_R)$, and so $E(R_R)$ is an R_P-module. Further, λ is an R_P-isomorphism. Also, for $r \in R$,

$$y \circ r = \lambda^{-1}(\lambda(y)r) = \lambda^{-1}(\lambda(yr)) = yr.$$

Hence the R_P-module scalar multiplication $\circ$ of $E(R_R)$ extends the R-module scalar multiplication of $E(R_R)$.

From the preceding argument, we have that

$$\gamma := \theta\,\lambda : E(R_R) \to E((R_P/PR_P)_{R_P})$$

is an R_P-isomorphism. Since R_P is an R-submodule of $E(R_R)$, R_P is an R_P-submodule of $E(R_R)$.

We show that R_P is a noetherian ring. For this, let $I \trianglelefteq R_P$. Then $I = (I \cap R)R_P$. Indeed, obviously $(I \cap R)R_P \subseteq I$. Conversely, take $\beta \in I$. Then $\beta = ac^{-1}$ with $a \in R$ and $c \in C = R \setminus P$. Therefore $a = \beta c \in I \cap R$, and thus $\beta = ac^{-1} \in (I \cap R)R_P$. Hence $I \subseteq (I \cap R)R_P$. So $I = (I \cap R)R_P$. Let

$$I_1 \subseteq I_2 \subseteq I_3 \subseteq \cdots$$

be a ascending chain of ideals of R_P. Then $I_1 \cap R \subseteq I_2 \cap R \subseteq I_3 \cap R \subseteq \cdots$ is an ascending chain of ideals of R. Since R is noetherian, there exists a positive integer n such that $I_n \cap R = I_{n+1} \cap R = I_{n+2} \cap R = \cdots$. Therefore

$$(I_n \cap R)R_P = (I_{n+1} \cap R)R_P = (I_{n+2} \cap R)R_P = \cdots .$$

Hence $I_n = I_{n+1} = I_{n+2} = \cdots$ because $(I \cap R)R_P = I$ for any ideal of I of R_P from the preceding argument. Therefore R_P is a noetherian ring.

By Theorem 4.11, $W := \gamma(R_P)$ is a quasi-injective R_P-module as W is an R_P-submodule of the injective hull $E((R_P/PR_P)_{R_P})$ of the R_P-module $(R_P/PR_P)_{R_P}$, where R_P is a commutative local noetherian ring with PR_P the maximal ideal. Hence R_P is quasi-injective as an R_P-module. Therefore R_P is self-injective, so $Q_{c\ell}(R)$ is self-injective because $R_P = Q_{c\ell}(R)$. As a consequence, by Step 1 and Step 2, for any commutative uniform noetherian ring R, $Q_{c\ell}(R)$ is self-injective.

Now in general, assume that R be a commutative noetherian ring. Let R/I be a uniform factor ring of R. Then the ring R/I is a uniform commutative noetherian ring, and hence $Q_{c\ell}(R/I)$ is self-injective by the preceding argument. Therefore R satisfies condition $(\star)$. □

From our previous results, we obtain the following interesting result for continuous modules.

Theorem 4.14. Let R be a commutative noetherian ring. Then every continuous R-module is quasi-injective.

Proof. The proof follows from Theorem 4.9 and Theorem 4.13. □

In contrast to Example 3.1, the next result shows that every module over a commutative noetherian ring has a continuous hull, which coincides with its quasi-injective hull.

Theorem 4.15. Let R be a commutative noetherian ring and M_R be a module. Then we have the following.

(i) M_R has a continuous hull.

(ii) The continuous hull of M_R is exactly the quasi-injective hull of M_R.

Proof. Let $H = \text{End}_R(E(M))$. Then by Theorem 2.9, HM is the quasi-injective hull of M. So HM is a continuous R-module. Let V be a continuous R-module such that $M \leq V \leq E(M)$. From Theorem 4.14, V is quasi-injective, and hence $HM \leq V$. Therefore HM is the continuous hull of M. □

Assume that R is a commutative finitary ring satisfying condition $(\star)$. Then every continuous R-module is quasi-injective by Theorem 4.9. Therefore every R-module M has a continuous hull which coincides with the quasi-injective hull of M from the argument in the proof of Theorem 4.15. Thus we have the following theorem related to Theorem 4.15.

Theorem 4.16. Assume that R is a commutative finitary ring with $(\star)$. Let M_R be a module. Then we have the following.

(i) M_R has a continuous hull.

(ii) The continuous hull of M_R is exactly the quasi-injective hull of M_R.

In Theorem 3.2, every uniform *cyclic* module over a commutative ring has a continuous hull (see also Theorem 4.4 and Theorem 4.5). Related to this result, we obtain the following.

Theorem 4.17. Assume that R is a commutative ring with condition $(\star)$. Then every uniform module M_R has a continuous hull which is exactly the quasi-injective hull of M_R.

Proof. Let $H = \text{End}(E(M))$. Then Theorem 2.8 yields that HM is the quasi-injective hull of M. Thus HM is a continuous overmodule of M. Let V be a continuous module such that $M \leq V \leq E(M)$. Since V is uniform, Theorem 4.6 yields that V is quasi-injective. Thus $HM \leq V$. Hence HM is the continuous hull of M. □

In the following Example 4.18, there exits a commutative ring R with condition $(\star)$ which is not finitary. However, the ring R is semiprimary with $J(R)^2 = 0$, where $J(R)$ is the Jacobson radical of R. (Recall that a ring R is called semiprimary if $R/J(R)$ is (right) artinian and $J(R)^n = 0$ for some positive integer n.) Hence from [**14**, Corollary 10], every continuous R-module is quasi-injective.

By using the argument used in the proof of Theorem 4.15, every R-module M over a commutative noetherian ring R has a continuous hull which coincides with the quasi-injective hull of M.

Example 4.18. By Theorem 4.13, any commutative noetherian ring satisfies condition $(\star)$. In the following, there exists a commutative ring R with condition $(\star)$ but not finitary.

Let $R = T(\mathbb{Q}, \mathbb{Q}[x])$, the trivial extension of $\mathbb{Q}$ by $\mathbb{Q}[x]$, that is,

$$R = \left\{ \begin{bmatrix} q & f(x) \\ 0 & q \end{bmatrix} \mid q \in \mathbb{Q} \text{ and } f(x) \in \mathbb{Q}[x] \right\}$$

with usual matrix addition and multiplication as a subring of $\text{Mat}_2(\mathbb{Q}[x])$.

Note that R is a commutative local ring and $R/J(R) \cong \mathbb{Q}$ with $J(R)^2 = 0$ (hence R is semiprimary). We let I be a proper ideal of R such that R/I is uniform. Then there exists $\mathbb{Q}$-subspace V of $\mathbb{Q}[x]$ such that

$$I = \begin{bmatrix} 0 & V \\ 0 & 0 \end{bmatrix}.$$

Then $\mathbb{Q}[x] = V \oplus W$ for some $\mathbb{Q}$-subspace W of $\mathbb{Q}[x]$.

If $\dim(W) > 1$ (where $\dim(W)$ is the $\mathbb{Q}$-vector space dimension), then there exist nonzero $w_1, w_2 \in W$ such that $\mathbb{Q}w_1 \oplus \mathbb{Q}w_2 \subseteq W$. Consider

$$\alpha := \begin{bmatrix} 0 & w_1 \\ 0 & 0 \end{bmatrix} + I \in R/I \text{ and } \beta := \begin{bmatrix} 0 & w_2 \\ 0 & 0 \end{bmatrix} + I \in R/I.$$

Then $\alpha(R/I) \neq 0$, $\beta(R/I) \neq 0$, and $\alpha(R/I) \cap \beta(R/I) = 0$. So R/I is not uniform. Hence if R/I is uniform, $\dim(W) = 0$ or $\dim(W) = 1$.

If $\dim(W) = 0$, then $V = \mathbb{Q}[x]$. Hence $R/I \cong \mathbb{Q}$, and thus R/I is uniform and $Q_{c\ell}(R/I) \cong \mathbb{Q}$ is self-injective.

Next assume that $\dim(W) = 1$. Then $W = \mathbb{Q}w$ for some $0 \neq w \in W$. Then

$$R/I = \left\{ \begin{bmatrix} a & bw \\ 0 & a \end{bmatrix} + I \mid a, b \in \mathbb{Q} \right\}.$$

Define $\theta : R/I \to T(\mathbb{Q}, \mathbb{Q})$, the trivial extension of $\mathbb{Q}$ by $\mathbb{Q}$, by

$$\theta \left(\begin{bmatrix} a & bw \\ 0 & a \end{bmatrix} + I \right) = \begin{bmatrix} a & b \\ 0 & a \end{bmatrix}.$$

Then θ is a ring isomorphism.

Now $T(\mathbb{Q}, \mathbb{Q})$ is a commutative uniform noetherian ring. Hence Theorem 4.13 yields that $T(\mathbb{Q}, \mathbb{Q})$ satisfies condition $(\star)$, so $Q_{c\ell}(T(\mathbb{Q}, \mathbb{Q}))$ is self-injective as $T(\mathbb{Q}, \mathbb{Q})$ is uniform (in fact, $Q_{c\ell}(T(\mathbb{Q}, \mathbb{Q})) = T(\mathbb{Q}, \mathbb{Q})$). See also Remark 4.19 for $T(\mathbb{Q}, \mathbb{Q})$. Therefore $Q_{c\ell}(R/I)$ is self-injective. Therefore R satisfies condition $(\star)$.

For a positive integer n, put

$$K_n = \left\{ \begin{bmatrix} 0 & ax^n \\ 0 & 0 \end{bmatrix} \mid a \in \mathbb{Q} \right\}.$$

Then each K_n is a nonzero right ideal of R, and $(K_1 \oplus K_2 \oplus \cdots) \leq R_R$. Thus R_R has no finite uniform dimension. So R is not finitary.

Remark 4.19. In Example 4.18, additionally we also note that the ring $T(\mathbb{Q}, \mathbb{Q})$ is QF (self-injective and artinian). Indeed, note that $T(\mathbb{Q}, \mathbb{Q}) \cong \mathbb{Q}[x]/x^2\mathbb{Q}[x]$ by corresponding

$$\begin{bmatrix} a & b \\ 0 & a \end{bmatrix} \in T(\mathbb{Q}, \mathbb{Q}) \text{ to } a + bx + x^2\mathbb{Q}[x] \in \mathbb{Q}[x]/x^2\mathbb{Q}[x].$$

As $\mathbb{Q}[x]$ is a commutative PID (hence Dedekind), the ring $\mathbb{Q}[x]/x^2\mathbb{Q}[x]$ is QF. Therefore the ring $T(\mathbb{Q}, \mathbb{Q})$ is QF.

Recall that a module is called *directly finite* if it is not isomorphic to a proper direct summand of itself. Let M be a continuous module. Then $M = U \oplus V$, where U is directly finite and V is purely infinite by [**10**, Proposition 6]. Thus U and V are continuous, U is V-injective, and V is U-injective from [**10**, Theorem 13].

Since V is continuous and purely infinite, V is quasi-injective by Corollary 3.38. Hence U is quasi-injective if and only if the continuous module $M = U \oplus V$ is quasi-injective (see [**2**, Theorem 2.1.7]). Therefore we may raise the following question for a continuous module to be quasi-injective so that a certain module has a continuous hull.

Question 4.20. *Under what condition, does every directly finite continuous module become quasi-injective?*

If Question 4.20 does hold true under a certain condition, then any continuous module is quasi-injective. So every module has a continuous hull, and further the continuous hull and the quasi-injective hull coincide by using the argument in the proof of Theorem 4.15.

Added in Proof. Related to Theorem 3.23, Theorem 3.24, and Corollary 3.39, it is shown in [**13**] that every finitely generated projective module over an abelian regular ring has a continuous hull.

Acknowledgments. The authors are very thankful to the referee for his/her nice comments and helpful suggestions for the improvement of this work. The authors are grateful for the partial grant support received from The Ohio State University at Lima and Mathematics Research Institute, The Ohio State University. The authors also thank each others' institutions for the hospitality received during the research work on this paper.

References

[1] K. Beidar and R. Wisbauer, *Strongly and properly semiprime modules and rings*, In: Ring Theory (S.K. Jain and S.T. Rizvi (eds.)), Proc. Ohio State-Denison Conf., World Scientific, Singapore (1993), 58–94.

[2] G.F. Birkenmeier, J.K. Park, and S.T. Rizvi, *Extensions of Rings and Modules*, Research Monograph, Birkhäuser/Springer, New York-Heidelberg-Dordrecht-London, 2013.

[3] B. Eckmann and A. Schopf, *Über injektive Moduln*, Arch. Math. **4** (1953), 75–78.

[4] V.K. Goel and S.K. Jain, *π-injective modules and rings whose cyclics are π-injective*, Comm. Algebra **6** (1978), 59–73.

[5] K.R. Goodearl, *Von Neumann Regular Rings*, Krieger, Malabar, 1991.

[6] R. Gordon and J.C. Robson, *Krull Dimension*, Memoirs **133**, Amer. Math. Soc., Providence, 1973.

[7] R. E. Johnson and E. T. Wong, *Quasi-injective modules and irreducible rings*, J. London Math. Soc. **36** (1961), 260–268, DOI 10.1112/jlms/s1-36.1.260. MR131445

[8] Eben Matlis, *Injective modules over Noetherian rings*, Pacific J. Math. **8** (1958), 511–528. MR99360

[9] Bruno J. Müller and S. Tariq Rizvi, *On the existence of continuous hulls*, Comm. Algebra **10** (1982), no. 17, 1819–1838, DOI 10.1080/00927878208822805. MR674693

[10] Bruno J. Müller and S. Tariq Rizvi, *On injective and quasicontinuous modules*, J. Pure Appl. Algebra **28** (1983), no. 2, 197–210, DOI 10.1016/0022-4049(83)90090-7. MR697666

[11] Kiyoichi Oshiro, *On torsion free modules over regular rings*, Math. J. Okayama Univ. **16** (1973), 107–114. MR469962

[12] Kiyoichi Oshiro and S. Tariq Rizvi, *The exchange property of quasi-continuous modules with the finite exchange property*, Osaka J. Math. **33** (1996), no. 1, 217–234. MR1381627

[13] J.K. Park, S.T. Rizvi, and C.S. Roman, *Module Hulls via Idempotents*, Manuscript, in preparation.

[14] S. Tariq Rizvi, *Commutative rings for which every continuous module is quasi-injective*, Arch. Math. (Basel) **50** (1988), no. 5, 435–442, DOI 10.1007/BF01196504. MR942540

[15] D. W. Sharpe and P. Vámos, *Injective modules*, Cambridge Tracts in Mathematics and Mathematical Physics, No. 62, Cambridge University Press, London-New York, 1972. MR0360706

[16] Kenjiro Shoda, *Zur Theorie der algebraischen Erweiterungen* (German), Osaka Math. J. **4** (1952), 133–143. MR52403

[17] Yuzo Utumi, *On quotient rings*, Osaka Math. J. **8** (1956), 1–18. MR78966

Department of Mathematics, Busan National University, Busan 46241, South Korea
Email address: `jkp1128@yahoo.com`

Department of Mathematics, The Ohio State University, Lima, OH 45804-3576, U.S.A.
Email address: `rizvi.1@osu.edu`

SELECTED PUBLISHED TITLES IN THIS SERIES

785 **A. Leroy and S. K. Jain, Editors,** Algebra and Coding Theory, 2023

783 **Alla Detinko, Michael Kapovich, Alex Kontorovich, Peter Sarnak, and Richard Schwartz, Editors,** Computational Aspects of Discrete Subgroups of Lie Groups, 2023

782 **Galina Filipuk, Alberto Lastra, and Sławomir Michalik, Editors,** Recent Trends in Formal and Analytic Solutions of Diff. Equations, 2023

781 **Emanuel Indrei, Diego Marcon, and Levon Nurbekyan, Editors,** Geometric and Functional Inequalities and Recent Topics in Nonlinear PDEs, 2023

780 **Erik Koelink, Stefan Kolb, Nicolai Reshetikhin, and Bart Vlaar, Editors,** Hypergeometry, Integrability and Lie Theory, 2022

779 **Samuele Anni, Valentijn Karemaker, and Elisa Lorenzo García, Editors,** Arithmetic, Geometry, Cryptography, and Coding Theory 2021, 2022

778 **Carlos Galindo, Alejandro Melle Hernández, Julio José Moyano-Fernández, and Wilson A. Zúñiga-Galindo, Editors,** p-Adic Analysis, Arithmetic and Singularities, 2022

777 **Bang-Yen Chen, Nicholas D. Brubaker, Takashi Sakai, Bogdan D. Suceavă, Makiko Sumi Tanaka, Hiroshi Tamaru, and Mihaela B. Vajiac, Editors,** Differential Geometry and Global Analysis, 2022

776 **Aaron Wootton, S. Allen Broughton, and Jennifer Paulhus, Editors,** Automorphisms of Riemann Surfaces, Subgroups of Mapping Class Groups and Related Topics, 2022

775 **Fernando Galaz-García, Cecilia González-Tokman, and Juan Carlos Pardo Millán, Editors,** Mexican Mathematicians in the World, 2021

774 **Randall J. Swift, Alan Krinik, Jennifer M. Switkes, and Jason H. Park, Editors,** Stochastic Processes and Functional Analysis, 2021

773 **Nicholas R. Baeth, Thiago H. Freitas, Graham J. Leuschke, and Victor H. Jorge Pérez, Editors,** Commutative Algebra, 2021

772 **Anatoly M. Vershik, Victor M. Buchstaber, and Andrey V. Malyutin, Editors,** Topology, Geometry, and Dynamics, 2021

771 **Nicolás Andruskiewitsch, Gongxiang Liu, Susan Montgomery, and Yinhuo Zhang, Editors,** Hopf Algebras, Tensor Categories and Related Topics, 2021

770 **Stéphane Ballet, Gaetan Bisson, and Irene Bouw, Editors,** Arithmetic, Geometry, Cryptography and Coding Theory, 2021

769 **Kiyoshi Igusa, Alex Martsinkovsky, and Gordana Todorov, Editors,** Representations of Algebras, Geometry and Physics, 2021

768 **Dražen Adamović, Andrej Dujella, Antun Milas, and Pavle Pandžić, Editors,** Lie Groups, Number Theory, and Vertex Algebras, 2021

767 **Moshe Jarden and Tony Shaska, Editors,** Abelian Varieties and Number Theory, 2021

766 **Paola Comparin, Eduardo Esteves, Herbert Lange, Sebastián Reyes-Carocca, and Rubí E. Rodríguez, Editors,** Geometry at the Frontier, 2021

765 **Michael Aschbacher,** Quaternion Fusion Packets, 2021

764 **Gabriel Cunningham, Mark Mixer, and Egon Schulte, Editors,** Polytopes and Discrete Geometry, 2021

763 **Tyler J. Jarvis and Nathan Priddis, Editors,** Singularities, Mirror Symmetry, and the Gauged Linear Sigma Model, 2021

762 **Atsushi Ichino and Kartik Prasanna,** Periods of Quaternionic Shimura Varieties. I., 2021

761 **Ibrahim Assem, Christof Geiß, and Sonia Trepode, Editors,** Advances in Representation Theory of Algebras, 2021